Maintenance of Forest Biodiversity

Maintenance of Forest Biodiversity

Editors

Runguo Zang
Yi Ding

MDPI • Basel • Beijing • Wuhan • Barcelona • Belgrade • Manchester • Tokyo • Cluj • Tianjin

Editors

Runguo Zang
Chinese Academy of Forestry
Beijing
China

Yi Ding
Chinese Academy of Forestry
Beijing
China

Editorial Office
MDPI
St. Alban-Anlage 66
4052 Basel, Switzerland

This is a reprint of articles from the Special Issue published online in the open access journal *Forests* (ISSN 1999-4907) (available at: https://www.mdpi.com/journal/forests/special_issues/maintenance_biodiversity).

For citation purposes, cite each article independently as indicated on the article page online and as indicated below:

LastName, A.A.; LastName, B.B.; LastName, C.C. Article Title. *Journal Name* **Year**, *Volume Number*, Page Range.

ISBN 978-3-0365-8442-3 (Hbk)
ISBN 978-3-0365-8443-0 (PDF)

Cover image courtesy of Yi Ding

Contents

About the Editors

Runguo Zang

Prof. Runguo Zang is a Research Fellow at the Ecology and Nature Conservation Institute, Chinese Academy of Forestry. He is interested in forest biodiversity conservation, population ecology, and forest ecosystem management. He has many experiences in the investigation, research, and extension of forest biodiversity conservation and management in major forest types from tropical to cold temperate regions in China. Dr. Runguo Zang has participated in the planning and management of many forest projects as a technical consultant. He has made great contributions to theory and practice in the area of natural disturbance regimes and maintenance of biodiversity dynamics in China.

Yi Ding

Prof. Yi Ding is a Research Fellow at the Ecology and Nature Conservation Institute, Chinese Academy of Forestry. Yi is interested in restoration mechanisms and restoration technology of degraded natural forests, as well as forest dynamics and community assembly theory. The main goals of his current research are to provide theoretical and technical support for accelerating the quality and efficiency improvement and carbon sequestration of degraded natural secondary forests, and establish an important foundation for the dynamic long-term monitoring of vegetation in typical natural forests in China for biodiversity conservation.

Preface

Among terrestrial ecosystems, forests play a dominant role in maintaining biodiversity and providing ecosystem functioning, such as nutrient cycling, carbon sequestration, and ecosystem stability. Over the past several decades, however, global forests are facing unprecedented pressure from climate change and anthropogenic disturbances, resulting in a high rate of biodiversity loss due to deforestation, fragmentation, and degradation. Determining how to maintain biodiversity is extremely urgent if we want to achieve the aims of the Convention on Biological Diversity (CBD) and the Sustainable Development Goals (SDGs). The complex structure, species interaction, heterogenic environment, and regional species pool all affect the spatial and temporal patterns of forest species, but it is still unclear how these effects maintain long-term forest biodiversity as forests undergo rapid variations in habitat features under global change. This book includes sixteen published articles related to the dynamics of forest spatial structure, community assembly, forest recovery patterns after disturbance, spatial distribution of species diversity, effects of abiotic and biotic variables on understory species diversity, and above- and below-ground biotic interactions. These studies cover different forest types across varied spatial and temporal scales, as well as different biological groups (plants, birds, and microbes). We acknowledge all authors who contributed to this special issue "Maintenance of Forest Biodiversity" of Forests. We hope this book can improve our understanding of the underlying mechanisms of forest biodiversity maintenance.

Runguo Zang and Yi Ding
Editors

forests

MDPI

Article

Spatial Structure Dynamics and Maintenance of a Natural Mixed Forest

Chaofan Zhou [1,2], Di Liu [1], Keyi Chen [3], Xuefan Hu [4], Xiangdong Lei [1], Linyan Feng [1], Yuchao Zhang [5] and Huiru Zhang [1,6,*]

1 Institute of Forest Resource Information Techniques, Chinese Academy of Forestry, Beijing 100091, China; cfzhou2021@163.com (C.Z.); liudi19920109@163.com (D.L.); xdlei@ifrit.ac.cn (X.L.); linyan_feng@caf.ac.cn (L.F.)
2 Ecology and Nature Conservation Institute, Chinese Academy of Forestry, Beijing 100091, China
3 Research Institute of Forestry Policy and Information, Chinese Academy of Forestry, Beijing 100091, China; lowrychen@sina.com
4 Beijing Institute of Landscape Architecture, Beijing 100102, China; hufanzi@163.com
5 Academy of Inventory and Planning, National Forestry and Grassland Administration, Beijing 100714, China; georgemacfee@163.com
6 Experimental Center of Forestry in North China, Chinese Academy of Forestry, Beijing 102300, China
* Correspondence: huiru@caf.ac.cn; Tel.: +86-10-69836348

Abstract: Spatial structure dynamics play a major role in understanding the mechanisms of forest structure and biodiversity formation. Recently, researches on the spatial structure dynamics utilizing multi-period data have been published. However, these studies only focused on comparative analyses of the spatial structure of multi-period living trees, without an in-depth analysis of the change processes. In this study, we propose a new comprehensive analysis method for dynamic change of the spatial structure at the individual level, which includes three processes (living trees' flow, mortality process and recruitment process) that have not been considered in previous researches. Four spatial structural parameters (SSSPs, Uniform angle index, Mingling, Dominance and Crowding) and a natural spruce-fir-broadleaf mixed forest with two-phase data were taken as an example to find out the laws of the spatial structure dynamics. All types of dynamic change were named and their proportions were analyzed. The proportion of changes in the SSSPs of individuals was relatively high, even though the mean values of the stand did not change considerably. The five values (0, 0.25, 0.5, 0.75, 1) of the SSSPs are in mutual flow, and the flows are typically one-step, with three-steps and four-steps changes being uncommon. The processes of mortality and recruitment have a higher influence on the spatial structure than the flow of living trees. The dynamic change of spatial structure analysis method created in this study can capture more features not discovered in earlier approaches, as well as guiding forest management in some ways. Understanding the nuances of these changes is a critical part of reasonable spatial structure and biodiversity maintenance, and should be the focus of future research efforts.

Keywords: spatial structure parameters; dynamic changes; the Sankey diagram; mortality process; recruitment process; natural mixed forests

Citation: Zhou, C.; Liu, D.; Chen, K.; Hu, X.; Lei, X.; Feng, L.; Zhang, Y.; Zhang, H. Spatial Structure Dynamics and Maintenance of a Natural Mixed Forest. *Forests* **2022**, *13*, 888. https://doi.org/10.3390/f13060888

Academic Editor: Roberto Molowny-Horas

Received: 14 April 2022
Accepted: 4 June 2022
Published: 7 June 2022

1. Introduction

On long-term temporal and geographical dimensions, forest structure is the outcome of the combined activity of a range of ecological processes [1]. Complex forest structures form by diverse species, which in turn promotes the coexistence of multiple species [2]. Affected by the systems theory that "structure determines function" and the goal of diversity conservation, an increasing number of studies begin to focus on forest structure [3]. Spatial and non-spatial characteristics can be used to define the forest structure [3]. The indexes of non-spatial structure such as tree species composition [4], basal area [5], tree DBH diversity [6,7] and stand density [8] represent the mean stand characteristics by neglecting the spatial information, while the spatial structure approaches take the distribution pattern of individuals and the spatial arrangement of their features into consideration; this includes stand crowding degree [9], species segregation index [10], DBH differentiation degree [11] and aggregation index [12]. It has been proved that spatial structure indicators play a greater role than non-spatial structure in stand reconstruction [13], describing the interaction between trees and thinning decision-making [9,14]. Especially the stand spatial structure parameters (SSSPs) [15], based on a structural unit constituted by a reference tree and its four nearest neighbors [13], have already been applied to the rationality evaluation of structural state [16,17], selective thinning [18,19] and close-to-nature planting for plantation [20].

The SSSPs, including Uniform angle index (*W*), Mingling (*M*), Dominance (*U*) and Crowding (C), can comprehensively analyze the micro spatial condition of individuals and the overall spatial structure of the stand [3,21]. These four parameters may be combined to generate a zero-variate (mean value), univariate, bivariate, trivariate and quadrivariate distribution, which can be used to evaluate forest structural properties at various resolutions [22]. Since it is simpler to understand and evaluate, zero-variate distribution, univariate distribution and bivariate distribution of parameters have been employed more commonly among them [18,23–29]. However, most of this research is concentrated on the spatial structure of forests in a specific time point. Response changes in spatial structure over time under anthropogenic and natural disturbances did not get the attention it deserves [30].

Fortunately, some scholars have noticed this, and several studies have been carried out. Deng and Katoh [31], Zhao et al. [32] and Xue et al. [33], and Zhang et al. [34] have analyzed the change of spatial structure characteristics on various natural forests and plantation in a short period (5 to 10 years) by the zero-variate, univariate and bivariate distribution methods, respectively. Wan et al. [35] used the zero-variate distribution and bivariate distribution of SSSPs to investigate the impacts of four treatments on the spatial structure dynamic of *Quercus aliena* var. *acuteserrata* forest in the Xiaolongshan mountain. These researchers found that the variations in the spatial structure are more affected by anthropogenic disturbances, especially structure-based forest management [35], than natural disturbances [36]. Unfortunately, the distribution dimensions (zero-variate to bivariate) of the spatial structure only had a slight effect on increasing the explanation of its change. All these approaches can examine changes in the spatial structure of living trees from early-stage to later-stage, though, it is unable to explain the origins of such changes adequately. As there are not only residual trees, but also felled trees, dead trees and recruitment trees during forest management or forest development [30,36], a simple examination of the difference in the spatial structure of live trees between the two phases cannot identify the causes of this change. This is not conducive to maintaining reasonable spatial structure through artificial measures. Therefore, a new method is needed to break the bottleneck.

Sankey diagrams represent the movement of information to and from various nodes in a network and are most commonly used to analyze energy and material flows [37]. These fluxes are represented by arrows or directed lines whose thickness corresponds to the size of the flow [38]. These diagrams are frequently used in industrial ecology to represent product lifetime evaluations, as well as in engineering to quickly visualize energy efficiency [37]. Sankey diagrams emphasize the quantity and direction of flows within a system, and they have been used in various geographic and human-environment study contexts due to their versatility [38]. The ability of Sankey diagrams to trace material movements has made them useful tools for estimating major greenhouse gas emissions [39,40], partitioning the global terrestrial water fluxes [41] and vegetation cover type conversion [42]. Although the use of the Sankey diagram in the study of spatial structure dynamics is innovative, it is ideally suited to tackling issues that cannot be answered using multidimensional distributions because it can accurately depict the changes between two periods. This study took the spruce-fir-broadleaf mixed forest as an example, disentangling its spatial structure changes into "living trees' flow", "mortality process" and "recruitment process", aimed to: (1) develop a comprehensive methodology for spatial structure dynamics analysis based on SSSPs; (2) understand the laws in dynamic changes of spatial structure for spruce-fir-broadleaf mixed forest; (3) emphasize the important role of mortality process and recruitment process in the dynamic changes of spatial structure.

2. Materials and Methods

2.1. Study Area and Experimental Design

The research area is located in the Jingouling Forest Farm (130°5′–130°20′ E, 43°17′–43°25′ N), Wangqing Forestry Bureau, Jilin Province, China (Figure 1). It belongs to the middle and low mountains of Changbai Mountain, with an altitude between 300 and 1200 m and a slope between 5 and 25 degrees. The region has a temperate continental monsoon climate, with four distinct seasons of long winter and short summer, and the coexistence of rain and heat. The average annual temperature is 4 °C, and the annual average rainfall ranges from 600 to 700 mm. The area is dominated by grayish-brown soil with a moist and loose granular structure, which has an acidic pH and high fertility. Existing forest types are coniferous forest, broad-leaved forest and mixed forest.

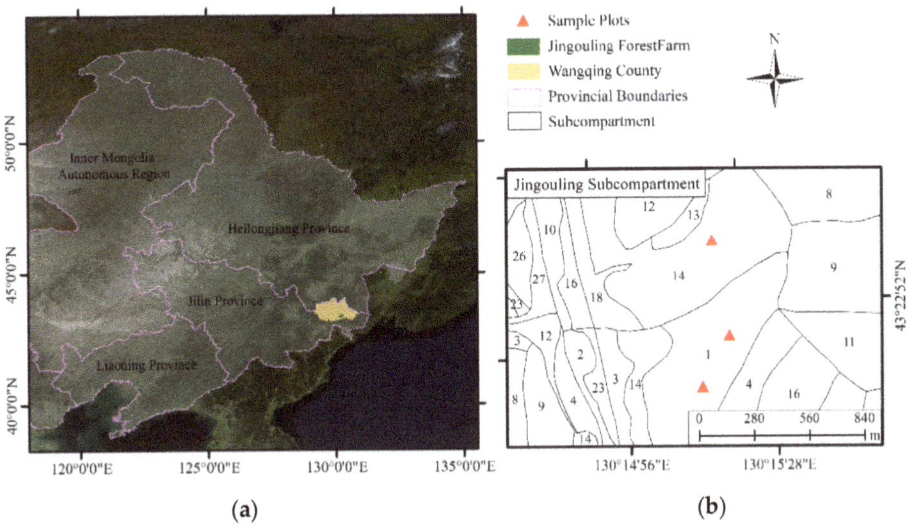

Figure 1. Location of the study area: Wangqing Forest Bureau (**a**) in northeast China and spatial distribution of 3 sample plots (**b**).

Three one-hectare sampling plots (100 m × 100 m each) were established in the natural spruce-fir-broadleaf mixed forest of the Jingouling Forest Farm in July 2013 (Figure 1). The main tree species in spruce-fir-broadleaf mixed forest include spruce (*Picea jezoensis* var. *microsperma* (Lindl.) Cheng et L.K. Fu), fir (*Abies nephrolepis* (Trautv.) Maxim.), larch (*Larix olgensis* Henry), Korean pine (*Pinus koraiensis* Siebold et Zucc.), white birch (*Betula platyphylla* Suk.), poplar (*Populus ussuriensis* Kom.), ribbed birch (*Betula costata* Trautv.), linden (*Tilia amurensis* Rupr.), elm (*Ulmus laciniata* (Trautv.) Mayr), maple (*Acer pictum* subsp. *mono* (Maxim.) H. Ohashi) and ash (*Fraxinus mandshurica* Rupr.). Trees with diameters at breast height (DBH) more than 5 cm were recorded, and species, DBH, tree height (H), crown diameter, relative coordinates and living state were measured (Table 1). In August 2018, these indicators were remeasured and recruitment trees (DBH of regeneration reach or exceed the threshold of DBH (5 cm) during the period of 2013–2018) were considered.

Table 1. Stand characteristics of 3 plots in 2013.

Plot Code	Trees/ ha	Average DBH/cm	Basal Area/ (m^2/ha)	Stock Volume/ (m^3/ha)	Canopy Density	Composition of Tree Species
YLK1	996	17.63	24.30	199.97	0.85	2Bc2Ta1An1Pj1Pk1Lo1Pu1Am
YLK2	1024	18.23	26.72	216.00	0.86	2Bc2Ta1Lo1An1Pk1Am1Pj1Os
YLK3	1018	17.38	24.15	182.75	0.63	3Lo1Bc1Fm1Pk1An1Bp1Pj1Ta

Note: Bp stands for *B. platyphylla;* Ta stands for *T. amurensis;* Bc stands for *B. costata;* Pk stands for *P. koraiensis;* An stands for *A. nephrolepis;* Lo stands for *L. olgensis;* Am stands for *A. pictum* subsp. *mono;* Fm stands for *F. mandschurica;* Pu stands for *P. ussuriensis;* Pj stands for *P. jezoensis;* Os stands for Other species.

2.2. Stand Spatial Structure Parameters

Four Stand Spatial Structure Parameters (SSSPs) including Uniform angle index, Mingling, Dominance and Crowding [3] were used to analyze the dynamics of the spatial structure of the spruce-fir-broadleaf mixed forest. Four parameters are calculated as follows:

1. Uniform angle index (W);

$$W_i = \frac{1}{4}\sum_{j=1}^{4} z_{ij} \quad z_{ij} = \begin{cases} 1, & if \alpha_j < \alpha_0 \\ 0, & otherwise \end{cases} \tag{1}$$

In the Equation (1), α_j stands for the angle shown in and α_0 stands for the standard angle 72°.

2. Mingling (M);

$$M_i = \frac{1}{4}\sum_{i=1}^{4} v_{ij} \quad v_{ij} = \begin{cases} 1, & if species_j \neq species_i \\ 0, & otherwise \end{cases} \tag{2}$$

In the Equation (2), $species_j$ and $species_i$ denote the species of *jth* neighboring tree and reference tree *i*, respectively.

3. Dominance (U);

$$U_i = \frac{1}{4}\sum_{i=1}^{4} k_{ij} \quad k_{ij} = \begin{cases} 1, & if DBH_j \geq DBH_i \\ 0, & otherwise \end{cases} \tag{3}$$

In the Equation (3), DBH_j and DBH_i denote the diameter at breast height of *jth* neighboring tree and reference tree *i*, respectively.

4. Crowding (C).

$$C_i = \frac{1}{4}\sum_{j=1}^{4} y_{ij} \quad y_{ij} = \begin{cases} 1, & if c_j + c_i > dist_{ij} \\ 0, & otherwise \end{cases} \tag{4}$$

In the Equation (4), c_j and c_i stand for the crown radius of the *jth* neighboring tree and reference tree *i*, respectively. The $dist_{ij}$ denotes the distance between the *jth* neighboring tree and reference tree *i*. In a structural unit, these four SSSPs looked at the spatial and attribute correlations between the reference tree *i* and its nearest neighbor *j*. Even though the four SSSPs focus on distinct parts of the spatial structure, they all have five possible values: 0, 0.25, 0.5, 0.75 and 1, which indicate the various spatial structure statuses (Figure 2). The quantization and analyses of SSSPs in this study were conducted using R version 4.1.1 [43].

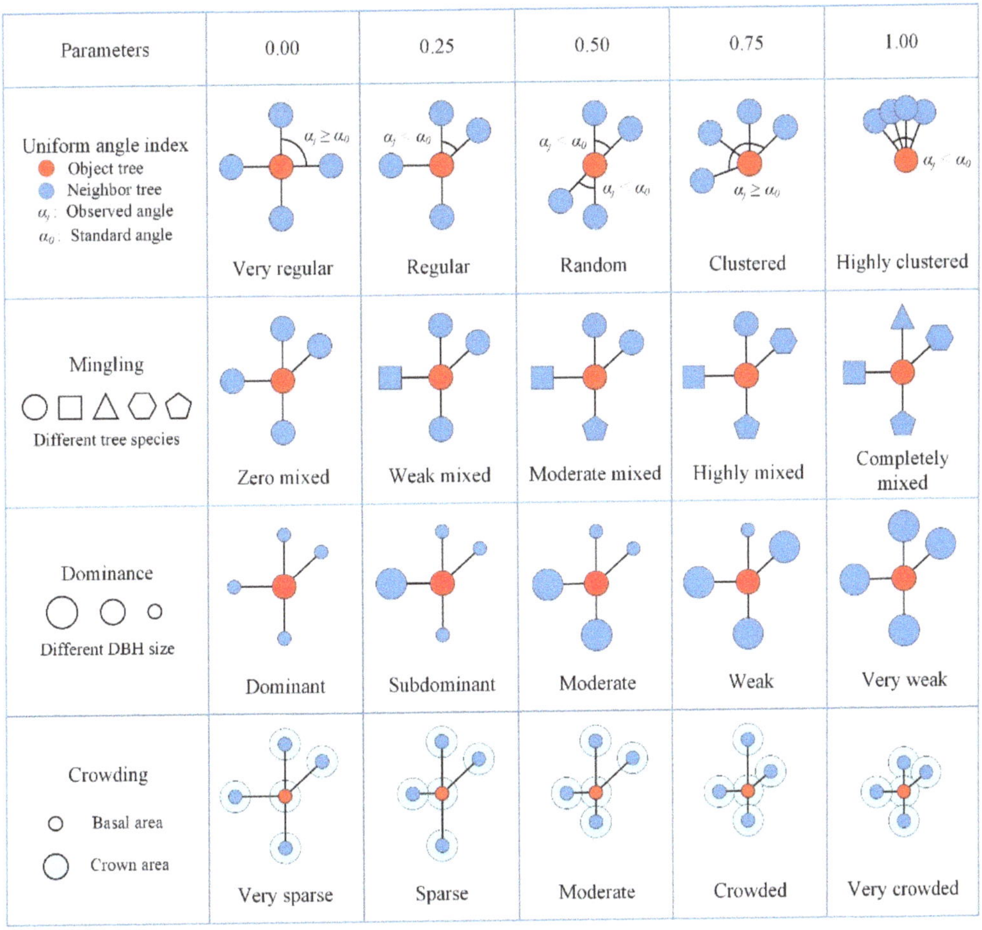

Figure 2. Schematic diagram of four stand spatial structural parameters. The Uniform angle index compared the size of observed angle α_j formed by two nearest neighbors and the reference tree to the standard angle α_0 (72°); the Mingling, Dominance and Crowding compared whether the tree species are the same, the size of DBH and whether the crowns overlap between the reference tree and its four neighboring trees, respectively.

2.3. Disentangle the Dynamic Changes of Spatial Structure

In natural forests, the processes of mortality and recruitment are the primary causes of dynamic changes [30,36]. By incorporating these two processes into the dynamic change analysis of SSSPs, the source of changes can be better revealed. We used the ggalluvial [44] and ggplot2 [45] package in the R program [43] to create a Sankey diagram to show the flow change of SSSPs from early-stage to later-stage (Figure 3a), and all "change types" have been shown and named (Figure 3b), followed by statistics to further analyze the change law of SSSPs. The change law is evaluated primarily from three perspectives: the flow from living trees to living trees (*Flow*), the change from living trees to deadwood (*Dead*) and the conversion from regeneration to living trees (*Reg*), which respectively abbreviated as "living trees' flow", "mortality process" and "recruitment process".

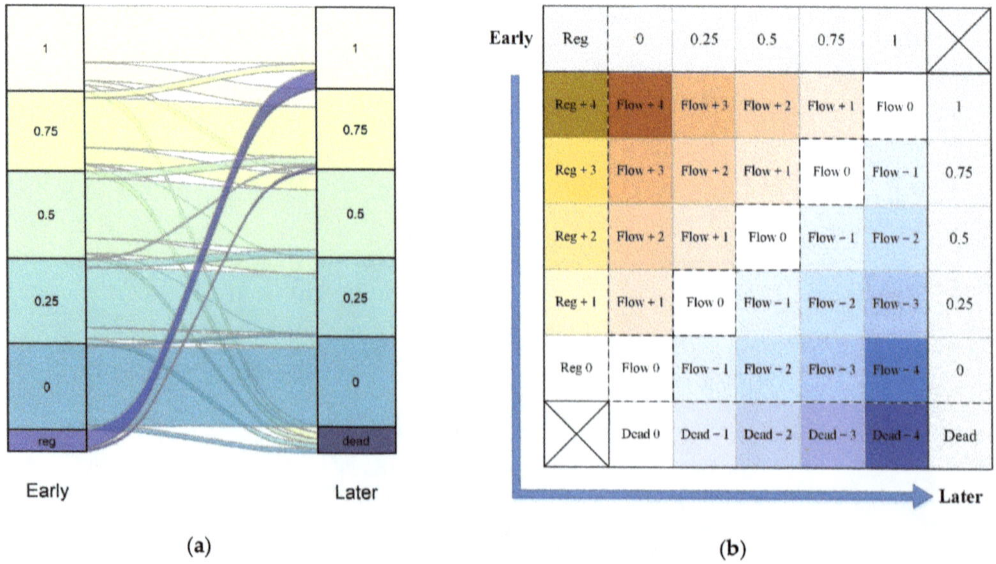

Figure 3. The Sankey diagram (**a**) and Change types (**b**) of stand spatial structural parameters between two periods. The "reg" means the regeneration in early stage and growing into a living tree (DBH $\geq$ 5 cm) in later stage. The "dead" means the living trees in early stage, that have died in later stage. The flows in (**a**) connect the structural attributes of trees between the early and later periods. Each flow represents a change type shown in (**b**), and the width of the flow represents the relative proportion of this change type. The dotted lines in (**b**) divide 35 change types into recruitment process, ascending flow, stable flow, descending flow and mortality process with the colors of gold, orange, white, light blue and dark blue, respectively. The darker the color, the greater the change range of structure attributes. These change types in the ascending flow, stable flow and descending flow change can be further combined into nine change types according to the change range (the depth of color, see Section 2.3.1 for more details).

2.3.1. The Flow from Living Tree to Living Tree

Each SSSPs have five possible values: 0, 0.25, 0.5, 0.75 and 1. There are three flow directions of values of living trees from the early-stage to the later-stage when the values are placed from top to bottom in descending order as in Figure 3a: "ascending", "stable" and "descending". A shift of one-step, such as from 0 to 0.25 or 1 to 0.75, is referred to as a "one-step change". When combining flow direction and step, the case like the value changes from 0 and 1 is called "ascending four-steps". On the contrary, when called "descending four-steps", while the value that does not change is called "stable". The naming method of other value changes are analogized.

Based on the naming method, there are nine types of living trees flow of SSSPs, namely "ascending four-steps (Flow + 4)", "ascending three-steps (Flow + 3)", "ascending two-steps (Flow + 2)", "ascending one-step (Flow + 1)", "stable (Flow 0)", "descending one-step (Flow − 1)", "descending two-steps (Flow − 2)", "descending three-steps (Flow − 3)" and "descending four-steps (Flow − 4)". The number of corresponding types increased from 1 to 5 and then decreased to 1 (Figure 3b). As shown by the dashed line in Figure 3b, the nine types of living trees' flow can be divided into three parts, the "ascending flow ($Flow_{Up}$)" in the upper left triangle (Flow + 4, Flow + 3, Flow + 2, Flow + 1), the "stable flow ($Flow_{Stable}$)" in the middle diagonal (Flow 0), and the "descending flow ($Flow_{Down}$)" in the lower right triangle. (Flow − 1, Flow − 2, Flow − 3, Flow − 4).

2.3.2. The Change from Living Tree to Deadwood

When trees die for a reason internal or external, then their SSSPs will change from five possible values to no value (just like the impact of value 0 when computing the mean value of SSSPs). As a result, the naming type for the mortality process is identical to that of 2.3.1, with names like "Dead 0", "Dead − 1", "Dead − 2", "Dead − 3" and "Dead − 4" (see Figure 3b).

2.3.3. The Conversion from Regeneration to Living Trees

The regeneration (seeding and sapling) first contributed to the computation of spatial structure when its DBH is more than 5 cm (becoming a living tree), then their SSSPs will change from no value to five possible values (opposite to the mortality process). Accordingly, the type naming for the recruitment process is analogical to that of 2.3.1, with names like "Reg 0", "Reg + 1", "Reg + 2", "Reg + 3", "Reg + 4" (see Figure 3b).

By calculating the proportion of each change type in these three processes, we can comprehend the impact of each process on the overall spatial structure and analyze the stability of SSSPs at the individual, species and stand level.

2.4. Equation of Changes in Spatial Structure

By disentangling the process of spatial structure change, we assume that the relationship between the mean values of SSSPs of early-stage and later-stage conforms to the following equation:

$$\omega_{Early} \times N_{Early} + Flow_{Up} + Flow_{Stable} + Flow_{Down} + Dead + Reg = \omega_{Later} \times N_{Later} \quad (5)$$

where, ω_{Early}, ω_{Later}, N_{Early}, N_{Later} stand for the mean values of SSSPs and the number of trees in the early and late periods, respectively. $Flow_{Up}$, $Flow_{Stable}$, $Flow_{Down}$, $Dead$ and Reg denote the changes of SSSPs caused by ascending flow, stable flow, descending flow, mortality process and recruitment process, respectively. The calculation formulas are as follows:

$$\begin{cases} Flow_{Up} = 1 \times N_{Flow+4} + 0.75 \times N_{Flow+3} + 0.5 \times N_{Flow+2} + 0.25 \times N_{Flow+1} \\ Flow_{Stable} = 0 \times N_{Flow0} \\ Flow_{Down} = -0.25 \times N_{Flow-1} - 0.5 \times N_{Flow-2} - 0.75 \times N_{Flow-3} - 1 \times N_{Flow-4} \\ Reg = 1 \times N_{Reg+4} + 0.75 \times N_{Reg+3} + 0.5 \times N_{Reg+2} + 0.25 \times N_{Reg+1} + 0 \times N_{Reg0} \\ Dead = -0 \times N_{Dead0} - 0.25 \times N_{Dead-1} - 0.5 \times N_{Dead-2} - 0.75 \times N_{Dead-3} - 1 \times N_{Dead-4} \end{cases} \quad (6)$$

N_i stands for the number of trees of the i-th change type mentioned above, and the relationship between N_{Early} and N_{Later} is as follows:

$$\begin{cases} N_{Early} + N_{Reg} - N_{Dead} = N_{Later} \\ N_{Reg} = N_{Reg+4} + N_{Reg+3} + N_{Reg+2} + N_{Reg+1} + N_{Reg0} \\ N_{Dead} = N_{Dead0} + N_{Dead-1} + N_{Dead-2} + N_{Dead-3} + N_{Dead-4} \end{cases} \quad (7)$$

Both the ascending flow ($Flow_{Up}$) and the recruitment process (Reg) will increase the sum of the SSSPs' values ($\omega_{Later} \times N_{Later}$) in the later-stage, while the descending flow ($Flow_{Down}$) and the mortality process ($Dead$) are the opposite, and the stable flow ($Flow_{Stable}$) does not change it. The difference between N_{Reg} and N_{Dead} determines whether N_{Later} increases or decreases compared with N_{Early}, which in turn affects the mean values of SSSPs of later-stage. To validate the feasibility of the correlation formula between the mean values of the SSSPs early and later, it must be extensively tested with actual samples (see below).

3. Results

3.1. The Changing Process of Stand Spatial Structure in Early- and Later-Period

The distribution proportions of the other three SSSPs rarely changed from 2013 to 2018, except for modest variations in the distribution proportions of Crowding (Figure 4).

The changing process of stand spatial structure will be analyzed in the following three (Sections 3.1.1–3.1.3), respectively.

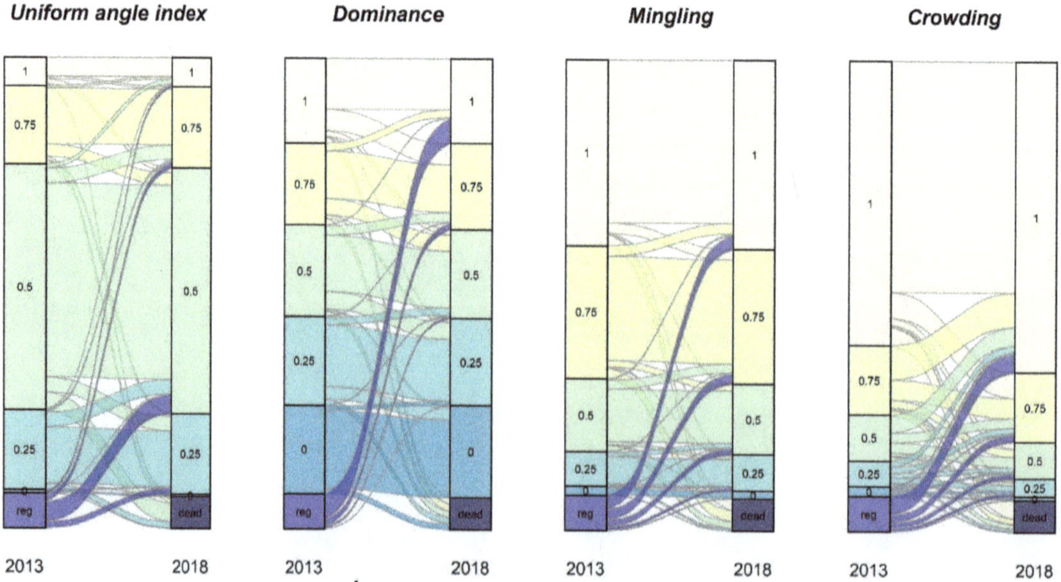

Figure 4. Changes in values distribution of different stand spatial structural parameters from 2013 to 2018 of all three plots. The "reg" means the regeneration in 2013 and growing into a living tree (DBH ≥ 5 cm) in 2018. The "dead" means the living trees in 2013, that have died in 2018. The flows in the figure connect the structural attributes of trees between 2013 and 2018. Each flow represents a change type of structure attributes, and the width of the flow represents the relative proportion of this change type.

3.1.1. The Flow of Living Trees

During the natural development of a spruce-fir-broadleaf mixed forest, there is almost no "four-steps change" (0–1.3%), and very little "three-steps change" (0–5%) and "two-steps change" (0–15.3%), while the "one-step change" are relatively large (3.7%–27%), second only to the case where the value does not change (46.7%–89.2%) (Figure 4 and Table 2). In most circumstances, changes between two values are not unidirectional, but rather flow in both directions (ascending and descending coexist), however, the quantity of mutual flow may be unequal.

Table 2. The proportion statistics of change types in living trees' flow (%).

Types	Uniform Angle Index	Dominance	Mingling	Crowding
Flow + 4	0.0 ± 0.0 (0.0–0.0)	0.0 ± 0.0 (0.0–0.0)	0.0 ± 0.0 (0.0–0.0)	0.4 ± 0.4 (0.0–1.3)
Flow + 3	0.1 ± 0.0 (0.0–0.1)	0.0 ± 0.0 (0.0–0.0)	0.0 ± 0.0 (0.0–0.0)	1.7 ± 1.4 (0.0–5.0)
Flow + 2	2.0 ± 0.3 (1.5–2.9)	0.6 ± 0.2 (0.3–0.9)	0.6 ± 0.0 (0.5–0.7)	6.2 ± 3.7 (1.4–15.3)
Flow + 1	8.3 ± 0.3 (7.8–8.9)	6.7 ± 0.3 (6.1–7.3)	6.6 ± 0.4 (5.7–7.2)	12.5 ± 5.9 (3.7–27.0)
Flow 0	80.3 ± 1.2 (77.7–82.5)	73.6 ± 2.3 (68.9–78.8)	87.3 ± 0.9 (85.4–89.2)	66.7 ± 8.2 (46.7–76.9)
Flow − 1	7.4 ± 0.7 (5.7–8.4)	16.2 ± 1.9 (13.1–20.7)	4.9 ± 0.6 (3.9–6.4)	9.6 ± 2.2 (4.5–13.6)
Flow − 2	1.9 ± 0.4 (1.2–2.7)	2.7 ± 0.5 (1.7–3.6)	0.5 ± 0.3 (0.0–1.1)	2.3 ± 1.0 (0.0–4.2)
Flow − 3	0.0 ± 0.0 (0.0–0.0)	0.2 ± 0.1 (0.0–0.4)	0.0 ± 0.0 (0.0–0.1)	0.5 ± 0.2 (0.1–0.9)
Flow − 4	0.0 ± 0.0 (0.0–0.0)	0.0 ± 0.0 (0.0–0.0)	0.0 ± 0.0 (0.0–0.0)	0.0 ± 0.0 (0.0–0.1)

Note: Values are presented as the mean ± standard error (minimum–maximum).

Analysis of the proportion of Flow 0 (value does not change) of each SSSPs shows that Mingling, followed by Uniform angle index and Dominance, is the steadiest, while

Crowding is the least stable (Table 2). For Uniform angle index, Dominance, Mingling and Crowding, the most stable and the least stable tree species are larch and poplar, white birch and fir, poplar and fir and poplar and larch, respectively (Table A1 in Appendix A).

When the fraction of ascending and descending change types is compared, it is clear that the overall value of Uniform angle index, Mingling and Crowding grew while Dominance fell throughout the living trees' flow (Table 2). The shifting trends of parameters for most tree species are the same as those of stands, with the exception of the following exceptions. Species including white birch, linden, ribbed birch, and poplar declined in the Uniform angle index; birch increased in Dominance; fir, maple, ash, and spruce reduced in Mingling; and fir decreased in Crowding (Table A1).

3.1.2. The Process of Mortality

Compared with the corresponding values in 2013, the proportion of change types in Uniform angle index (Dead − 2 and Dead − 3), Dominance (Dead 0, Dead − 1 and Dead − 2), Mingling (Dead 0, Dead − 1 and Dead − 4), Crowding (Dead − 4) is lower (Table 3), indicating that the mortality process encourages an increase in the number of reference trees with the following situation: random and aggregated of Uniform angle index (W = 0.5 and 0.75), dominant, subdominant and moderate of Dominance (U = 0, 0.25 and 0.5), zero mixed, weak mixed and extremely highly mixed of Mingling (M = 0, 0.25 and 1), very crowded of Crowding (C = 1).

Table 3. The proportion statistics of change types in mortality process (%).

Types	Uniform Angle Index	Dominance	Mingling	Crowding
Dead 0	1.0 ± 0.4 (0.0 − 1.7)	17.6 ± 1.5 (15.0–21.1) ↓	2.3 ± 0.6 (1.4–3.7) ↓	2.3 ± 1.4 (0.0–5.6)
Dead − 1	22.6 ± 5.2 (9.9–29.6)	19.9 ± 0.6 (18.5–21.1) ↓	7.5 ± 2.5 (1.4–11.1) ↓	8.4 ± 5.6 (1.4–22.2)
Dead − 2	56.4 ± 2.1 (51.9–60.6) ↓	20.0 ± 2.1 (16.7–25.0) ↓	22.5 ± 3.6 (14.8–30.0)	14.6 ± 3.9 (8.3–24.1)
Dead − 3	13.2 ± 2.4 (8.3–18.3) ↓	21.2 ± 0.6 (19.7–22.2)	34.2 ± 0.7 (32.4–35.2)	17.9 ± 3.3 (12.7–25.9)
Dead − 4	6.8 ± 1.3 (5.0–9.9)	21.3 ± 1.9 (18.3–25.9)	33.6 ± 4.5 (23.3–42.3) ↓	56.8 ± 14.1 (22.2–75.0) ↓

Note: Values are presented as the mean ± standard error (minimum–maximum). "↓" indicated that the proportion of this change type is smaller than corresponding values in 2013.

Some tree species exhibit significant variances with the stand in change patterns of four parameters throughout the mortality process, as shown in Table A2: White birch, linden, ash, poplar, and elm have increased in Uniform angle index (W = 1), whereas fir and spruce have increased in Dominance (U = 0.75 and 1), respectively. The Mingling of linden, larch, poplar, and spruce, zero mixed and weak mixed (M = 0 and 0.25) have not risen; very crowded (C = 1) have not increased in Crowding of linden and elm.

3.1.3. The Process of Recruitment

Compared with the corresponding values in 2018, the proportion of change types in Uniform angle index (Reg + 1, Reg + 3 and Reg + 4), Dominance (Reg + 3 and Reg + 4), Mingling (Reg + 1, Reg + 2 and Reg + 3) and Crowding (Reg + 1, Reg + 2, Reg + 3 and Reg + 4) is larger (Table 4), showing that the recruitment process promotes a rise in the number of reference trees with the following situation: regular, clustered and very clustered of Uniform angle index (W = 0.25, 0.75 and 1), weak and very weak of Dominance (U = 0.75 and 1), weak mixed, moderate mixed and highly mixed of Mingling (M = 0.25, 0.5 and 0.75), very sparse, sparse, moderate and crowded of Crowding (C = 0, 0.25, 0.5 and 0.75).

The white birch, ribbed birch, poplar, Larch, and Ash have no or only one tree in the recruitment process, but the fir accounts for more than half of the population in the recruitment. Although the change patterns of four stand characteristics are nearly equivalent to fir, several tree species differ significantly in the following circumstances (Table A3): random (W = 0.5) has grown in Uniform angle index of Korean pine, maple, elm, and spruce; highly mixed or completely mixed (M = 0.75 and 1) has increased in Mingling of Korean pine, maple, ash, elm, and spruce; and very crowded ($C = 1$) has increased in Crowding of elm and spruce.

Table 4. The proportion statistics of change types in recruitment process (%).

Types	Uniform Angle Index	Dominance	Mingling	Crowding
Reg 0	0.0 ± 0.0 (0.0–0.0)	0.0 ± 0.0 (0.0–0.0)	2.2 ± 1.4 (0.0–5.6)	1.2 ± 0.5 (0.0–2.2)↑
Reg + 1	19.2 ± 3.4 (11.0–23.6)↑	0.5 ± 0.4 (0.0–1.4)	6.8 ± 3.3 (0.0−13.9)↑	7.1 ± 0.8 (5.1–8.3)↑
Reg + 2	55.2 ± 2.8 (48.6–60.4)	6.0 ± 3.0 (0.0–12.5)	16.3 ± 6.0 (2.6–27.8)↑	13.6 ± 3.0 (9.7–20.9)↑
Reg + 3	19.2 ± 0.7 (17.9–20.8)↑	20.8 ± 0.6 (19.8–22.2)↑	32.2 ± 5.1 (22.2–43.6)↑	22.5 ± 1.1 (20.5–25.0)↑
Reg + 4	6.5 ± 1.7 (2.6–9.9)↑	72.7 ± 3.8 (63.9–79.5)↑	42.4 ± 5.5 (30.6–53.8)	55.6 ± 4.0 (47.3–64.1)

Note: Values are presented as the mean ± standard error (minimum–maximum). "↑" indicated that the proportion of this change type is larger than corresponding values in 2018.

3.2. The Variation and Disentangle of Mean Values of Stand Spatial Structure Parameters

The shifting patterns of the four SSSPs of the three plots were not consistent from 2013 to 2018 (Figure 5). The mean values of Uniform angle index and Mingling in all three plots tended to rise, indicating that the degree of tree aggregation and intermixing of tree species in the plots has increased. The mean values of Crowding in YLK1 and YLK2 decreased with time, showing these plots became less crowded, whereas the converse was true for plot YLK3. The mean of Dominance changes is complicated and has no fixed trend (Figure 5). In the average variation range, Crowding (0.090) > Uniform angle index (0.006) > Mingling (0.006) > Dominance (0.004).

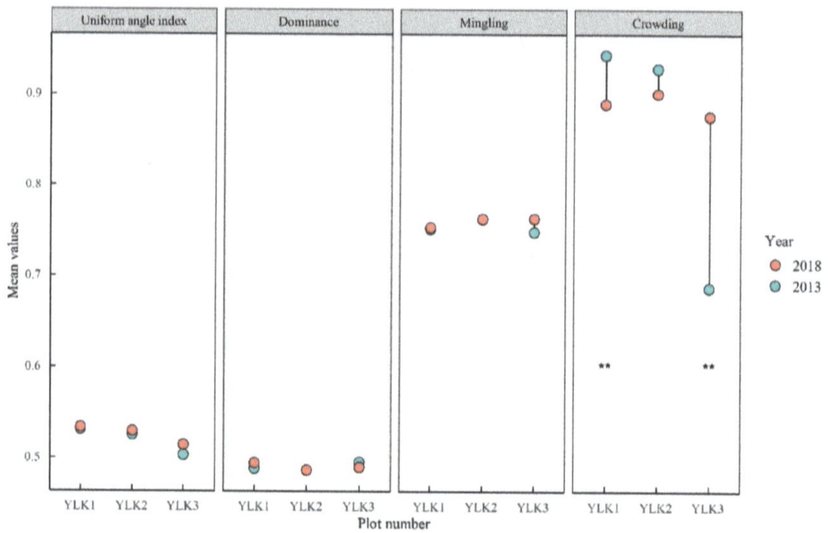

Figure 5. Changes in mean values of stand spatial structural parameters of plots from 2013 to 2018. The Chi-square test was used to check the significant difference between years for each of the four parameters of the five possible values (0, 0.25, 0.5, 0.75 and 1). Significant codes: $p < 0.01$ '**'.

The total value of the previous spatial structure (the product of the mean value of SSSPs and the number of trees in the early-stage ($\omega_{early} \times N_{early}$)), plus the sum of the five parts of SSSPs change processes at the individual level ($Flow_{up}$, $Flow_{stable}$, $Flow_{down}$, *Dead* and *Reg*), are exactly equal to the sum of the later spatial structure values (the product of the mean value of the SSSPs and the number of trees in later-stage($\omega_{later} \times N_{later}$)), irrespective of the category of SSSPs and sample plots (Table 5). The values of *Dead* or *Reg* are generally greater than that of $Flow_{up}$ or $Flow_{down}$ for the sum (absolute value) of the value changes of the SSSPs. In Dominance, the values of *Reg* are invariably greater than that of *Dead*, and the values of $Flow_{down}$ are always bigger than that of $Flow_{up}$, whereas other SSSPs have little difference or are inconclusive.

Table 5. Disentangle the changes in spatial structure. The values in the table equal to the sum of the product the type value and quantity of SSSPs (see equations in Section 2.4). Positive and negative values in the change processes indicate an increase and decrease in the total value of SSSPs from early-stage to later-stage, respectively.

Parameters	No.	Early-Stage	Change Processes					Later-Stage
		$\omega_{early} \times N_{early}$	$Flow_{up}$	$Flow_{stable}$	$Flow_{down}$	Dead	Reg	$\omega_{later} \times N_{later}$
Uniform angle index	YLK1	427.25	26.25	0.00	−25.50	−40.00	51.75	439.75
	YLK2	420.00	20.75	0.00	−21.25	−28.00	38.00	429.50
	YLK3	405.75	23.25	0.00	−15.25	−26.25	19.50	407.00
Dominance	YLK1	393.00	14.50	0.00	−49.25	−34.75	84.00	407.50
	YLK2	389.00	17.00	0.00	−42.75	−31.25	62.75	394.75
	YLK3	399.50	12.50	0.00	−31.25	−30.00	37.00	387.75
Mingling	YLK1	605.00	15.25	0.00	−13.25	−55.50	70.00	621.50
	YLK2	610.00	15.25	0.00	−12.00	−40.25	46.50	619.50
	YLK3	604.00	13.25	0.00	−8.25	−38.75	34.25	604.50
Crowding	YLK1	759.25	11.75	0.00	−43.75	−63.00	69.25	733.50
	YLK2	742.75	19.75	0.00	−35.00	−54.50	58.50	731.50
	YLK3	554.50	146.75	0.00	−9.25	−32.00	33.50	693.50

4. Discussion

4.1. The Dynamics of Spatial Structure and Its Implications for Management

Continuous spatial structure observations for individual trees were not carried out in previous studies, and ongoing research mainly focused more on changes at the stand level, such as the mean value or distribution of SSSPs. However, the changes in the mean value are generally slight. By following the change of the spatial structure at the individual level, the changes in the stand spatial structure can be disentangled into three categories to better understand the change process: the living trees' flow, the mortality process and the recruitment process. The formation of the spatial structure variation equation was validated during the process by comparing the mean values of the SSSPs of the sample plots in early-stage and later-stage, and the accuracy of disentangling process was confirmed. We were able to achieve several outcomes that were not found using typical approaches thanks to the methods used in this research:

(a) The proportion of changes in the spatial structure of individuals was relatively high, even though the overall mean values of the stand did not change considerably. This indicates that the changes in spatial structure at the individual level have a large counteracting effect when summarized to stand level. In order to further study the changes in spatial structure, it is necessary to start from the dynamics of spatial structure at the individual level.

(b) The five values of the SSSPs are in mutual flow, although the flow may be asymmetric. Flow changes are typically one-step, with three-steps and four-steps changes being uncommon. It is suggested that the change of spatial structure is a gradual process during the development of the stand, which has some guidance for imitating the natural change of stand spatial structure in forest management.

(c) In general, the processes of mortality and recruitment have a higher influence on the spatial structure than the flow of living trees. This is due to the dead woods and recruitment trees not only changing the stand spatial structure through their values, but also changing the spatial structure of other reference trees which take them as adjacent trees. A further thorough analysis of the influence mechanism on stand spatial structure can be explored through the mortality process and recruitment process.

(d) The impacts of the mortality and recruitment process on the direction of spatial structure variety are opposites, but their influences are equivalent in most cases. This suggests that the role of both cutting (mortality process) and replanting (recruitment process) measures should be emphasized in improving and maintaining the spatial structure [18,35].

We also found that the mean values of Uniform angle index and Mingling of the plots increased, whereas Crowding and Dominance both increased and decreased in five years. Our results share several similarities and differences with Deng and Katoh [31], Zhao et al. [32], Xue et al. [33] and Wan et al. [35], which may be due to the correlation between the change direction of the SSSPs in different forest communities and the initial spatial structure at the stand level.

According to Hui et al. [46], the mean value of Uniform angle index should remain at (0.475, 0.517), indicating an acceptable pattern of random distribution. From 2013 to 2018, the mean values of the Uniform angle index of three plots all rose, and YLK1 and YLK2 have already crossed the upper threshold of the tolerable interval, with YLK3 perhaps exceeding the upper threshold if the development trend continues ($W = 0.75$ and 1). Given this, in the reference unit with clustered and very clustered conditions, one or two clustered neighboring trees in the unit should be removed, and the opposite way should be checked for an adjacent tree to reintegrate the unit and decide the implementation of replanting [47].

In general, trees in a dominant condition ($U = 0$ and 0.25) have a better chance of surviving, and our results at the stand level back this up. However, other species, such as fir and spruce, are not affected in our study, dominant fir and spruce died more frequently than tiny ones (a weak state with $U = 0.75$ and 1). This might be explained by these species' shallow roots, which are readily blown down by the wind or broken down by the snow [48,49]. As a result, thinning should be done away from these species' huge trees to ensure that nearby trees provide adequate support.

The variety of tree species is frequently advocated, particularly spatial mixing, because pest and disease propagation can be reduced if the species are mixed [50]. In our study, the mean values of Mingling of spruce-fir-broadleaf mixed forest are steady at approximately 0.75 or slightly increased; most tree species show the same trend, although fir, maple, ash, and spruce have declined. One possible explanation is that the recruiting tree with the same species reintegrated the unit of the reference tree (such as the species often grow into a clump). Trees with Mingling descending flow and states of zero mixed, weak mixed, and moderate mixed ($M = 0$, 0.25, and 0.5) should be examined for thinning for enhancing Mingling, meanwhile the thinning should also guide by the health condition and dominance. [18].

Crowding has increased over time induced by tree canopy development in most species and plot YLK3, whereas reduced in fir and plots YLK1 and YLK2. This suggested that plots YLK1 and YLK2 had self-thinning due to the intense crown competition, and thinning procedures should be implemented as soon as feasible. It may also reveal that the nearby tree is more possible to die in the unit of reference tree of fir.

According to the aforementioned analysis, the fir should be improved first in the spruce-fir-broadleaf mixed forest of our study, followed by maple, ash and spruce; thinning should highlight Crowding, with the help of Mingling and Uniform angle index; and, if required, Mingling and Uniform angle index should also guide the species and the position of replanting.

4.2. The Reasons for the Change of Stand Spatial Structure Parameters

According to the construction method of the SSSPs, the reasons for the change of the spatial structure at the individual level can be attributed to:

(a) The reference tree dies or recruitment, causing the spatial structure value change from existence to nothing or in the opposite;

(b) The death or recruitment of the neighbors of the reference tree induces structural unit rearrangement [47], which alters the size, species, relative location, and canopy overlap relationships between the reference tree and its neighbors;
(c) A qualitative change has occurred in the connection between the size and canopy overlap of neighboring trees and the reference tree.

All four SSSPs could be affected by (a) and (b), whereas Dominance and Crowding are also affected by (c). The main reason for the change of the spatial structure is (c) when there is no change in the composition of a structural unit formed by the reference tree and neighboring trees. Otherwise, the main reason is (b), and (c) plays a supporting role. In a developing forest with tree mortality and recruitment, it can be shown that the change of spatial structure is generally dominated by tree mortality and recruitment processes, whilst the tree growth process just fine-tunes the change of spatial structure. The research on natural spruce-fir-broadleaf mixed forest backs this up.

The change of mean values at the stand level can be explained by the change in the values of the SSSPs at the individual level, according to the above theory and our study data. The stability of SSSPs at the individual level may be determined using the ratio of Flow 0 in the flow of living trees of a natural spruce-fir-broadleaf mixed forest. The most stable parameter is Mingling, which is followed by Uniform angle index, Dominance, and lastly Crowding.

The Dominance and Crowding in living trees' flow are affected by (b) and (c), so the stabilities are not as high as Mingling and Uniform angle index stability, which are solely affected by (b). The Dominance is more stable than Crowding, which could be explained as the replacement of adjacent trees of the reference tree caused by (b) has an equivalent effect on both the Dominance and Crowding. However, the change in canopy overlap relationship caused by (c) is greater than the change in tree size connection between reference tree and its neighbors, so the effect on Crowding is greater. The Mingling is more stable than the Uniform angle index, which might be owing to the fact that reference trees account for up to 71.4 percent of highly mixed and completely mixed ($M = 0.75$ and 1), and the replacement of neighboring tree species induced by (b) did not modify the state of mixing intensity of the reference tree, whereas the replacement of adjacent trees could cause a significant change in the positional relation between reference tree and its neighbors, resulting in a change in the Uniform angle index.

The mean values of Crowding fluctuate a lot, the Mingling and Uniform angle index are moderate, and the Dominance is most stable. According to Deng and Katoh [31], Zhao et al. [32], Xue et al. [33] and Wan et al. [35], the variation of the Mingling was larger than that of the Uniform angle index, while Deng and Katoh [31] and Xue et al. [33] calculated the mean of the Dominance and found that the variation of the Mingling was also larger than Dominance and Uniform angle index is equivalent with Dominance at most of the time.

The Dominance is determined by comparing the size relationships of the reference tree and its neighbors [3]. Since each tree in the core area is not only a reference tree, but also an adjacent tree of other trees, the Dominance will be around 0.5 when aggregated to the mean value at the stand level [51]. Changes in growth, mortality or recruitment at the individual level could hardly affect the stable of Dominance. The fact that the mortality or recruitment process impacts the spatial connection between reference trees and adjacent trees at the individual level, but the effect is reduced when computing the mean value of the stand, which may explain why the mean of the Uniform angle index is reasonably constant. When aggregated to the stand average, however, changes in Mingling and Crowding at the individual tree level will not be moderated but will double the effect of promotion, making their stability worse than the Uniform angle index and Dominance. The relationship that the Crowding is less stable than the Mingling at the individual level is naturally extended to the stand level.

4.3. The Advantages of the New Method and the Limitations in This Study

Based on the principle of Sankey diagrams, it has strong advantages in displaying the flow among various types or groups in multi-period data [38]. Theoretically, these types or groups can source from any classification variable as long as there are some possible flows between them, like values of SSSPs in this research, trees' group size in a simulated fire study [52], or other attributes in individual level. Therefore, Sankey diagrams can not only show land use change at a large scale, but also have great advantages in displaying stand dynamics at a single tree scale. To further analyze the laws of the stand dynamic changes, each type of flow between two-period was named in this study. The results are greatly helpful to understand the formation mechanism of spatial structure. The analytical method for spatial structure dynamics used in this paper should be popularized in future studies for precise management of the forest structure [53].

We used field data from dynamic monitoring to examine changes in SSSPs at the individual and stand levels, and we attempted to explain the changes using the creative process of each parameter. Although the findings of the study illustrate the dynamic changes laws of the spatial structure of this forest community in this setting, various elements, such as stand type and beginning condition, can impact the changing laws of stand spatial structure [35], the presence of tree pests or diseases can also affect the outcomes of tree mortality if the same species are clustered [50]. The generality of the findings must be verified in other forest communities or simulated by computer programs.

5. Conclusions

Multi-period data can evaluate the dynamics of stand spatial structure, whereas previous researches solely focused on comparative assessments of the spatial structure of multi-period live trees, without an in-depth investigation of the change processes. This study attempts to disentangle the dynamics of spatial structure into the flow of live trees, the process of mortality and the process of recruitment at the individual level. The statistics of each "change type" in the three processes were analyzed. Our results show the proportion of changes in the SSSPs of individuals was relatively high, even though the mean values of the stand did not change considerably. The five values (0, 0.25, 0.5, 0.75, 1) of the SSSPs are in mutual flow, and the flows are typically one-step, with three-steps and four-steps changes being uncommon. The spatial structure of fir, maple, ash and spruce in the spruce-fir-broadleaf mixed forest of our study should be improved based on the Crowding, Mingling and Uniform angle index. Our findings also emphasize the impact of the mortality and recruitment processes on spatial structure. The time series can incorporate the spatial structure of a tree as a chain value (change type) rather than limiting it to a single value. These types are highly useful for investigating the changes and stability of SSSPs at the individual, species, and stand level, as well as guiding forest management in some ways. The dynamic change of spatial structure analysis method created in this study can capture more features not discovered in earlier approaches. Understanding the nuances of these changes is an important aspect of maintaining an acceptable spatial structure and biodiversity, and it should be the focus of future study efforts.

Author Contributions: Conceptualization, C.Z. and H.Z.; methodology, C.Z. and X.L.; software, C.Z. and L.F.; validation, D.L., K.C. and X.H.; formal analysis, D.L., Y.Z. and C.Z.; investigation, C.Z., K.C. and X.H.; resources, H.Z.; data curation, C.Z. and X.L.; writing—original draft preparation, C.Z. and D.L.; writing—review and editing, K.C., X.H., Y.Z. and H.Z.; visualization, C.Z. and L.F.; supervision, H.Z.; project administration, H.Z.; funding acquisition, H.Z. All authors have read and agreed to the published version of the manuscript.

Funding: This research was funded by the Thirteenth Five-year Plan Pioneering project of the High Technology Plan of the National Department of Technology (No.2017YFC0504101).

Institutional Review Board Statement: Not applicable.

Informed Consent Statement: Not applicable.

Data Availability Statement: Data sharing is not applicable.

Acknowledgments: We would like to thank all people contributing to sample plot survey for this study. Moreover, we would like to thank the anonymous reviewers and editor for their relevant suggestions on the manuscript.

Conflicts of Interest: The authors declare no conflict of interest.

Appendix A

Table A1. The proportion statistics of change types of main tree species in living trees' flow (%).

Parameters	Types	White Birch	Linden	Ribbed Birch	Korean Pine	Fir	Larch	Maple	Ash	Poplar	Elm	Spruce
Uniform Angle Index	Flow + 4	0.0	0.0	0.0	0.0	0.0	0.0	0.0	0.0	0.0	0.0	0.0
	Flow + 3	0.0	0.0	0.0	0.0	0.0	0.0	0.0	0.0	0.0	0.0	0.0
	Flow + 2	1.9	1.6	2.4	1.2	1.9	1.4	3.5	2.3	0.0	3.1	1.1
	Flow + 1	0.0	6.0	7.7	12.0	9.4	6.2	11.2	10.5	16.0	7.7	7.7
	Flow 0	81.5	84.0	79.0	78.4	78.3	86.3	72.9	80.2	64.0	76.9	81.3
	Flow − 1	13.0	6.3	8.6	7.2	8.2	5.3	8.8	5.8	12.0	12.3	9.9
	Flow − 2	3.7	2.2	2.4	1.2	2.2	0.8	3.5	1.2	8.0	0.0	0.0
	Flow − 3	0.0	0.0	0.0	0.0	0.0	0.0	0.0	0.0	0.0	0.0	0.0
	Flow − 4	0.0	0.0	0.0	0.0	0.0	0.0	0.0	0.0	0.0	0.0	0.0
Dominance	Flow + 4	0.0	0.0	0.0	0.0	0.0	0.0	0.0	0.0	0.0	0.0	0.0
	Flow + 3	0.0	0.0	0.0	0.0	0.0	0.0	0.0	0.0	0.0	0.0	0.0
	Flow + 2	0.0	0.8	0.6	0.6	0.3	0.8	0.6	0.0	0.0	0.0	1.1
	Flow + 1	5.6	4.9	8.0	8.4	5.7	5.9	9.4	5.8	0.0	4.6	6.6
	Flow 0	92.6	73.4	78.7	74.9	62.3	80.4	69.4	77.9	84.0	72.3	83.5
	Flow − 1	1.9	17.7	11.5	15.0	25.8	12.0	14.7	14.0	16.0	16.9	8.8
	Flow − 2	0.0	3.3	1.2	0.6	5.7	0.8	5.3	2.3	0.0	6.2	0.0
	Flow − 3	0.0	0.0	0.0	0.6	0.3	0.0	0.6	0.0	0.0	0.0	0.0
	Flow − 4	0.0	0.0	0.0	0.0	0.0	0.0	0.0	0.0	0.0	0.0	0.0
Mingling	Flow + 4	0.0	0.0	0.0	0.0	0.0	0.0	0.0	0.0	0.0	0.0	0.0
	Flow + 3	0.0	0.0	0.0	0.0	0.0	0.0	0.0	0.0	0.0	0.0	0.0
	Flow + 2	1.9	1.1	0.0	0.0	0.6	1.1	0.0	0.0	0.0	0.0	0.0
	Flow + 1	1.9	5.2	5.9	4.2	10.7	8.7	2.9	2.3	0.0	6.2	1.1
	Flow 0	96.3	87.8	90.8	92.2	76.1	87.1	88.8	94.2	100.0	90.8	96.7
	Flow − 1	0.0	5.4	3.3	3.6	10.1	2.8	8.2	3.5	0.0	3.1	2.2
	Flow − 2	0.0	0.5	0.0	0.0	2.2	0.3	0.0	0.0	0.0	0.0	0.0
	Flow − 3	0.0	0.0	0.0	0.0	0.3	0.0	0.0	0.0	0.0	0.0	0.0
	Flow − 4	0.0	0.0	0.0	0.0	0.0	0.0	0.0	0.0	0.0	0.0	0.0
Crowding	Flow + 4	0.0	0.0	0.0	0.6	0.0	2.5	0.0	0.0	0.0	0.0	0.0
	Flow + 3	0.0	0.3	0.3	1.8	0.3	7.8	0.6	1.2	0.0	1.5	1.1
	Flow + 2	5.6	2.2	3.8	3.6	3.8	16.8	6.5	9.3	0.0	9.2	5.5
	Flow + 1	16.7	6.5	10.9	12.0	12.6	22.4	11.2	16.3	8.0	13.8	9.9
	Flow 0	70.4	82.9	77.2	65.9	64.2	39.5	62.9	72.1	88.0	61.5	67.0
	Flow − 1	7.4	6.0	5.9	11.4	13.8	9.2	14.1	1.2	4.0	9.2	12.1
	Flow − 2	0.0	1.6	1.5	4.2	4.7	1.1	2.9	0.0	0.0	4.6	4.4
	Flow − 3	0.0	0.3	0.3	0.6	0.6	0.6	1.8	0.0	0.0	0.0	0.0
	Flow − 4	0.0	0.3	0.0	0.0	0.0	0.0	0.0	0.0	0.0	0.0	0.0
Number of Trees		54	368	338	167	318	357	170	86	25	65	91

Note: The proportions of each species are calculated by the sum of all three plots. Main species: White birch (*B. platyphylla*); Linden (*T. amurensis*); Ribbed birch (*B. costata*); Korean pine (*P. koraiensis*); Fir (*A. nephrolepis*); Larch (*L. olgensis*); Maple (*A. pictum* subsp. *mono*); Ash (*F. mandschurica*); Poplar (*P. ussuriensis*); Elm (*U. laciniata*); Spruce (*P. jezoensis*). The same is below.

Table A2. The proportion statistics of change types of main tree species in mortality process (%).

Parameters	Types	White Birch	Linden	Ribbed Birch	Korean Pine	Fir	Larch	Maple	Ash	Poplar	Elm	Spruce
Uniform Angle Index	Dead 0	25.0	0.0↓	0.0↓	0.0↓	0.0↓	5.3	0.0↓	0.0↓	0.0	0.0	0.0
	Dead − 1	50.0	30.8	12.5↓	15.4↓	15.7↓	21.1	0.0↓	40.0	0.0↓	25.0	33.3
	Dead − 2	25.0↓	61.5	75.0	61.5	58.8↓	36.8↓	66.7	60.0	50.0↓	66.7	46.7↓
	Dead − 3	0.0↓	7.7↓	6.3↓	7.7↓	17.6↓	26.3	0.0↓	0.0↓	50.0	8.3↓	13.3↓
	Dead − 4	0.0↓	0.0↓	6.3	15.4	7.8	10.5	33.3	0.0↓	0.0↓	0.0↓	6.7
Dominance	Dead 0	0.0↓	0.0↓	0.0↓	7.7↓	21.6	21.1↓	0.0↓	40.0	25.0↓	0.0↓	60.0
	Dead − 1	0.0↓	7.7↓	18.8↓	7.7↓	23.5	36.8	0.0↓	0.0↓	50.0	16.7↓	6.7↓
	Dead − 2	75.0	30.8	6.3↓	15.4↓	23.5	15.8↓	0.0↓	0.0↓	0.0↓	16.7↓	26.7
	Dead − 3	0.0↓	30.8	50.0	23.1	11.8↓	21.1	33.3	20.0	25.0	33.3	6.7↓
	Dead − 4	25.0	30.8	25.0	46.2	19.6↓	5.3↓	66.7	40.0	0.0↓	33.3↓	0.0↓
Mingling	Dead 0	0.0↓	7.7	6.3	0.0	0.0↓	10.5	0.0↓	0.0	0.0	0.0	0.0
	Dead − 1	0.0↓	23.1	0.0↓	0.0↓	5.9	21.1	0.0↓	0.0↓	0.0	0.0↓	0.0
	Dead − 2	25.0	23.1	25.0	15.4	33.3	21.1	0.0↓	20.0	0.0↓	41.7	0.0↓
	Dead − 3	25.0	38.5	37.5	38.5	37.3	21.1↓	66.7	40.0	0.0↓	16.7↓	20.0
	Dead − 4	50.0↓	7.7↓	31.3↓	46.2↓	23.5↓	26.3↓	33.3↓	40.0↓	100.0	41.7↓	80.0

Table A2. *Cont.*

Parameters	Types	White Birch	Linden	Ribbed Birch	Korean Pine	Fir	Larch	Maple	Ash	Poplar	Elm	Spruce
Crowding	Dead 0	0.0↓	0.0↓	6.3	0.0↓	3.9	5.3↓	0.0↓	0.0↓	0.0↓	0.0↓	0.0↓
	Dead − 1	25.0	0.0↓	12.5	7.7	2.0↓	31.6	0.0↓	0.0↓	0.0↓	8.3	6.7
	Dead − 2	0.0↓	7.7	31.3	7.7↓	13.7	36.8	0.0↓	40.0	0.0↓	8.3↓	6.7↓
	Dead − 3	25.0	7.7↓	12.5↓	23.1	21.6	0.0↓	33.3	20.0	50.0	16.7↓	40.0
	Dead − 4	50.0↓	84.6	37.5↓	61.5↓	58.8↓	26.3↓	66.7↓	40.0↓	50.0↓	66.7	46.7↓
Number of Trees		4	13	16	13	51	19	3	5	4	12	15

Note: "↓" indicated that the proportion of this change type is smaller than corresponding values in 2013.

Table A3. The proportion statistics of change types of main tree species in recruitment process (%).

Parameters	Types	White Birch	Linden	Ribbed Birch	Korean Pine	Fir	Larch	Maple	Ash	Poplar	Elm	Spruce
Uniform Angle Index	Reg 0	0.0	0.0	0.0	0.0	0.0	0.0	0.0	0.0	0.0	0.0	0.0
	Reg + 1	0.0	18.8↑	0.0	33.3↑	18.8↑	100.0↑	13.0	0.0	0.0	11.1	0.0
	Reg + 2	0.0	56.3	0.0	66.7↑	50.0	0.0	60.9↑	0.0	0.0	66.7↑	100.0↑
	Reg + 3	0.0	25.0↑	0.0	0.0	21.4↑	0.0	17.4	100.0↑	0.0	22.2↑	0.0
	Reg + 4	0.0	0.0	0.0	0.0	9.8↑	0.0	8.7	0.0	0.0	0.0	0.0
Dominance	Reg 0	0.0	0.0	0.0	0.0	0.0	0.0	0.0	0.0	0.0	0.0	0.0
	Reg + 1	0.0	0.0	0.0	0.0	0.9	0.0	0.0	0.0	0.0	0.0	0.0
	Reg + 2	0.0	6.3	0.0	0.0	10.7	0.0	0.0	0.0	0.0	0.0	0.0
	Reg + 3	0.0	25.0↑	0.0	11.1	24.1↑	0.0	17.4	0.0	0.0	33.3↑	0.0
	Reg + 4	0.0	68.8↑	0.0	88.9↑	64.3↑	100.0↑	82.6↑	100.0↑	0.0	66.7↑	100.0↑
Mingling	Reg 0	0.0	0.0	0.0	0.0	3.6↑	0.0	4.3	0.0	0.0	0.0	0.0
	Reg + 1	0.0	25.0↑	0.0	0.0	9.8↑	0.0	4.3	0.0	0.0	0.0	0.0
	Reg + 2	0.0	18.8	0.0	0.0	25.0↑	100.0↑	4.3	0.0	0.0	0.0	0.0
	Reg + 3	0.0	31.3	0.0	33.3↑	31.3	0.0	30.4↑	100.0↑	0.0	44.4↑	0.0
	Reg + 4	0.0	25.0	0.0	66.7↑	30.4	0.0	56.5↑	0.0	0.0	55.6	100.0↑
Crowding	Reg 0	0.0	0.0	0.0	0.0	2.7↑	0.0	0.0	0.0	0.0	0.0	0.0
	Reg + 1	0.0	0.0	0.0	0.0	8.9↑	0.0	13.0↑	0.0	0.0	0.0	33.3↑
	Reg + 2	0.0	12.5↑	0.0	0.0	19.6↑	0.0	8.7	100.0↑	0.0	0.0	0.0
	Reg + 3	0.0	12.5↑	0.0	44.4↑	19.6	100.0↑	30.4↑	0.0	0.0	22.2↑	0.0
	Reg + 4	0.0	75.0	0.0	55.6	49.1	0.0	47.8	0.0	0.0	77.8↑	66.7↑
Number of Trees		0	16	0	9	112	1	23	1	0	9	3

Note: "↑" indicated that the proportion of this change type is larger than corresponding values in 2018.

References

1. Legendre, P.; Mi, X.; Ren, H.; Ma, K.; Yu, M.; Sun, I.-F.; He, F. Partitioning beta diversity in a subtropical broad-leaved forest of China. *Ecology* **2009**, *90*, 663–674. [CrossRef] [PubMed]
2. Staudhammer, C.L.; LeMay, V.M. Introduction and evaluation of possible indices of stand structural diversity. *Can. J. For. Res.* **2001**, *31*, 1105–1115. [CrossRef]
3. Hui, G.; Zhang, G.; Zhao, Z.; Yang, A. Methods of Forest Structure Research: A Review. *Curr. For. Rep.* **2019**, *5*, 142–154. [CrossRef]
4. Gao, T.; Hedblom, M.; Emilsson, T.; Nielsen, A.B. The role of forest stand structure as biodiversity indicator. *For. Ecol. Manag.* **2014**, *330*, 82–93. [CrossRef]
5. Ali, A. Forest stand structure and functioning: Current knowledge and future challenges. *Ecol. Indic.* **2019**, *98*, 665–677. [CrossRef]
6. Gao, W.; Lei, X.; Gao, D.; Li, Y. Mass-ratio and complementarity effects simultaneously drive aboveground biomass in temperate Quercus forests through stand structure. *Ecol. Evol.* **2021**, *11*, 16806–16816. [CrossRef]
7. Lei, X.; Wang, W.; Peng, C. Relationships between stand growth and structural diversity in spruce-dominated forests in New Brunswick, Canada. *Can. J. For. Res.* **2009**, *39*, 1835–1847. [CrossRef]
8. Baran, J.; Pielech, R.; Kauzal, P.; Kukla, W.; Bodziarczyk, J. Influence of forest management on stand structure in ravine forests. *For. Ecol. Manag.* **2020**, *463*, 118018. [CrossRef]
9. Hui, G.; Lianjin, Z.; Hu, Y.; Wang, H.; Zhang, G. Stand crowding degree and its application. *J. Beijing For. Univ.* **2016**, *38*, 1–6. [CrossRef]
10. Pielou, E. Segregation and Symmetry in Two-Species Populations as Studied by Nearest- Neighbour Relationships. *J. Ecol.* **1961**, *49*, 255–269. [CrossRef]
11. Pommerening, A. Approaches to quantifying forest structures. *Forestry* **2002**, *75*, 305–324. [CrossRef]
12. Clark, P.J.; Evans, F.C. Distance to Nearest Neighbor as a Measure of Spatial Relationships in Populations. *Ecology* **1954**, *35*, 445–453. [CrossRef]
13. Pommerening, A.; Stoyan, D. Reconstructing spatial tree point patterns from nearest neighbour summary statistics measured in small subwindows. *Can. J. For. Res.* **2008**, *38*, 1110–1122. [CrossRef]
14. Pommerening, A.; Sánchez Meador, A. Tamm review: Tree interactions between myth and reality. *For. Ecol. Manag.* **2018**, *424*, 164–176. [CrossRef]

15. Li, Y.; Yang, H.; Wang, H.; Ye, S.; Liu, W. Assessing the influence of the minimum measured diameter on forest spatial patterns and nearest neighborhood relationships. *J. Mt. Sci.* **2019**, *16*, 2308–2319. [CrossRef]
16. Zhang, G.; Hui, G.; Yang, A.; Zhao, Z. A simple and effective approach to quantitatively characterize structural complexity. *Sci. Rep.* **2021**, *11*, 1326. [CrossRef] [PubMed]
17. Xu, J.; Zhang, G.; Zhao, Z.; Hu, Y.; Liu, W.; Yang, A.; Hui, G. Effects of Randomized Management on the Forest Distribution Patterns of Larix kaempferi Plantation in Xiaolongshan, Gansu Province, China. *Forests* **2021**, *12*, 981. [CrossRef]
18. Li, Y.; Ye, S.; Hui, G.; Hu, Y.; Zhao, Z. Spatial structure of timber harvested according to structure-based forest management. *For. Ecol. Manag.* **2014**, *322*, 106–116. [CrossRef]
19. Dong, L.; Bettinger, P.; Liu, Z. Optimizing neighborhood-based stand spatial structure: Four cases of boreal forests. *For. Ecol. Manag.* **2022**, *506*, 119965. [CrossRef]
20. Zhang, G.; Hui, G.; Hu, Y.; Zhao, Z.; Guan, X.; Gadow, K.V.; Zhang, G. Designing near-natural planting patterns for plantation forests in China. *For. Ecosyst.* **2019**, *6*, 28. [CrossRef]
21. Aguirre, O.; Hui, G.; Gadow, K.V.; Jiménez, J. An analysis of spatial forest structure using neighbourhood-based variables. *For. Ecol. Manag.* **2003**, *183*, 137–145. [CrossRef]
22. Zhang, G.; Hui, G.; Zhang, G.; Zhao, Z.; Hu, Y. Telescope method for characterizing the spatial structure of a pine-oak mixed forest in the Xiaolong Mountains, China. *Scand. J. For. Res.* **2019**, *34*, 751–762. [CrossRef]
23. Pommerening, A. Evaluating structural indices by reversing forest structural analysis. *For. Ecol. Manag.* **2006**, *224*, 266–277. [CrossRef]
24. Li, Y.; Hui, G.; Zhao, Z.; Hu, Y. The bivariate distribution characteristics of spatial structure in natural Korean pine broad-leaved forest. *J. Veg. Sci.* **2012**, *23*, 1180–1190. [CrossRef]
25. Li, Y.; Hui, G.; Zhao, Z.; Hu, Y.; Ye, S. Spatial structural characteristics of three hardwood species in Korean pine broad-leaved forest—Validating the bivariate distribution of structural parameters from the point of tree population. *For. Ecol. Manag.* **2014**, *314*, 17–25. [CrossRef]
26. Chai, Z.; Sun, C.; Wang, D.; Liu, W.; Zhang, C. Spatial structure and dynamics of predominant populations in a virgin old-growth oak forest in the Qinling Mountains, China. *Scand. J. For. Res.* **2016**, *32*, 19–29. [CrossRef]
27. Li, Y.; Hui, G.; Wang, H.; Zhang, G.; Ye, S. Selection priority for harvested trees according to stand structural indices. *iForest Biogeosci. For.* **2017**, *10*, 561–566. [CrossRef]
28. Zhang, L.; Hui, G.; Hu, Y.; Zhao, Z. Spatial structural characteristics of forests dominated by Pinus tabulaeformis Carr. *PLoS ONE* **2018**, *13*, e0194710. [CrossRef]
29. Nguyen, H.H.; Hai; Erfanifard, Y.; Tran, B.; Petritan, A.M.; Hien Mai, T.; Petritan, I. Phylogenetic Community and Nearest Neighbor Structure of Disturbed Tropical Rain Forests Encroached by Streblus macrophyllus. *Forests* **2020**, *11*, 722. [CrossRef]
30. Pretzsch, H. Forest dynamics, growth, and yield: A review, analysis of the present state, and perspective. In *Forest Dynamics, Growth and Yield*; Springer: Berlin/Heidelberg, Germany, 2009; pp. 1–39.
31. Deng, S.; Katoh, M. Change of Spatial Structure Characteristics of the Forest in Oshiba Forest Park in 10 years. *J. For. Plan.* **2011**, *17*, 9–19. [CrossRef]
32. Zhao, Z.; Liu, W.; Shi, X.; Li, A.; Guo, X.; Zhang, G.; Hui, G. Structure Dynamic of Quercus aliena var. acuteserrata Natural Forest on Xiaolongshan. *For. Res.* **2015**, *28*, 759–766. [CrossRef]
33. Xue, W.; Guo, Q.; Ai, X.; Huang, Y.; Li, W.; Luo, X. Study on the Dynamic Changes of Main Tree Species Composition and Stand Spatial Structure of Natural Forest in Southwest Hube. *Acta Bot. Boreali-Occident. Sin.* **2021**, *41*, 1051–1061. [CrossRef]
34. Zhang, L.; Feng, H.; Du, M.; Wang, Y.; Lai, G.; Guo, J. Dynamic Effects of Structure-Based Forest Management on Stand Spatial Structure in a Platycladus orientalis Plantation. *Forests* **2022**, *13*, 852. [CrossRef]
35. Wan, P.; Zhang, G.; Wang, H.; Zhao, Z.; Hu, Y.; Zhang, G.; Hui, G.; Liu, W. Impacts of different forest management methods on the stand spatial structure of a natural Quercus aliena var. acuteserrata forest in Xiaolongshan, China. *Ecol. Inform.* **2019**, *50*, 86–94. [CrossRef]
36. Gadow, K.v.; Zhang, C.Y.; Wehenkel, C.; Pommerening, A.; Corral-Rivas, J.; Korol, M.; Myklush, S.; Hui, G.Y.; Kiviste, A.; Zhao, X.H. Forest structure and diversity. In *Continuous Cover Forestry*; Springer: Berlin/Heidelberg, Germany, 2012; pp. 29–83.
37. Schmidt, M. The Sankey diagram in energy and material flow management: Part II: Methodology and current applications. *J. Ind. Ecol.* **2008**, *12*, 173–185. [CrossRef]
38. Cuba, N. Research note: Sankey diagrams for visualizing land cover dynamics. *Landsc. Urban Plan.* **2015**, *139*, 163–167. [CrossRef]
39. Bachmaier, J.; Effenberger, M.; Gronauer, A. Greenhouse gas balance and resource demand of biogas plants in agriculture. *Eng. Life Sci.* **2010**, *10*, 560–569. [CrossRef]
40. Schnitzer, H.; Brunner, C.; Gwehenberger, G. Minimizing greenhouse gas emissions through the application of solar thermal energy in industrial processes. *J. Clean. Prod.* **2007**, *15*, 1271–1286. [CrossRef]
41. Curmi, E.; Fenner, R.; Richards, K.; Allwood, J.M.; Bajželj, B.; Kopec, G.M. Visualising a stochastic model of Californian water resources using Sankey diagrams. *Water Resour. Manag.* **2013**, *27*, 3035–3050. [CrossRef]
42. Perez, G.; Comiso, J.; Aragones, L.; Merida, H.C.; Ong, P. Reforestation and Deforestation in Northern Luzon, Philippines: Critical Issues as Observed from Space. *Forests* **2020**, *11*, 1071. [CrossRef]
43. R Core Team. *R: A Language and Environment for Statistical Computing*; Version 4.1.1; R Foundation for Statistical Computing: Vienna, Austria, 2021.

44. Brunson, J.C. Ggalluvial: Layered grammar for alluvial plots. *J. Open Source Softw.* **2020**, *5*, 2017. [CrossRef]
45. Wickham, H. *ggplot2: Elegant Graphics for Data Analysis*; Springer: New York, NY, USA, 2016.
46. Hui, G.; Gadow, K.v.; Albert, M. The neighbourhood pattern—A new structure parameter for describing distribution of forest tree position. *Sci. Silvae Sin.* **1999**, *35*, 37–42. [CrossRef]
47. Zhang, G.; Hui, G. Random Trees Are the Cornerstones of Natural Forests. *Forests* **2021**, *12*, 1046. [CrossRef]
48. Hallinger, M.; Johansson, V.; Schmalholz, M.; Sjöberg, S.; Ranius, T. Factors driving tree mortality in retained forest fragments. *For. Ecol. Manag.* **2016**, *368*, 163–172. [CrossRef]
49. Kharuk, V.I.; Im, S.T.; Petrov, I.A.; Dvinskaya, M.L.; Fedotova, E.V.; Ranson, K.J. Fir decline and mortality in the southern Siberian Mountains. *Reg. Environ. Change* **2017**, *17*, 803–812. [CrossRef]
50. Hui, G.; Hu, Y. Measuring species spatial isolation in mixed forests. *For. Res.* **2001**, *14*, 23–27. [CrossRef]
51. Zhao, Z.; Hui, G.; Hu, Y.; Li, Y.; Wang, H. Method and application of stand spatial advantage degree based on the neighborhood comparison. *J. Beijing For. Univ.* **2014**, *36*, 78–82. [CrossRef]
52. Ziegler, J.; Hoffman, C.; Collins, B.; Knapp, E.; Mell, W. Pyric tree spatial patterning interactions in historical and contemporary mixed conifer forests, California, USA. *Ecol. Evol.* **2020**, *11*, 820–834. [CrossRef] [PubMed]
53. Zhang, H.; Lei, X.; Zhang, C.; Zhao, X.; Hu, X. Research on theory and technology of forest quality evaluation and precision improvement. *J. Beijing For. Univ.* **2019**, *41*, 1–18. [CrossRef]

forests

MDPI

Article

Community Assembly of Forest Vegetation along Compound Habitat Gradients across Different Climatic Regions in China

Liangjin Yao [1,2,3], Yue Xu [2,3], Chuping Wu [1], Fuying Deng [4], Lan Yao [5], Xunru Ai [5] and Runguo Zang [2,3,*]

1 Zhejiang Academy of Forestry, Hangzhou 310023, China
2 Key Laboratory of Forest Ecology and Environment of National Forestry and Grassland Administration, Ecology and Nature Conservation Institute, Chinese Academy of Forestry, Beijing 100091, China
3 Co-Innovation Center for Sustainable Forestry in Southern China, Nanjing Forestry University, Nanjing 210037, China
4 College of Resources Environment and Earth Science, Yunnan University, Kunming 650091, China
5 School of Forestry and Horticulture, Hubei University for Nationalities, Enshi 445000, China
* Correspondence: zangrung@caf.ac.cn

Abstract: Community assembly research has mostly focused on areas with single vegetation types; however, the abiotic and biotic factors affecting community assembly act across regions. Integrating biotic and abiotic factors into "compound" habitats has gained attention as an emerging strategy to analyze spatial and temporal patterns of biodiversity. We used a compound habitat approach to explore the relative roles of habitat filtering, biotic competition, and stochastic processes in the forest community assembly of four climatic zones (tropical, subtropical, temperate, and cold temperate forests). Specifically, we combined biotic and abiotic factors in four compound ecological gradients by principal component analysis (PCA), which we used to assess the geographic and phylogenetic distribution of multiple woody plant functional traits. We found that forest functional and phylogenetic diversity shifted from clustered to overdispersed along the first compound habitat gradient (PC1) across climate zones. This finding indicates that competitive exclusion strongly affected the community assembly in tropical and subtropical forests, while habitat filtering played a key role in cold temperate forests; these mechanisms may both exist and interact in temperate forests. We also found that both habitat filtering and biotic competition affected forest community assembly across climatic regions in China. Our results elucidate the underlying mechanisms driving geographical differentiation of forest vegetation across climatic zones, and bolster empirical evidence for the conservation of forest biodiversity in China. Further research is also needed to explore whether the patterns found in this paper are prevalent in different locations in different climatic zones in China.

Keywords: phylogenetic signal; community assembly; habitat filtering; intra- and interspecific interactions

Citation: Yao, L.; Xu, Y.; Wu, C.; Deng, F.; Yao, L.; Ai, X.; Zang, R. Community Assembly of Forest Vegetation along Compound Habitat Gradients across Different Climatic Regions in China. *Forests* **2022**, *13*, 1593. https://doi.org/10.3390/f13101593

Academic Editor: Cate Macinnis-Ng

Received: 19 July 2022
Accepted: 21 September 2022
Published: 29 September 2022

1. Introduction

A major focus of ecology is to understand community assembly and maintenance across geographic scales [1]. Two main theories have been used to explain community assembly: the deterministic processes of niche theory and the stochastic processes of neutral theory [2,3]. Niche theory holds that each species has unique space and resource demands in an ecosystem; therefore, niche differentiation and resource allocation allow stable species coexistence in a community, with habitat filtering and interspecific competition shaping community composition [3–5]. The neutral theory posits that all species in a community have equal ecological opportunities, and emphasizes the importance of stochastic processes such as colonization opportunities, genetic variation, random extinction, and ecological drift [2]. Both theories are needed to describe community assembly, as niche and neutral processes are not mutually exclusive, and similar communities may arise under various combinations of underlying mechanisms [2]. Some studies suggest that competition and diffusion affect community assembly simultaneously [6–8], and that forest community assembly at a regional scale is a continuous dynamic process from niche- to neutral-driven. Therefore, the integration of niche and neutral theories, and their compound implications for community assembly, may continue to drive ecological research [9–11].

Combining ecological and evolutionary information has facilitated the understanding of biodiversity patterns and species assembly processes [6–9,11]. Specifically, the incorporation of a phylogenetic signal—a measurement of the tendency for closely related species to resemble each other more than they resemble species drawn at random—in functional traits and ecological factors has become a popular approach [12]. The phylogenetic signal (or lack thereof) in functional traits can be used to assess the degree to which traits are evolutionarily conserved or driven by habitat factors [13,14]. However, ecological and phylogenetic similarities often are not correlated, and therefore, phylogenetic signals alone should not be used to conclude ecological phenomena, such as niche conservatism [12]. Null-model-based approaches using functional and/or phylogenetic diversity have been used to quantify the relative importance of community assembly processes, including neutral effects, habitat filtering, and biotic competition [7,8,15,16]. Recent studies have linked patterns of phylogenetic clustering (i.e., co-occurring species more closely related than expected) to habitat filtering and phylogenetic divergence (i.e., sympatric species less closely related than expected) to competitive exclusion [7,14,17]. Mori et al. [18] compared the observed patterns of functional diversity of soil faunal communities with patterns expected from a given regional species pool, and found that species in harsh environments (e.g., little rain and high altitude) showed trait convergence due to habitat filtering, while the strong interaction between species resulted in different functional traits and different strategies to obtain resources and growth space in superior habitats. Both habitat filtering and intra- and interspecific interactions can result in convergent or divergent community functional and phylogenetic diversity structures [19–24].

Functional and phylogenetic diversity in forest communities with different environmental factors and disturbance levels can jointly explain community assembly mechanisms [7]. Ryo et al. [25] found a shift from phylogenetic overdispersion to clustering with the increase in slope and decrease in soil depth in a forest community; they interpreted this as evidence for non-random community assembly [7,23]. Plant diversity along environmental gradients is determined by both abiotic factors and biotic interactions, such as competition, at the local scale [26]. Swenson [7] demonstrated the importance of biotic interactions on phylogenetic or functional aggregation [16,27,28]. Community assembly research has mostly focused on coexistence and community assembly mechanisms along habitat gradients in forests of different climatic zones [29–32]. However, studies across climatic zones or at regional scales generally focus on latitudinal gradients associated with changes in climate, soil, and anthropogenic disturbances [33].

In this study, we bridge this gap in scales of community ecology to understand how abiotic and biotic factors influence community assembly at regional scales. We combined forest survey and plant functional trait data from different climatic regions in China to quantify functional and phylogenetic diversity across broad geographic and habitat gradients, and to assess the relative importance of niche and neutral processes in community assembly. We focused on the following questions: (1) Is there a phylogenetic signal in the functional traits of forest vegetation in China? (2) What is the geographical pattern of functional and phylogenetic diversity of forest vegetation in China? (3) What roles do habitat filtering and species interactions play in forest community assembly across climatic regions in China?

2. Materials and Methods

2.1. Study Site

We investigated seven forest communities of four typical forest types in different climatic regions of China (Table 1), including three tropical rainforests (Jianfengling, Bawangling, Xishuangbanna), two subtropical evergreen deciduous broad-leaved mixed forests

(Mulinzi, Xingdoushan), one warm temperate mixed coniferous and broad-leaved forest (Xiaolongshan), and one temperate coniferous forest (Kanasi) (Table 1). All sites were well-preserved old-growth forests that had not been disturbed by human activity for at least 100 years.

Table 1. Geographic, sampling, biodiversity, climate, and tree size information of forest plots across four climatic zones in China.

Sites		Jianfengling (JFL)	Bawangling (BWL)	Xishuangbanna (XSBN)	Mulinzi (MLZ)	Xingdoushan (XDS)	Xiaolongshan (XLS)	Kanasi (KNS)
Climatic zone		Tropical			Subtropical		Temperate	Cold temperate
Number of plots		50	50	40	44	25	50	50
number of species		247	236	208	130	94	111	7
Elevation		726–937	865–1023	715–864	1368–1652	1521–1627	1853–1867	1624–1810
Latitude		18.71–18.80	19.02–19.16	21.58–21.63	29.89–30.10	30.00–30.07	34.29–34.43	48.69–48.70
Longitude		108.84–108.92	109.07–109.17	101.56–101.60	109.22–110.12	100.21–110.20	106.08–106.10	86.94–86.95
pH	Soil pH	4.81 ± 0.29	4.37 ± 0.35	4.91 ± 0.56	4.41 ± 0.25	4.49 ± 0.31	5.70 ± 0.53	5.54 ± 0.20
SOM	Soil organic matter (g/kg)	15.62 ± 5.62	6.39 ± 1.83	3.19 ± 0.76	8.40 ± 2.05	9.21 ± 2.20	9.69 ± 2.92	3.38 ± 1.03
TN	Soil total N content (g/kg)	1.32 ± 0.43	2.11 ± 0.81	1.34 ± 0.44	5.90 ± 1.72	6.54 ± 1.51	3.58 ± 1.17	1.06 ± 0.34
TP	Soil total P content (g/kg)	0.14 ± 0.08	0.33 ± 0.19	0.31 ± 0.07	0.50 ± 0.16	0.68 ± 0.33	0.38 ± 0.11	0.64 ± 0.13
AN	Soil available K (mg/kg)	183.22 ± 54.10	193.00 ± 46.70	156.69 ± 27.93	299.38 ± 73.7	355.99 ± 102.1	266.03 ± 82.37	64.29 ± 19.53
AP	Soil available N (mg/kg)	192.14 ± 136.72	179.90 ± 123.20	84.49 ± 28.69	34.59 ± 11.99	23.25 ± 12.47	60.37 ± 27.78	37.50 ± 16.61
AK	Soil available P (mg/kg)	125.54 ± 53.66	150.18 ± 43.78	130.60 ± 58.25	142.84 ± 34.04	134.00 ± 33.78	186.19 ± 53.12	227.84 ± 58.83
MAT	Annual mean temperature (°C)	205.18 ± 5.42	211.52 ± 15.59	217.40 ± 3.88	96.52 ± 54.39	95.20 ± 34.91	81.18 ± 2.79	−9.90 ± 2.83
MAP	Mean annual precipitation (mm)	1633.00 ± 49.81	1579.08 ± 145.0	1584.43 ± 27.00	1213.91 ± 363.3	13,854.00 ± 301.50	671.60 ± 5.16	316.34 ± 4.77
CVPS	Precipitation seasonality (coefficient of variation) (%)	79.80 ± 0.45	80.44 ± 0.50	72.93 ± 0.47	71.91 ± 23.45	63.44 ± 18.53	78.62 ± 0.49	56.28 ± 0.81
DBH	Diameter at breast height (DBH, cm)	6.34 ± 1.38	5.62 ± 0.84	6.98 ± 2.94	11.06 ± 3.79	10.68 ± 1.21	7.03 ± 1.91	15.39 ± 2.87
TH	Tree height (m)	18.43 ± 4.65	25.90 ± 5.34	37.25 ± 14.14	20.00 ± 3.19	18.46 ± 3.14	15.72 ± 1.58	23.76 ± 1.8
BA	Basal area (≥1 cm DBH) (m^2/ha)	34.93 ± 11.94	45.77 ± 14.90	51.75 ± 20.03	35.14 ± 10.09	28.88 ± 9.53	33.86 ± 6.60	46.84 ± 10.66
CVDBH	Coefficient of variation of DBH (%)	125.55 ± 24.39	158.26 ± 23.22	164.40 ± 51.23	87.15 ± 28.56	54.74 ± 12.23	110.48 ± 20.22	88.43 ± 15.24
CVH	Coefficient of variation of tree height (%)	59.91 ± 12.14	79.96 ± 8.36	95.37 ± 28.94	50.65 ± 15.42	32.69 ± 9.33	62.70 ± 11.10	67.10 ± 9.03

2.2. *Vegetation Survey and Functional Trait Sampling*

At each of the seven forest communities, we selected 25–50 plots, each with an area of 20 m × 20 m, for a total of 309 dynamic monitoring plots (Table 1). In each old-growth forest type, we ensured that the distance among each of the 20 m × 20 m plot was more than 100 m. Thus, each of the plots in each forest type were randomly distributed, and they had no spatial autocorrelation. In addition, we conducted an autocorrelation analysis for each plots in each of the four forest types, the results of which showed that there were no spatial autocorrelations among the plots within each forest type (the distribution of sample sites is shown in Figure A1). In an initial vegetation survey, we identified all individual woody plants with a diameter at breast height (DBH, cm) of ≥1 cm. Species names, DBH, tree height (TH, m), and relative coordinates of individual trees in the plot were recorded. We measured six key functional traits that represent major axes of resource use and allocation strategies [34]: specific leaf area (SLA, mm^2/mg), leaf dry matter content (LDMC, mm^2/mg^2), leaf nitrogen content (LNC, mg/g), leaf phosphorus content (LPC, mg/g), wood density (WD, g/cm^3), and leaf nitrogen/phosphorus ratio (N:P, %). We sampled at least 10 individuals of each species for functional traits; for species with more than 10 individuals, samples were taken from 10 random trees. Five healthy and intact mature sun leaves were collected from every individual, weighed fresh, and then dried in an oven at 60 °C for 72 h. SLA was calculated as the ratio of fresh leaf area to leaf dry mass, and LDMC was the leaf dry mass divided by leaf fresh mass. The collected leaves were brought back to the laboratory to measure LNC and LPC. To avoid harming trees, WD was calculated based on measurements of branches with diameters between 1 cm and 2 cm, rather than from tree cores. The bark was removed from branches before measuring the fresh volume, and branches were dried in an oven at 105 °C for 72 h to measure the dry mass. WD was calculated as the ratio of dry mass to fresh branch volume. The same methods were used to collect and measure functional traits of forest communities in all climatic regions.

2.3. Data Collection of Biotic and Abiotic Factors

We measured five tree size metrics to represent overall productivity and canopy density in each plot: average DBH (DBH), average tree height (TH), basal area (BA), coefficient of variation of DBH (CVDBH), and coefficient of variation of tree height (CVTH).

The longitude and latitude coordinates were recorded for each plot during the survey, and were used to extract the 19 WorldClim2 bioclimatic variables for each plot using climate raster layers with a resolution of 30″ (https://www.worldclim.org/, founded on 24 June 2005). Removing highly correlated predictor variables can increase model performance; therefore, we calculated pairwise Pearson correlation coefficients for each pair of variables. We removed those variables with a high load (correlation greater than 0.8) for two or more factors. Finally, three climatic variables (MAP, MAT, and CVPS) were selected. These climatic factors have been shown to significantly influence plant diversity and help verify ecological hypotheses in other ecosystems [4].

In each plot, measurements were taken at the center point of the quadrat and along two diagonals 14.14 m away from the center (0–20 cm depth). Subsequently, we thoroughly mixed the samples to create one bulk sample per plot for analysis. The samples were analyzed for pH, soil organic matter (SOM), total nitrogen (TN), total phosphorus (TP), available nitrogen (AN), available phosphorus (AP), and available potassium (AK) (Table 1).

2.4. Construction of a Compound Habitat Gradient of Biotic and Abiotic Factors

Principal component analysis (PCA) was performed on the 15 climatic, soil, and biotic factors to reduce the redundancy of the variables using the "factoextra" package in the software R [35]. We took the first four principal components (PC1–PC4) as compound habitat gradients, which explained 73.1% of the variation of the 15 variables (Figure 1). The first axis (PCA1) explained 31.8% of the variation, and climate variables (MAP and MAT) and CVDBH were significantly positively correlated with PCA1, while DBH was significantly negatively correlated with PCA1. The second axis (PCA2) explained 22.0% of the variation, and the biotic variables (CVH, BA, and TH) were significantly positively correlated with PCA2, while TN and AN were significantly negatively correlated with PCA2. The overall contributions of the third (PCA3) and fourth (PCA4) axes to the explanatory variation of all variables were relatively low, and mainly related to biotic and soil variables (Figure 1).

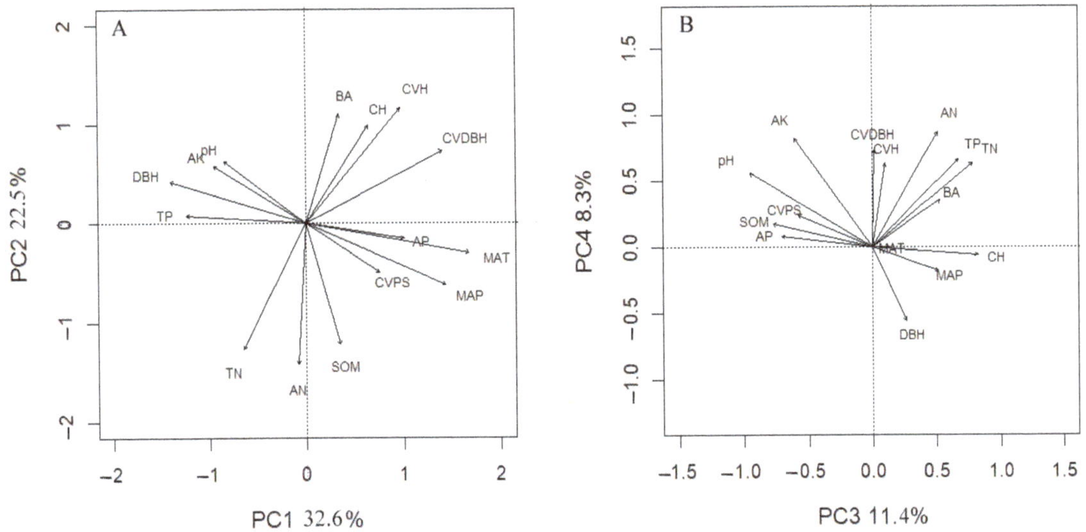

Figure 1. Loadings of biotic and abiotic factors of all forest plots on the first four principal components. The variable abbreviations are listed in Table 1. Note: (**A**) PC1 and PC2 of the compound habitat gradient axis; (**B**) PC3 and PC4 of the compound habitat gradient axis.

2.5. Phylogenetic Signal in Functional Traits

We identified a total of 46,280 woody plants (788 species) in our survey. We used the AWK version of Phylomatic to generate a phylogenetic tree for all species. To reduce redundant information and excessive fitting, we used the first three principal components as comprehensive functional trait factors. Then the trait matrix was transformed into a distance matrix by calculating Euclidean distance. Finally, hierarchical aggregation was used to construct trait trees.

We used Blomberg's K [36] to assess the phylogenetic signal. Values of K close to 1 indicate that species' traits are distributed as expected under a model of Brownian evolution and $K > 1$ implies strong phylogenetic conservatism; K values near 0 imply that traits are phylogenetically independent.

2.6. Phylogenetic Diversity of Communities

The net relatedness index (NRI) describes the phylogenetic clustering or dispersion in a community by quantifying the phylogenetic distance between all species in the community. The nearest taxon index (NTI) describes the degree of relatedness between the most similar species by quantifying the mean phylogenetic distance between pairs of the closest species in the community [19]. NTI and NRI values are calculated based on null models to assess whether communities are more clustered or dispersed than expected. NTI and NRI values greater than 0 indicate that species within a community are more closely related than expected, which provides evidence of community assembly by habitat filtering; negative values indicate that species are less closely related than expected, and suggest community assembly by competitive exclusion [23].

2.7. Functional Diversity Structure of Communities

We first tested for group differences in forest vegetation across different climatic regions in China. We then used all species from the 309 plots as the species pool in null models of functional diversity of a random community to assess the assembly processes of the observed communities. Specifically, we calculated the standardized effect size (SES) of functional richness (FRic) and functional dispersion (FDis) to identify habitat filtering and interactions between species as follows:

$$\text{SES.FRic} = \frac{FRic_{obs} - FRic_{null}}{SD(FRic_{null})} \tag{1}$$

$$\text{SES.FDis} = \frac{FDis_{obs} - FDis_{null}}{SD(FDis_{null})} \tag{2}$$

where $FRic_{obs}$ and $FDis_{obs}$ are the observed FRis and FDis values of the communities, respectively; $FRic_{null}$ and $FDis_{null}$ are the mean FRis and FDis values calculated from 999 random communities, respectively; and $SD(FRic_{null})$ and $SD(FDis_{null})$ are the standard deviations of $FRic_{null}$ and $FDis_{null}$, respectively. We interpreted the resulting values as follows: SES.FRic values significantly less than 0 indicate that the observed community functional trait space is smaller than that of random communities, implying the niche processes of habitat filtering and interspecies interactions; SES.FDis values significantly greater than 0 indicate neutral processes of community assembly; and if there is no significant difference between ses.FDis and 0, community assembly was dominated by stochastic processes [37].

2.8. Statistical Testing and Data Analysis

We used linear regression to test the variation of functional and phylogenetic diversity with compound habitat gradients and chi-square tests to analyze the correlation between functional and phylogenetic diversity structure. All statistical analyses were performed in R 3.2.5 [35].

3. Results

3.1. Phylogenetic Signal in Functional Traits

Tests of the phylogenetic signal revealed that four functional traits (SLA, LDMC, LNC, and WD) had K values significantly ($p < 0.05$) less than 1 for all species across the 309 forest vegetation plots (Tables 2 and A1).

Table 2. Phylogenetic signal, as measured by Blomberg's K, in functional traits of forest vegetation across climatic regions in China. * $p < 0.05$; ** $p < 0.01$; NS, non-significant.

	SLA	LDMC	LNC	LPC	N:P	WD
K	0.09 **	0.07 **	0.08 **	0.04 (NS)	0.04 (NS)	0.05 *

3.2. Functional Diversity along Compound Habitat Gradients

Generally, SES.FRic and SES.FDis were significantly negatively correlated with PC1 and significantly positively correlated with PC2; however, values differed across forests after testing for group differences in forest vegetation across different climatic regions in China (Figure 1). In the cold temperate zone with low temperature and little rain, the SES.FRic and SES.FDis of the forest community was greater than 0, and the functional space of the community was larger than that of the random community. On the contrary, the SES.FRic and SES.FDis of tropical forest communities in China was less than 0, and the community structure was lower than stochastic (Figure 2). There were no significant relationships between SES.FRic or SES.FDis and PCA3 or PCA4 (Figure A2).

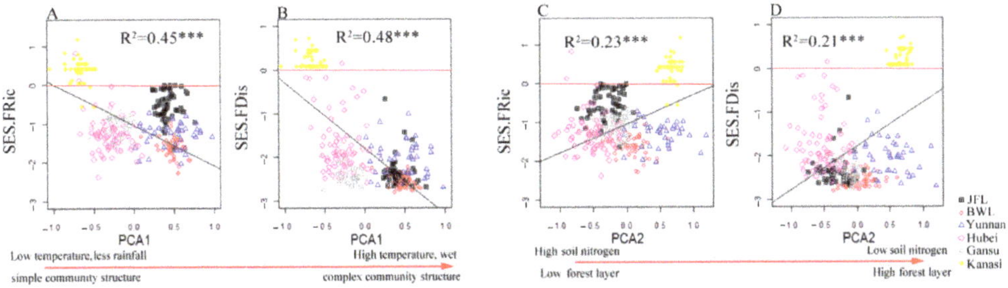

Figure 2. The relationships between functional diversity, SES.FRic (**A**,**C**) and SES.FDis (**B**,**D**), and compound habitat gradients, PC1 (**A**,**B**) and PC2 (**C**,**D**), of forest vegetation communities in China. Different color–marker combinations represent each forest plot. The abbreviations for each variable are listed in Table 1. Significance values: ***, $p < 0.001$.

3.3. Phylogenetic Diversity along Compound Habitat Gradients

Generally, phylogenetic diversity was significantly negatively correlated with the first two compound habitat gradients (PC1 and PC2; Figure 3). In both cold and warm temperate climate zones, the NRI and NTI of forest vegetation were greater than 0, and the community structure was higher than stochastic. In tropical forest communities, the NRI and NTI were less than 0, and the community structure was lower than stochastic (Figure 2). There were no significant relationships between NRI or NTI and PCA3 or PCA4 (Figure A2).

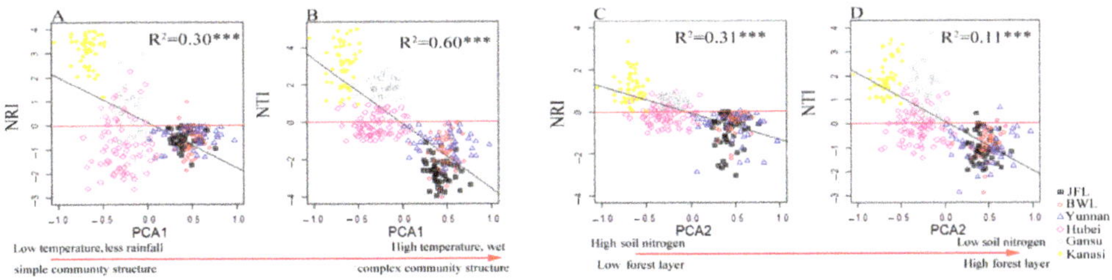

Figure 3. The relationships between phylogenetic diversity, NRI (**A**,**C**) and NTI (**B**,**D**), and compound habitat gradients, PC1 (**A**,**B**) and PC2 (**C**,**D**), of forest vegetation communities in China. Different color–marker combinations represent each forest plot. The abbreviations for each variable are listed in Table 1. Significance values: ***, $p < 0.001$.

4. Discussion

4.1. Phylogenetic Signal in Functional Traits

We found that K values of functional traits across climate regions of China were all much lower than 1, and therefore showed phylogenetic independence (Table A1). Therefore, in this study, phylogeny was not a good predictor of functional diversity. Four functional traits (SLA, LDMC, LPC, and WD) had significant phylogenetic signals ($p < 0.05$); however, the lack of signal in LNC and N:P suggests that these two functional traits may be more affected by factors such as habitat [38]. Community phylogenies tend to exhibit stronger correlations with functional traits at larger taxonomic scales [39]. Secondly, the phylogenetic relationship cannot completely reflect all the information on functional traits of species in the community, and is a necessary but not sufficient condition required to study the similarity of functional traits of species in the community. The phylogenetic relationship can only be used as one of the indirect indicators, and cannot be substituted for functional traits in the study [7]. Finally, the factors and conditions that form the random phylogenetic relationship in the community are compound and diverse; some traits may show random divergence, while other traits show a non-random conserved state in the whole community [7,14].

4.2. Climate Influences Forest Community Construction across Climatic Regions

Habitat filtering, biotic competition, and interspecific interactions are three major ecological processes driving community assembly [40,41]. In this study, we show that functional diversity tended to converge along a compound habitat gradient composed of climate, soil, and biotic factors across multiple climatic regions of China (Figure 3). Our results provide evidence that niche processes (i.e., habitat filtering and interspecific interaction) had greater effects than neutral processes on the maintenance of forest vegetation diversity across climatic regions of China. This may be mainly because the species in these communities are not functionally equivalent, and therefore, random processes may play a role at the regional scale [42]. We found that the functional diversity of cold temperate forest vegetation was higher than that of random communities, and that the functional diversity of tropical and subtropical forest vegetation tended to be divergent. The high degree of functional dispersion in tropical forest communities confirms that interspecific interactions lead to higher functional differences [19]. In the cold temperate forest community, habitat filtering dominated the community assembly process due to the influence of extreme climates, such as low temperature and little rain, which improved species similarity and reduced the range of functional traits [4,19,43].

We also found that cold temperate forest vegetation tended to be phylogenetically clustered, while tropical and subtropical forest vegetation was overdispersed. Phylogenetic clustering may result from habitat filtering, which we interpret as niche conservatism. For example, closely related species may possess similar adaptions to environmentally challenging conditions [34]. Phylogenetic overdispersion in subtropical and tropical forest communities may indicate competitive exclusion of species with different survival strategies [5]. Competition for habitat resources, such as light, soil nutrients, and water, increases with species richness [44]. The limited similarity between species may lead to niche divergence and phylogenetic overdispersion. Therefore, it is possible that habitat filtering and biotic interactions (e.g., competition and exclusion) combined to affect phylogenetic patterns of trait values in forest vegetation in China across different climatic regions.

4.3. Biotic Interactions and Soil Characteristics Affect Community Construction

We found that interspecific interactions gradually intensified from cold temperate to tropical forest communities, which may be due to the differentiation of functional traits in species during long-term evolution and phenotypic plasticity [34]. Functional trait divergence can reduce niche overlap, therefore allowing species to be more evenly distributed along resource axes [19] and improving resource utilization [42]. Our results show that the overall similarity of functional traits among all species in cold temperate forest communities was higher, which may be because the community structure of the cold temperate forest is single and the dominant species are obvious, while the functional traits in the community are determined by the species with greater abundance.

In contrast, the overall similarity of functional traits of all tropical forest species was higher, which may result from local habitat filtering and biotic competition [44,45]. In tropical and subtropical forest communities, limited similarity plays a more important role in increasing variability in community functional traits [34]. We confirmed that habitat modification only acted on the functional traits or relatedness of the species rather than the species itself [4,19,42].

In our study, compound habitat gradients acted as the abiotic and biotic filters, jointly influencing the phylogenetic diversity of forest communities. We found that niche processes based on habitat filtering and biotic competition played important roles in the formation of forest functional diversity across climatic regions in China, and in determining the mechanism of community assembly [29,46]. Furthermore, whether the patterns found in this paper are prevalent in different locations in different climatic zones in China will be the focus of future research.

5. Conclusions

We found that phylogenetic and function diversity shifted from clustered to overdispersed temperate to tropical forest communities. Our results suggest that habitat filtering had greater effects in temperate forests, while competition had greater effects in tropical forests. Phylogenetic signal in functional traits was generally low across climatic regions. We further uncovered an interaction between habitat selection and biotic competition across climatic regions in China. Our results support the joint effects of habitat filtering and competition in community assemblies, and we often found more than one mechanism influencing community structure and function. In future studies, we need more plots and more accurate environmental data to verify whether the patterns found in this paper are prevalent in different locations in different climatic regions of China.

Author Contributions: R.Z. and C.W. designed this study and improved the English language and grammatical editing. L.Y. (Liangjin Yao) wrote the first draft of the manuscript and performed the data analysis. L.Y. (Lan Yao), F.D. and Y.X. did the fieldwork. X.A. gave guidance and methodological advice. All the coauthors contributed to the discussion, revision, and improvement of the manuscript. All authors have read and agreed to the published version of the manuscript.

Funding: This work was supported by the Fundamental Research Funds for the Central Non-profit Research Institution of the Chinese Academy of Forestry (CAFYBB2020ZA002-2), the National Natural Science Foundation of China (41671047, 41771059), and the Zhejiang Provincial Scientific Research Institute Special Project (2022F1068-2).

Institutional Review Board Statement: Not applicable.

Informed Consent Statement: Not applicable.

Data Availability Statement: The data presented in this study are available in the article.

Acknowledgments: We would like to thank Elizabeth Tokarz at Yale University for her assistance with the English language and grammatical editing of the manuscript, and the support of all staff of Zhejiang Hangzhou Urban Forest Ecosystem Research Station.

Conflicts of Interest: This is the first submission of this manuscript, and no parts of this manuscript are being considered for publication elsewhere. All authors have read and approved the content of the manuscript. No financial, contractual, or other interest conflicts exist for the study.

Appendix A Appendix

Table A1. The phylogenetic signal of functional traits in forest communities occupying different climatic zones in China.

Indexes	Jianfengling		Bawangling		Xishuangbanna		Mulinzi		Xingdoushan		Xiaolongshan		Kanasi	
	K	P	K	P	K	P	K	P	K	P	K	P	K	P
SLA	0.19	0.029	0.21	0.001	0.24	0.149	0.008	0.87	0.006	0.64	0.42	0.001	1.38	0.022
LDMC	0.2	0.001	0.25	0.001	0.25	0.069	0.04	0.696	0.047	0.66	0.27	0.001	0.68	0.035
LNC	0.07	0.25	0.15	0.095	0.17	0.329	0.02	0.816	0.033	0.78	0.19	0.011	0.89	0.049
LPC	0.09	0.054	0.14	0.069	0.18	0.263	0.04	0.286	0.038	0.32	0.12	0.079	1.01	0.082
N:P	0.05	0.786	0.12	0.426	0.18	0.27	0.008	0.906	0.008	0.71	0.17	0.07	0.57	0.686
WD	0.06	0.545	0.11	0.358	0.23	0.157	0.03	0.745	0.041	0.65	0.27	0.001	0.59	0.034

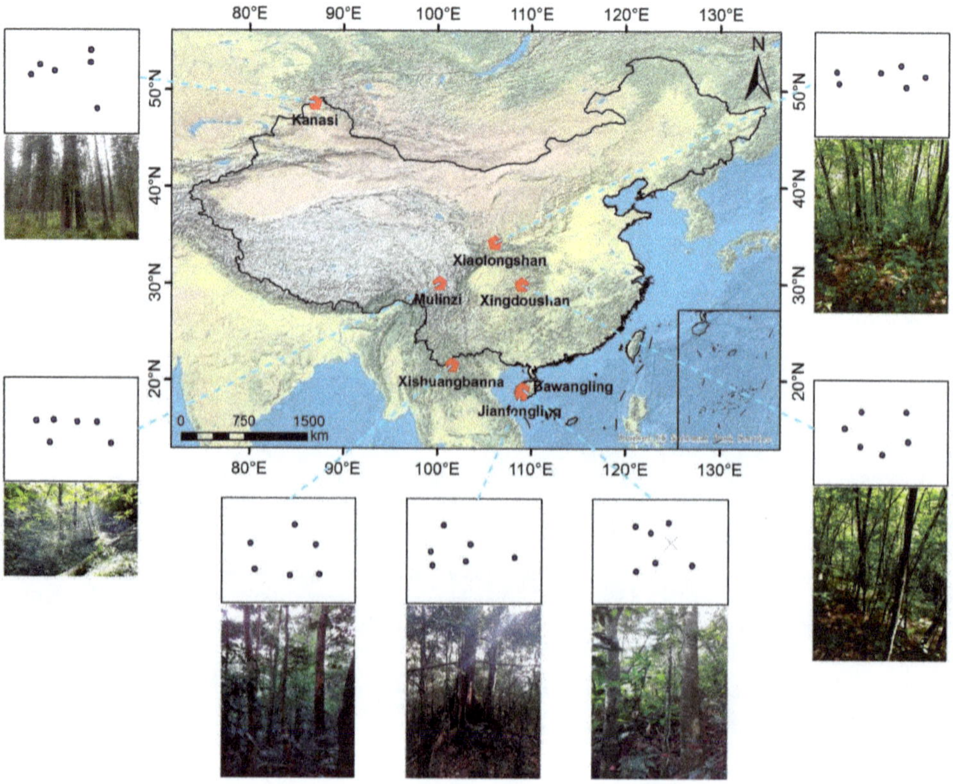

Figure A1. Geographical distribution of the 309 plots. Plots were taken from seven forest sites in China. At each site, plots with an area of 20 m × 20 m were randomly established.

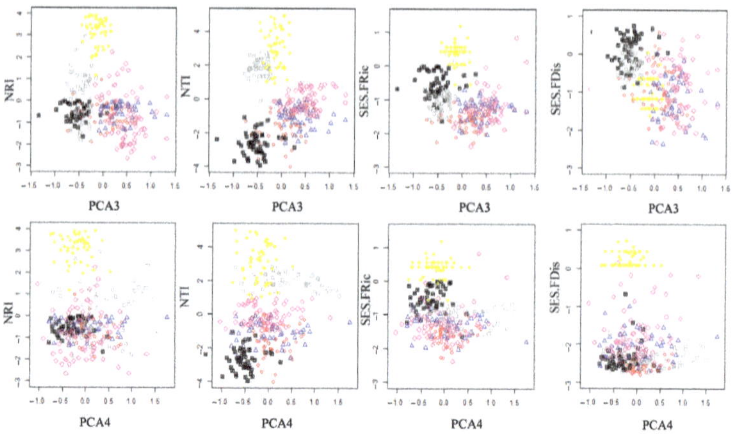

Figure A2. The relationships between functional and phylogenetic diversity and compound habitat gradients PC3 and PC4. Different color–marker combinations represent each forest plot. The abbreviations for each variable are listed in Table 1.

References

1. Willig, M.R.; Kaufman, D.M.; Stevens, R.D. Latitudinal Gradients of Biodiversity: Pattern, Process, Scale, and Synthesis. *Annu. Rev. Ecol. Evol. Syst.* **2003**, *34*, 273–309. [CrossRef]
2. Hubbell, S. Neutral Theory and the Evolution of Ecological Equivalence. *Ecology* **2006**, *87*, 1387–1398. [CrossRef]
3. Chase, J.M.; Myers, J.A. Disentangling the Importance of Ecological Niches from Stochastic Processes across Scales. *Philos. Trans. R. Soc. B Biol. Sci.* **2011**, *366*, 2351–2363. [CrossRef] [PubMed]
4. Yao, L.; Ding, Y.; Xu, H.; Deng, F.; Yao, L.; Ai, X.; Zang, R. Patterns of Diversity Change for Forest Vegetation across Different Climatic Regions—A Compound Habitat Gradient Analysis Approach. *Glob. Ecol. Conserv.* **2020**, *23*, e01106. [CrossRef]
5. Qian, H.; Field, R.; Zhang, J.-L.; Zhang, J.; Chen, S. Phylogenetic Structure and Ecological and Evolutionary Determinants of Species Richness for Angiosperm Trees in Forest Communities in China. *J. Biogeogr.* **2016**, *43*, 603–615. [CrossRef]
6. Gravel, D.; Canham, C.D.; Beaudet, M.; Messier, C. Reconciling Niche and Neutrality: The Continuum Hypothesis. *Ecol. Lett.* **2006**, *9*, 399–409. [CrossRef]
7. Swenson, N.G. The Assembly of Tropical Tree Communities—The Advances and Shortcomings of Phylogenetic and Functional Trait Analyses. *Ecography* **2013**, *36*, 264–276. [CrossRef]
8. Dong, L.; Liang, C.; Li, F.Y.; Zhao, L.; Ma, W.; Wang, L.; Wen, L.; Zheng, Y.; Li, Z.; Zhao, C.; et al. Community Phylogenetic Structure of Grasslands and Its Relationship with Environmental Factors on the Mongolian Plateau. *J. Arid. Land* **2019**, *11*, 595–607. [CrossRef]
9. Alonso, D.; Etienne, R.; Mckane, A. The Merits of Neutral Theory. *Trends Ecol. Evol.* **2006**, *21*, 451–457. [CrossRef]
10. Wennekes, P.L.; Rosindell, J.; Etienne, R.S. The Neutral—Niche Debate: A Philosophical Perspective. *Acta Biotheor.* **2012**, *60*, 257–271. [CrossRef]
11. Adler, P.B.; HilleRisLambers, J.; Levine, J.M. A Niche for Neutrality. *Ecol. Lett.* **2007**, *10*, 95–104. [CrossRef]
12. Losos, J.B. Phylogenetic Niche Conservatism, Phylogenetic Signal and the Relationship between Phylogenetic Relatedness and Ecological Similarity among Species. *Ecol. Lett.* **2008**, *11*, 995–1003. [CrossRef]
13. Freckleton, R.P.; Cooper, N.; Jetz, W. Comparative Methods as a Statistical Fix: The Dangers of Ignoring an Evolutionary Model. *Am. Nat.* **2011**, *178*, E10–E17. [CrossRef]
14. Carvalho, F.; Brown, K.A.; Waller, M.P.; Razafindratsima, O.H.; Boom, A. Changes in Functional, Phylogenetic and Taxonomic Diversities of Lowland Fens under Different Vegetation and Disturbance Levels. *Plant Ecol.* **2020**, *221*, 441–457. [CrossRef]
15. Legendre, P.; Mi, X.; Ren, H.; Ma, K.; Yu, M.; Sun, I.-F.; He, F. Partitioning Beta Diversity in a Subtropical Broad-leaved Forest of China. *Ecology* **2009**, *90*, 663–674. [CrossRef]
16. Yang, J.; Ci, X.; Lu, M.; Zhang, G.; Cao, M.; Li, J.; Lin, L. Functional Traits of Tree Species with Phylogenetic Signal Co-Vary with Environmental Niches in Two Large Forest Dynamics Plots. *J. Plant Ecol.* **2014**, *7*, 115–125. [CrossRef]
17. Dini-Andreote, F.; Stegen, J.C.; van Elsas, J.D.; Salles, J.F. Disentangling Mechanisms That Mediate the Balance between Stochastic and Deterministic Processes in Microbial Succession. *Proc. Natl. Acad. Sci. USA* **2015**, *112*, E1326–E1332. [CrossRef]
18. Mori, A.S.; Ota, A.T.; Fujii, S.; Seino, T.; Kabeya, D.; Okamoto, T.; Ito, M.T.; Kaneko, N.; Hasegawa, M. Biotic Homogenization and Differentiation of Soil Faunal Communities in the Production Forest Landscape: Taxonomic and Functional Perspectives. *Oecologia* **2015**, *177*, 533–544. [CrossRef]
19. Kraft, N.J.B.; Ackerly, D.D. Functional Trait and Phylogenetic Tests of Community Assembly across Spatial Scales in an Amazonian Forest. *Ecol. Monogr.* **2010**, *80*, 401–422. [CrossRef]
20. HilleRisLambers, J.; Adler, P.B.; Harpole, W.S.; Levine, J.M.; Mayfield, M.M. Rethinking Community Assembly through the Lens of Coexistence Theory. *Annu. Rev. Ecol. Evol. Syst.* **2012**, *43*, 227–248. [CrossRef]
21. Cardinale, B.J. Towards a General Theory of Biodiversity for the Anthropocene. *Elem. Sci. Anthr.* **2013**, *1*, 000014. [CrossRef]
22. Cavender-Bares, J.; Kozak, K.H.; Fine, P.V.A.; Kembel, S.W. The Merging of Community Ecology and Phylogenetic Biology. *Ecol. Lett.* **2009**, *12*, 693–715. [CrossRef] [PubMed]
23. Mayfield, M.M.; Bonser, S.P.; Morgan, J.W.; Aubin, I.; McNamara, S.; Vesk, P.A. What Does Species Richness Tell Us about Functional Trait Diversity? Predictions and Evidence for Responses of Species and Functional Trait Diversity to Land-Use Change. *Glob. Ecol. Biogeogr.* **2010**, *19*, 423–431. [CrossRef]
24. Kunstler, G.; Falster, D.; Coomes, D.A.; Hui, F.; Kooyman, R.M.; Laughlin, D.C.; Poorter, L.; Vanderwel, M.; Vieilledent, G.; Wright, S.J.; et al. Plant Functional Traits Have Globally Consistent Effects on Competition. *Nature* **2016**, *529*, 204–207. [CrossRef]
25. Kitagawa, R.; Mimura, M.; Mori, A.S.; Sakai, A. Topographic Patterns in the Phylogenetic Structure of Temperate Forests on Steep Mountainous Terrain. *AoB Plants* **2015**, *7*, plv134. [CrossRef]
26. Roslin, T.; Hardwick, B.; Novotny, V.; Petry, W.K.; Andrew, N.R.; Asmus, A.; Barrio, I.C.; Basset, Y.; Boesing, A.L.; Bonebrake, T.C.; et al. Higher Predation Risk for Insect Prey at Low Latitudes and Elevations. *Science* **2017**, *356*, 742–744. [CrossRef]
27. Kress, W.J.; Erickson, D.L.; Jones, F.A.; Swenson, N.G.; Perez, R.; Sanjur, O.; Bermingham, E. Plant DNA Barcodes and a Community Phylogeny of a Tropical Forest Dynamics Plot in Panama. *Proc. Natl. Acad. Sci. USA* **2009**, *106*, 18621–18626. [CrossRef]
28. Liu, X.; Swenson, N.G.; Zhang, J.; Ma, K. The Environment and Space, Not Phylogeny, Determine Trait Dispersion in a Subtropical Forest. *Funct. Ecol.* **2013**, *27*, 264–272. [CrossRef]
29. Jeffers, E.S.; Bonsall, M.B.; Froyd, C.A.; Brooks, S.J.; Willis, K.J. The Relative Importance of Biotic and Abiotic Processes for Structuring Plant Communities through Time. *J. Ecol.* **2015**, *103*, 459–472. [CrossRef]

30. Li, S.; Cadotte, M.W.; Meiners, S.J.; Hua, Z.; Jiang, L.; Shu, W. Species Colonisation, Not Competitive Exclusion, Drives Community Overdispersion over Long-Term Succession. *Ecol. Lett.* **2015**, *18*, 964–973. [CrossRef]
31. Ricklefs, R.E.; He, F. Region Effects Influence Local Tree Species Diversity. *Proc. Natl. Acad. Sci. USA* **2016**, *113*, 674–679. [CrossRef] [PubMed]
32. Poorter, L.; van der Sande, M.T.; Arets, E.J.M.M.; Ascarrunz, N.; Enquist, B.J.; Finegan, B.; Licona, J.C.; Martínez-Ramos, M.; Mazzei, L.; Meave, J.A.; et al. Biodiversity and Climate Determine the Functioning of Neotropical Forests. *Glob. Ecol. Biogeogr.* **2017**, *26*, 1423–1434. [CrossRef]
33. Chu, C.; Lutz, J.A.; Král, K.; Vrška, T.; Yin, X.; Myers, J.A.; Abiem, I.; Alonso, A.; Bourg, N.; Burslem, D.F.R.P.; et al. Direct and Indirect Effects of Climate on Richness Drive the Latitudinal Diversity Gradient in Forest Trees. *Ecol. Lett.* **2019**, *22*, 245–255. [CrossRef]
34. Yao, L.; Ding, Y.; Yao, L.; Ai, X.; Zang, R. Trait Gradient Analysis for Evergreen and Deciduous Species in a Subtropical Forest. *Forests* **2020**, *11*, 364. [CrossRef]
35. R Core Team. R: A Language and Environment for Statistical Computing. Computing 2011. Available online: https://www.r-project.org/ (accessed on 19 September 2022).
36. Bolmberg, S.P.; Garland, T.; Ives, A.R. Testing for Phylogenetic Signal in Comparative Data: Behavioral Traits Are More Labile. *Evolution* **2003**, *57*, 717–745. [CrossRef]
37. Cornwell, W.K.; Schwilk, D.W.; Ackerly, D.D. A Trait-Based Test for Habitat Filtering: Convex Hull Volume. *Ecology* **2006**, *87*, 1465–1471. [CrossRef]
38. Swenson, N.G.; Erickson, D.L.; Mi, X.; Bourg, N.A.; Forero-Montaña, J.; Ge, X.; Howe, R.; Lake, J.K.; Liu, X.; Ma, K.; et al. Phylogenetic and Functional Alpha and Beta Diversity in Temperate and Tropical Tree Communities. *Ecology* **2012**, *93*, S112–S125. [CrossRef]
39. Amaral, E.J.; Franco, A.C.; Rivera, V.L.; Munhoz, C.B.R. Environment, Phylogeny, and Photosynthetic Pathway as Determinants of Leaf Traits in Savanna and Forest Graminoid Species in Central Brazil. *Oecologia* **2021**, *197*, 1–11. [CrossRef]
40. Scherrer, D.; Mod, H.K.; Pottier, J.; Litsios-Dubuis, A.; Pellissier, L.; Vittoz, P.; Götzenberger, L.; Zobel, M.; Guisan, A. Disentangling the Processes Driving Plant Assemblages in Mountain Grasslands across Spatial Scales and Environmental Gradients. *J. Ecol.* **2019**, *107*, 265–278. [CrossRef]
41. Fang, S.; Cadotte, M.W.; Yuan, Z.; Lin, F.; Ye, J.; Hao, Z.; Wang, X. Intraspecific Trait Variation Improves the Detection of Deterministic Community Assembly Processes in Early Successional Forests, but Not in Late Successional Forests. *J. Plant Ecol.* **2019**, *12*, 593–602. [CrossRef]
42. Milardi, M.; Gavioli, A.; Soana, E.; Lanzoni, M.; Fano, E.A.; Castaldelli, G. The Role of Species Introduction in Modifying the Functional Diversity of Native Communities. *Sci. Total Environ.* **2020**, *699*, 134364. [CrossRef] [PubMed]
43. Mouchet, M.A.; Villéger, S.; Mason, N.W.H.; Mouillot, D. Functional Diversity Measures: An Overview of Their Redundancy and Their Ability to Discriminate Community Assembly Rules. *Funct. Ecol.* **2010**, *24*, 867–876. [CrossRef]
44. Orrock, J.L.; Dutra, H.P.; Marquis, R.J.; Barber, N. Apparent Competition and Native Consumers Exacerbate the Strong Competitive Effect of an Exotic Plant Species. *Ecology* **2015**, *96*, 1052–1061. [CrossRef] [PubMed]
45. Pakeman, R.J. Functional Diversity Indices Reveal the Impacts of Land Use Intensification on Plant Community Assembly. *J. Ecol.* **2011**, *99*, 1143–1151. [CrossRef]
46. Gerhold, P.; Cahill, J.F.; Winter, M.; Bartish, I.V.; Prinzing, A. Phylogenetic Patterns Are Not Proxies of Community Assembly Mechanisms (They Are Far Better). *Funct. Ecol.* **2015**, *29*, 600–614. [CrossRef]

forests

MDPI

Article

Variations in Functional Richness and Assembly Mechanisms of the Subtropical Evergreen Broadleaved Forest Communities along Geographical and Environmental Gradients

Caishuang Huang [1,2], Yue Xu [1,2] and Runguo Zang [1,2,*]

1 Key Laboratory of Forest Ecology and Environment of National Forestry and Grassland Administration, Institute of Forest Ecology, Environment and Nature Conservation, Chinese Academy of Forestry, Beijing 100091, China; huangcaishuang@caf.ac.cn (C.H.); xuyue@caf.ac.cn (Y.X.)
2 Co-Innovation Center for Sustainable Forestry in Southern China, Nanjing Forestry University, Nanjing 210037, China
* Correspondence: zangrung@caf.ac.cn; Tel.: +86-10-62889546

Abstract: Linking functional trait space and environmental conditions can help to understand how species fill the functional trait space when species increase along environmental gradients. Here, we examined the variations in functional richness (FRic) and their correlations with key environmental variables in forest communities along latitudinal, longitudinal, and elevational gradients, by measuring seven functional traits of woody plants in 250 forest plots of 0.04 ha across five locations in the subtropical evergreen broadleaved forests (SEBLF) of China. On this basis, we explored whether environmental filtering constrained the functional volume by using a null model approach. Results showed that FRic decreased with increasing elevation and latitude, while it increased with increasing longitude, mirroring the geographical gradients in species richness. FRic was significantly related to precipitation of driest quarter, soil pH, and total phosphorus. Negative SES.FRic was prevalent (83.2% of the communities) in most SEBLF communities and was negatively related to mean diurnal range. Our study suggested that the geographical variation in the functional space occupied by SEBLF communities was affected mainly by climate and soil conditions. The results of the null model revealed that niche packing was dominant in SEBLF communities, highlighting the importance of environmental filtering in defining functional volume within SEBLF communities.

Keywords: biogeography; subtropical evergreen broadleaved forest; functional richness; environmental variables; community assembly

Citation: Huang, C.; Xu, Y.; Zang, R. Variations in Functional Richness and Assembly Mechanisms of the Subtropical Evergreen Broadleaved Forest Communities along Geographical and Environmental Gradients. *Forests* **2022**, *13*, 1206. https://doi.org/10.3390/f13081206

Academic Editor: Alejandro A. Royo

Received: 7 July 2022
Accepted: 28 July 2022
Published: 31 July 2022

1. Introduction

Exploring and understanding spatial patterns of biodiversity is central to understanding mechanisms of species coexistence and community assembly processes in natural ecosystems [1,2]. Geographical gradients in species diversity are of widespread concern, and species richness in most plants and animals is generally explained by declining from the tropics to the poles and from low to high elevations [3,4]. Despite the generality of these patterns, it has been recognized that considering species diversity alone (e.g., species richness or taxonomic diversity) is not sufficient to understand the underlying processes that influence communities along geographical gradients [5–7]. Species richness does not explain fully the community structure as the differences among species in their evolutionary history and ecological roles are ignored [6,8]. Thus, recent research argued that the study of biodiversity theories should move beyond species diversity, especially focusing more explicitly on the functional aspects of diversity [5]. Functional diversity can not only mirror species richness along geographical gradients [9] but also complement species richness in explaining species coexistence and ecological processes [10].

Functional trait values of species within communities are a reflection of the ability of species to respond to their environment [11]. Therefore, the location occupied by a species within a multidimensional trait space can also shift across geographical and environmental gradients [12]. Climate is one of the most important drivers shaping the functional structure of forest ecosystems [13]. Specifically, climate harshness and seasonality were proven to be key factors controlling the variation in functional volume in subtropical forests [14]. Researchers predicted that environments that are more abiotically benign will allow the invasion and success of peripheral phenotypes, resulting in a large morphological volume, whereas harsher environments will limit the invasion and success of peripheral phenotypes, resulting in a small morphological volume [15]. For instance, abundant and evenly distributed precipitation and stable temperatures allow communities to contain a large number of species, while species that occur in those regions of drought and temperature instability are strongly limited [14,16]. Soil fertility is another important factor shaping the functional diversity of plant ecosystems [17]. Changes in soil resource availability can directly affect the size of the niche space occupied by species [18].

Studies argued that changes in the functional trait space along an environmental gradient in terms of species richness could be associated with the patterns of functional space occupancy [10]. As species richness increases, communities would change either by tighter packing of species into the niche space (niche packing) or by occupying unexploited portions of the niche space (niche expansion) to accommodate more species in a given environmental condition [19,20]. Niche space characterizes the phenotypic space occupied by a set of species, which can be quantified by assessing the multidimensional trait space [21]. Comparing changes in the occupancy of the multidimensional trait space across communities can allow inferences regarding the association of niche expansion and packing with environments [12]. To test the two patterns of functional trait space, it is necessary to use a functional diversity index that can directly measure the volume and occupancy pattern of the niche space by species and quantify the diversity in trait combinations in the research of biodiversity and community assembly [22]. Functional richness (FRic) is measured as the volume of the minimum convex hull that includes all the species of the community and represents the amount of functional space filled by the community, which has been recognized as the best candidate for this research [23]. It has been widely used in many recent papers to explore the patterns of niche space in different biological groups [12,19,24]. In addition, previous studies also showed that FRic was useful for tracking diversity variations along environmental gradients [25]. Therefore, FRic has reasonable power to detect the changes in the occupancy of the multidimensional trait space under different environmental conditions and reveal how species pack and fill the trait space [5].

Niche packing and expansion are not mutually exclusive and may occur simultaneously [12]. However, the relative importance of the two patterns remains unknown [19]. Quantifying the volume, packing, and expansion of functional trait space and their relative importance enables us to understand the ecological processes structuring functional diversity and ecological strategies [5,26]. In the absence of niche-based processes, the trait composition of a local assemblage is predicted to be a random subset of a shared regional species pool [5,12]. Therefore, comparing the observed value of FRic to a random expectation—that is, computing a standardized effect size (SES.FRic) [15]—allows us to understand whether niche-based processes occur in a given location [12]. Environmental filtering theory predicts that although FRic can also increase with species richness in more stressful environments, FRic will be smaller than the null expectation (negative SES.FRic), revealed by niche packing [5,27]. Conversely, competitive exclusion theory predicts that new species are most likely to fill an expanded variety of niche space; thus, FRic should always be larger than a sampling expectation (positive SES.FRic), revealed by niche expansion [5,28]. Determining which pattern of functional space occupancy is dominant in a given location helps us to understand the relative importance of environmental filtering and biotic competition. Substantial evidence showed that environmental filtering, driven by a set of environmental variables, is more important than biotic competition in shaping plant diversity [29,30]. Therefore, we expected that the pattern of niche packing would be dominant within the forest communities.

The subtropical evergreen broadleaved forest (SEBLF) is one of the main vegetation types around the world and is widely distributed in China [31,32]. The subtropical region in China hosts unique and rich biodiversity and encompasses different environmental gradients in topography and climate, providing a natural setting for studying patterns of biodiversity change along geographical gradients. While trait–environment relationships have been already explored in some separate sites of subtropical forest ecosystems [33], few studies have been carried out in detecting the patterns of functional trait space (niche packing/expansion) along geographical and environmental gradients. Here, we used a large trait and spatial distributional data set of SEBLF communities to describe the geographical patterns of functional richness, species richness, and their correlations with key environmental factors in the subtropical region of China. Specifically, to test whether niche packing is dominant in the SEBLF of China, we used a null model to test whether the environmental filtering constrained the functional trait space, thereby increasing the packing of species within the SEBLF communities.

2. Materials and Methods

2.1. Study Area

This study was conducted in the 250 permanent forest plots with an area of 0.04 ha (20 m × 20 m) (latitude: 27.58°–30.18° N, longitude: 102.95°–120.00° E; elevation: 200–1948 m) across five locations (Figure S1) in the natural old-growth subtropical evergreen broadleaved forests of China (SEBLF). Locations were selected at random within the study areas and represented an unbiased, representative sample of natural old-growth vegetation. The study areas have a subtropical monsoon climate, with a mean annual temperature of 15.0 °C and mean precipitation of 1391.1 mm. The plots have a rough terrain, with the slope ranging from 4° to 42°. The floristic composition of these plots is characteristic of evergreen broadleaved forests, with *Symplocos anomala* (Symplocaceae), *Camellia fraternal* (Theaceae), *Loropetalum chinense* (Hamamelidaceae), *Eurya muricata* (Theaceae), *Symplocos lucida* (Symplocaceae), and *Cyclobalanopsis glauca* (Fagaceae) being the dominant canopy species. All plots were established and investigated according to the standard of the Center for Tropical Forest Science (CTFS) [34] during the summer of 2018 and 2019. For woody plant species, all individuals with a diameter at breast height ≥1 cm were tagged, mapped, and identified to species level with the help of local botanists. The abundances of species were determined by calculating the number of individuals of a certain species in each plot. In total, 292 woody plant species (48,680 individuals) belonging to 59 families and 134 genera were collected.

2.2. Trait Data

We selected seven functional traits that represent the variation in plant form and function. These traits included specific leaf area (SLA, $cm^2\ g^{-1}$), leaf nitrogen concentration (LNC, $g\ kg^{-1}$), leaf phosphorus concentration (LPC, $g\ kg^{-1}$), wood density (WD, $g\ cm^{-3}$), leaf dry matter content (LDMC, $g\ g^{-1}$), maximum plant height (Hmax, m), and seed mass (SM, g). Selected traits correlate well with ecological characteristics and are good predictors of resource utilization, tolerance to drought, and mechanical damage [29]. Except for SM, all functional traits were measured according to standardized measurement protocols [35]. SM was compiled from multiple sources and databases, including FOC (http://www.efloras.org/ (accessed on 20 June 2021)), Seeds of Woody Plants in China [36], the TRY Plant Trait Database (https://www.try-db.org/de/ (accessed on 20 June 2021)) from the KEW Seed Information Database [37] and PLANTS Database [38], and the BIEN trait database (accessed using the BIEN package [39] in R). In each 20 m × 20 m plot, ten individuals of every species were sampled. For those species with less than ten individuals, we added additional individuals of the same species from surrounding areas. A detailed description of the measurement methods for these functional traits can be found in our previous works [40]. The mean functional trait values for each species were applied to all individuals.

2.3. Functional and Species Richness Calculation

Finding a suitable measure of functional space to appropriately quantify the diversity in trait combinations is challenging [22]. Here, we considered functional richness (FRic) as the best candidate measure for our research purposes [23]. FRic measures the extent of the functional volume of an assemblage as the smallest possible multidimensional convex hull volume that contains all species in an assemblage [22,41]. We quantified FRic using the package 'FD' in the program R [42]. To compare the geographical variation between FRic and species diversity and explore the patterns of FRic related to species richness, we also quantified species richness (SR). Species richness in each plot was estimated as the number of species.

There are numerical dependencies between functional and species richness. Functional richness is expected to be positively correlated with species richness. We thus accounted for such variations by computing a null model in which species identities (and therefore trait values) were shuffled 999 times while maintaining species richness and occupancy within each assemblage, and we recalculated the functional richness for each assemblage for each iteration. This generated a null distribution of 999 values for functional richness in each assemblage [15,24]. The observed values and this null distribution were used to calculate a standardized effect size for functional richness (SES.FRic) = [Observed FRic–mean expected FRic]/SD expected FRic. The SES.FRic shows how many SDs the observed FRic is above or below the mean expected value from the null models [15,43]. The direction of SES.FRic (higher or lower than the expected null) was correlated with different deterministic processes of community assembly. Negative SES.FRic indicates that the observed functional volume is less than expected by chance, resulting from environmental filtering. Conversely, positive SES.FRic indicates that the observed functional volume is larger than expected given the species richness, resulting from biotic competition [15]. If we predicted that the environmental filtering led to the dominance of niche packing in SEBLF communities, this could be supported by the following two observations. First, the proportions of communities with negative SES.FRic would be higher than those with positive SES.FRic. Second, and more importantly, differences between the observed and simulated FRic, measured as the mean of the associated SES.FRic across all communities, would be significantly lower than the expectation of zero when using a two-tailed Student's *t*-test [44,45]. Additionally, the magnitude of SES.FRic was interpreted as the strength of the signal of deterministic processes on the assemblages [15]. More negative SES values indicate stronger environmental filtering [46,47].

2.4. Environmental Variables

To estimate the influences of environmental variables on community-level functional richness, we collected 19 climatic and 7 edaphic variables to represent environmental conditions. A detailed description of the data collection and measurement methods for these environmental variables can be found in our previous works [40]. We performed correlations to assess possible multicollinearity and variable redundancy among the environmental variables. Pearson's correlation values ($R < 0.8$) between explanatory variables were used as a cutoff criterion to retain variables that were more relevant to the FRic (Table 1). Final models were run with precipitation of driest quarter (PDQ), precipitation of wettest quarter (PWQ), mean diurnal range (MDR), mean temperature of driest quarter (TDQ), soil pH (SpH), and soil total phosphorus (STP).

Table 1. Correlations between environmental variables. Significance levels: *, $p < 0.05$; **, $p < 0.01$; ***, $p < 0.001$.

Variables	MDR	TDQ	PDQ	PWQ	STP
TDQ	−0.42 ***				
PDQ	0.22 ***	0.70 ***			
PWQ	0.26 ***	0.25 ***	0.53 ***		
STP	0.16 *	−0.72 ***	−0.70 ***	−0.18 **	
SpH	0.04	0.15 *	0.19 **	0.57 ***	0.11

Abbreviations: mean temperature of driest quarter (TDQ), precipitation of driest quarter (PDQ), precipitation of wettest quarter (PWQ), mean diurnal range (MDR), soil pH (SpH), soil total phosphorus (STP).

2.5. Data Analysis

To assess geographical variation in functional volume and species richness, linear mixed models (LMMs) were used to test shifts in FRic, SES.FRic, and SR along latitude, longitude, and elevation. We also used the LMMs to test the effects of climate and soil on FRic, SES.FRic, and SR. LMMs were conducted using FRic, SES.FRic, and SR as the response variables, and the three geographical variables and the six climatic and edaphic variables as fixed predictors. The site was included as a random effect, given that we could not exclude the potential role of unmeasured, spatially autocorrelated environmental factors in our analysis. Linear mixed models were built using the "lme4" package [48] as implemented in the R statistical software [42]. The proportion of variance explained by the LMMs was assessed using the marginal R^2m (variance explained by the fixed predictors) and the conditional R^2c (variance explained by both the fixed and random effects) [49]. The direction and magnitude of selected predictors were assessed from the sign and values of standardized coefficients. The parameter estimates, t-statistics, and p values were obtained using Satterthwaite's method for denominator degrees of freedom in the "lmerTest" package [50]. The spatial autocorrelations in model residuals were tested using Moran's I, which was conducted using the "spdep" package [51,52].

Finally, we performed hierarchical partitioning (HP) to determine the relative importance and independent contribution of each explanatory factor to FRic SES.FRic, and SR [53], as the multicollinearity between the factors in linear mixed models could not be completely excluded (VIFs of all variables < 10) despite it being controlled in the previous analysis [54]. HP can solve the multicollinearity problems effectively by averaging the contributions of an independent variable in models.

In order to fit assumptions about the uniformity of data and the homoscedasticity of errors, all functional traits (and thus functional richness) and environmental variables, and species richness, were log-transformed prior to analysis.

3. Results

3.1. Geographical Patterns of Functional and Species Richness in the Subtropical Evergreen Broadleaved Forests

We observed significant geographical patterns in FRic and SR (Table 2, Figure 1) along the gradients of latitude, longitude, and elevation. FRic was higher along increasing longitude and lower along increasing latitude and elevation, mirroring the geographical gradients in species richness (Figure 1A–C,G–I). Apparently, the patterns of SR along latitude and elevation were not significant, although they had similar trends to the FRic (Figure 1G,I). SESE.FRic was overall negative, with values of SES.FRic in most communities below null expectations, although they did not shift significantly along geographical gradients (Figure 1D,E).

Table 2. Summary of the mixed linear models analyzing the effects of geographic gradients on the functional richness (FRic), standardized effect of the functional richness (SES.FRic), and (log-transformed) species richness [log(SR)]. The marginal (R^2m) and conditional (R^2c) coefficients of determination are presented. Values in bold indicate p values < 0.05.

Model Factors	Latitude	Longitude	Elevation	AIC	R^2m	R^2c
FRic	**−2.47**	0.070	**−0.002**	1358.14	0.37	0.41
SES.FRic	−0.15	−0.004	−0.0002	247.10	0.09	0.58
SR	−0.05	**0.020**	−0.0001	−15.45	0.10	0.48

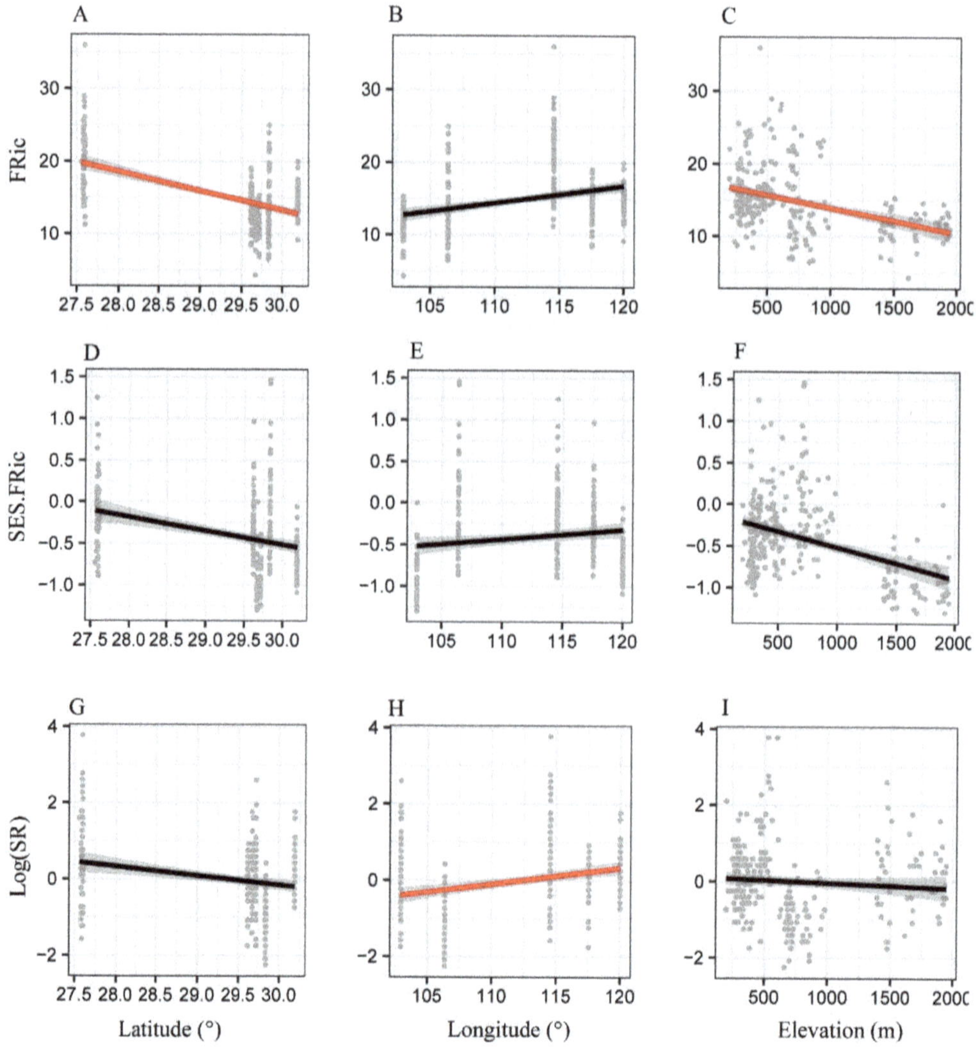

Figure 1. (**A–I**) Shifts in the functional richness (FRic), standardized effect of the functional richness (SES.FRic), and (log-transformed) species richness [log(SR)] along geographical gradients. (**A–C**) FRic, (**D–F**) SES.FRic, as well as (**G–I**) log(SR) across latitude, elevation, and longitude. The significance was tested by mixed linear models (see Table 2 for details on the models). Red lines represent significant linear regressions.

3.2. Correlations of Functional and Species Richness with Environmental Variables

Linear mixed effects models showed significant correlations of FRic, SESE.FRic, and SR with environmental factors (Table 3). The total percentage of variance explained by the LMM was higher (R^2 conditional = 0.50, 0.50, 0.47, respectively), while the variance explained by fixed predictors taken separately was relatively lower (R^2 marginal = 0.27, 0.50, 0.12, respectively). This indicated an important contribution of the random factor (i.e., the variability sources related to specific characteristics of sites) that was not reflected by fixed predictors. Results showed that FRic was significantly and positively correlated with precipitation of driest quarter (PDQ), soil pH (SpH), and soil total phosphorus (STP). Similarly, SES.FRic showed significant and positive associations with PDQ and SpH, but a negative association with mean diurnal range (MDR). SR was significantly and positively related to PDQ and MDR.

Table 3. Summary of the mixed linear models analyzing the effects of environmental variables on the functional richness (FRic), standardized effect of the functional richness (SES.FRic), and (log-transformed) species richness [log(SR)]. Estimate values are the standardized regression coefficients of predictors. The marginal (R^2m) and conditional (R^2c) coefficients of determination are presented. Values in bold indicate p values < 0.05.

	PDQ	PWQ	TDQ	MDR	SpH	STP	AIC	R^2m	R^2c	Moran's I
FRic	**3.40**	−0.97	0.10	−0.98	**0.81**	**0.86**	242.94	0.27	0.5	0.19
SES.Fric	**0.17**	−0.01	0.06	**−0.28**	**0.15**	0.06	1352.73	0.50	0.5	0.04
SR	**2.48**	−0.33	−1.08	**1.78**	0.01	−0.19	1537.93	0.12	0.47	0.18

Abbreviations: precipitation of driest quarter (PDQ), precipitation of wettest quarter (PWQ), mean temperature of driest quarter (TDQ), mean diurnal range (MDR), soil pH (SpH), soil total phosphorus (STP), variance explained by the fixed effects (R^2m), variance explained by both fixed and random effects (R^2c), Akaike Information Criterion (AIC), and Moran's I of linear mixed effects models (Moran's I).

The analysis of hierarchical variation partitioning showed that PDQ and MDR explained most of the variance in the Fric, SES.Fric, and SR (Figure 2). Results indicated the most important effects of PDQ and MDR on the variation in functional trait space and species richness in subtropical forest communities.

3.3. The Assembly Mechanisms of Forest Communities along Geographical Gradients

FRic was positively related to species richness and the relationship was always stronger with increasing species richness (Figure 3A; $p < 0.01$). However, the FRic was significantly different from the null random expectation after controlling species richness within each plot, indicating that the SEBLF community assembly was primarily governed by deterministic processes based on the FRic. Specifically, observed FRic in 83.2% of communities was lower than expected given species richness (i.e., negative SES.FRic, Figure 3B). Consistent with the prevalence of lower than expected FRic in each forest community, when FRic was evaluated as the mean of SES.FRic across all communities, we also found a significant and negative SES.FRic (Figure 3C). The result indicated that the average FRic was also lower than expected by chance across the study sites.

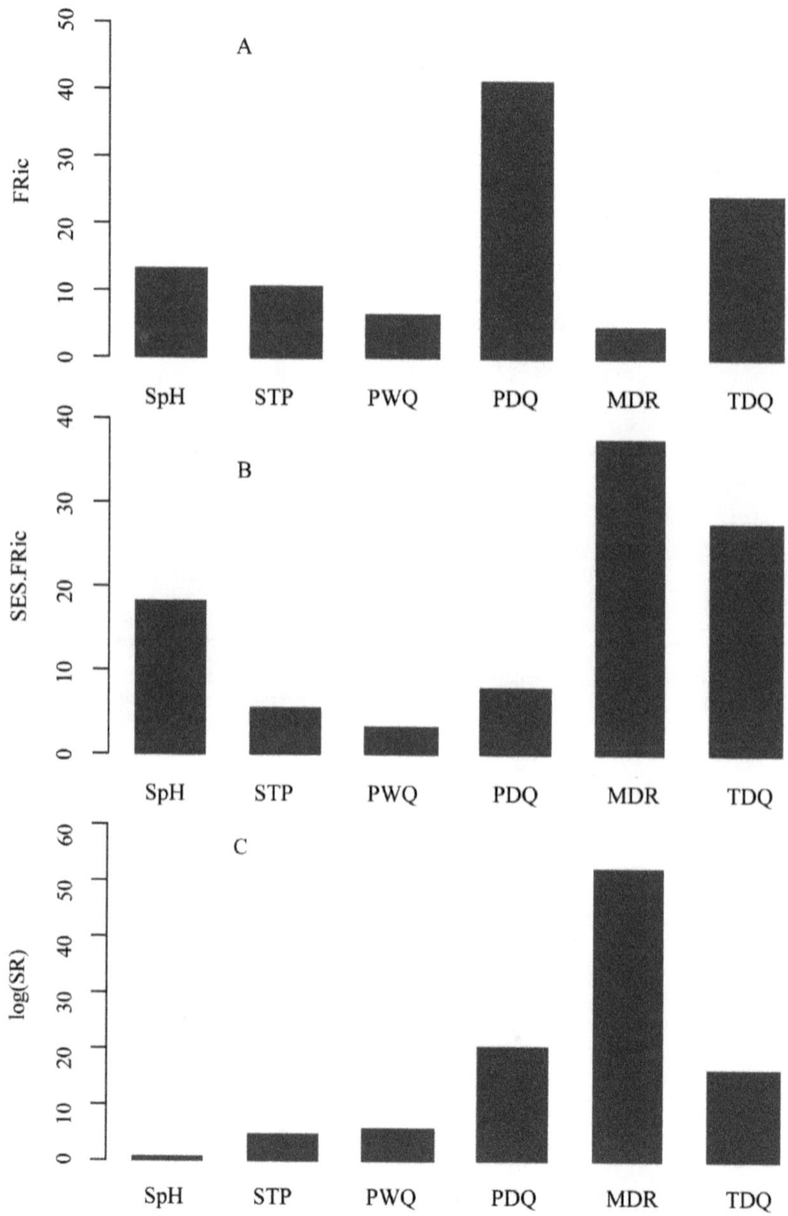

Figure 2. (**A–C**) Summary of hierarchical partitioning results for functional richness (FRic), the standardized effect of the functional richness (SES.FRic), and (log-transformed) species richness [log(SR)]. Each bar is the percentage of the variance obtained by a separate HP analysis for (**A**) FRic, (**B**) SES.FRic, and (**C**) log(SR) by using consistent environmental variables. Abbreviations: precipitation of driest quarter (PDQ), precipitation of wettest quarter (PWQ), mean diurnal range (MDR), mean temperature of driest quarter (TDQ), soil pH (SpH), soil total phosphorus (STP).

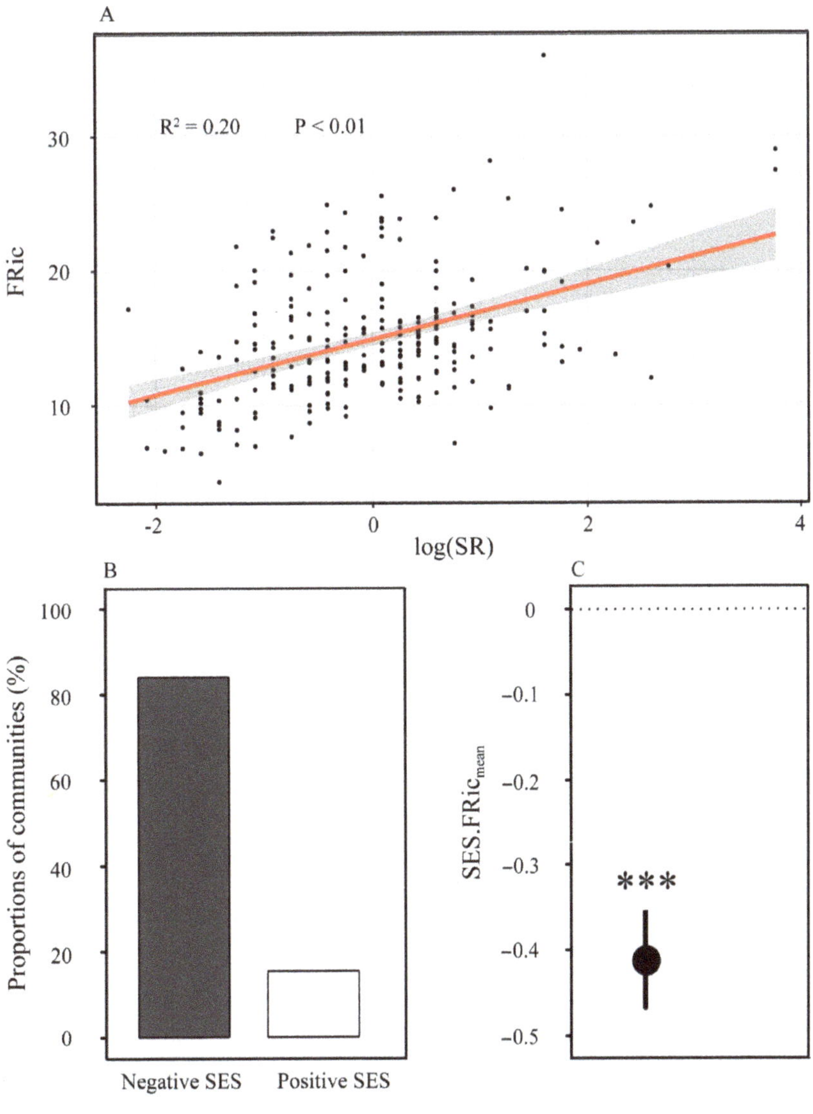

Figure 3. (**A**) Correlations between functional diversity (FRic) and (log-transformed) species richness [log(SR)]. (**B**) The relative prevalence of negative SES.FRic versus positive SES.FRic in forest communities, evaluated based on standardized effect size of the FRic. (**C**) Mean (mean ±95% confidence interval) of standardized effect size of the FRic across all communities (n = 250). ***, $p < 0.001$.

4. Discussion

We showed significant and similar geographical patterns of functional space and species richness from lower to higher latitudes, longitudes, and elevations, with FRic decreasing with increasing latitude and elevation and increasing with increasing longitude (Table 2, Figure 1). This was consistent with previous studies in which the geographical gradients in niche breadth mirrored the species richness [55]. In general, geographical gradients (latitude, longitude, and elevation) can affect functional space only indirectly through species richness; thus, a geographical pattern of functional space would be observed only when there is a geographical pattern of species richness and an effect of species richness on functional space [55]. However, it was notable that the functional volume (FRic) and species richness did not shift consistently and significantly along these geographical gradients. For instance, functional richness did not shift significantly as the species richness increased with increasing longitude. This means that the geographical variation in niche space might not mirror the geographical gradients in plant species richness completely. The results also indicated that there may be some limiting factors that restricted the expansion of niche space when the number of species increased. Therefore, the direct influences of environmental factors on functional volume should be considered in accounting for the geographical variation in functional trait space, given that the geographical gradients were closely related to environmental factors and strongly reflected resource use by plant species in our study areas [56,57].

Results showed that the geographical patterns of FRic were significantly and positively linked to precipitation of the driest quarter (PDQ), soil pH (SpH), and soil total phosphorus (STP). These results were consistent with previous studies wherein areas where functional volume was greater (i.e., lower elevations and latitudes) were significantly characterized by high amounts of rainfall during the driest quarter and higher soil nutrient content [58–60]. This suggested that precipitation and soil resources were closely related to the expansion of the niche space of subtropical forests. Greater availability of water and soil nutrients might create a wider niche space, allowing plant individuals with a larger range of viable functional strategies to coexist [58]. Significantly, PDQ outperformed other environmental variables that were not excessively co-linear with it, in accounting for the geographical variation in functional volume. The result was in line with a previous study conducted on BCI in Central Panama, which was dominated by evergreen tropical forests [61]. The significant effects of soil pH and phosphorus on the niche space occupied by plant species could be a result of the poor and acidic soil in subtropical regions of China [60,62,63].

Importantly, consistent with our expectation, the observed FRic was generally smaller than expected after controlling species richness, as indicated by the negative standardized effect sizes [24]. The results indicated that niche packing appeared to be prevalent and dominant in most forest assemblages across the entire study site. This implied that as species richness increased, new species were not preferentially added in unfilled portions of the trait space to expand the functional volume [5,19]. Instead, species tended to enter into the interior of the existing functional space, causing tighter packing of species (niche packing), since the increase in functional trait space was constrained within the SEBLF communities [19,20,64]. For instance, species richness increased significantly with increasing mean diurnal range (MDR), while FRic decreased (although not significant). The results suggested that the species with convergent trait combinations would increase with the increases in MDR, therefore increasing the overlap of niche space (niche packing) [12]. Presumably, the patterns of this species coexistence in our study were determined primarily by environmental filtering [65].

Similarly, we also found significant patterns of negative SES.FRic (niche packing) along environmental gradients, as indicated by significant associations between SES.FRic and environmental variables. The positive correlation of SES.FRic with PDQ and SpH provided evidence that the niche volume appeared to increase (i.e., niche packing decreased) [5] with increasing rainfall and soil pH. In contrast, a negative relationship between SES.FRic and mean diurnal range (MDR) was found, which indicated that temperature variability was the key factor in the prevalent pattern of niche packing. The result was in line with many previous studies. For instance, Thakur et al. reported that "temperature filtering" can result in a narrow niche [66]. Specifically, SES.FRic generally changed more quickly with the mean diurnal range (MDR) compared to other environmental variables. This result implied that the relative role of environmental filtering changed faster across MDR gradients in the subtropics, with stronger relevance in the assembly of communities with larger temperature fluctuations [40,67,68]. Less change for niche packing along gradients of precipitation and soil suggested that environmental filters are relatively homogeneous across water and soil conditions in subtropical forests, regardless of their importance [69]. Therefore, our results suggested that even if both niche patterns might accompany the increase in species richness, the niche expansion was insufficient to become the dominant pattern at the upper end of the precipitation and soil pH gradients. Instead, the temperature variability might play a stronger role in niche pattering than the actual increase in resources of water and soil nutrients, leading to more frequent niche packing in SEBLF communities [12]. Together, our study results suggested that climate and soil significantly affected the variation in functional volume along geographical gradients. The increase in functional trait space was constrained within the SEBLF communities. Environmental filtering, especially associated with temperature variability, probably defined the functional volume and caused the tighter packing of species into the niche space (niche packing).

A previous conceptual framework proposed that species first pass through an environmental filter at a regional scale, which constrains their functional trait space and thereby increases the packing of species. Next, the species pass through a biotic filter within a limited functional volume, which maximizes the possible functional diversity by competition or facilitation [24]. However, the importance of biotic interactions was not tested in our study. Many studies revealed that biotic interactions triggering niche partitioning in resource use [70,71] could explain the niche packing within a functional volume. Additionally, although the approach of the null model comparing the expected and observed patterns of functional diversity was widely used in many studies, it has fallen short in delivering generalizable conclusions [72]. For instance, clustering of species in functional traits could also emerge from hierarchical competition [73], in addition to environmental filtering. Therefore, it is necessary to build more mechanistic, dynamic models of community assembly to contrast the different ecological theories and processes [72,74,75] in future research.

Supplementary Materials: The following supporting information can be downloaded at: https://www.mdpi.com/article/10.3390/f13081206/s1, Figure S1: Geographical distribution of the 250 plots. Plots were taken from five forest sites in subtropical China. At each site, 50 plots with an area of 20 m × 20 m were randomly established.

Author Contributions: R.Z. conceived this project. C.H. and Y.X. conducted the field investigation and collected the data. C.H. performed the statistical analyses and wrote the first draft with Y.X. All the authors contributed to improving the quality of the manuscript. All authors have read and agreed to the published version of the manuscript.

Funding: This research was supported by the National Natural Science Foundation of China (41701055) and the Fundamental Research Funds for the Central Non-Profit Research Institution of the Chinese Academy of Forestry (CAFYBB2020ZA002-2).

Institutional Review Board Statement: Not applicable.

Informed Consent Statement: Not applicable.

Data Availability Statement: The climatic data used are available at https://www.worldclim.org/ (accessed on 20 June 2021) and other data about functional traits and local soil variables are available upon request from the authors.

Acknowledgments: We thank the graduate students from Sichuan Agricultural University, Anhui Agricultural University, and Hangzhou Normal University for their help in field sampling measurements, as well as the local staff who aided the field investigation work.

Conflicts of Interest: The authors declare no conflict of interest.

References

1. Swenson, N.G.; Enquist, B.J.; Pither, J.; Kerkhoff, A.J.; Boyle, B.; Weiser, M.D.; Elser, J.J.; Fagan, W.F.; Forero-Montana, J.; Fyllas, N.; et al. The biogeography and filtering of woody plant functional diversity in North and South America. *Glob. Ecol. Biogeogr.* **2012**, *21*, 798–808. [CrossRef]
2. Swenson, N.G.; Weiser, M.D. Plant geography upon the basis of functional traits: An example from eastern North American trees. *Ecology* **2010**, *91*, 2234–2241. [CrossRef] [PubMed]
3. Hillebrand, H. On the generality of the latitudinal diversity gradient. *Am. Nat.* **2004**, *163*, 192–211. [CrossRef] [PubMed]

4. Rahbek, C. The elevational gradient of species richness: A uniform pattern? *Ecography* **1995**, *18*, 200–205. [CrossRef]
5. Lamanna, C.; Blonder, B.; Violle, C.; Kraft, N.J.; Sandel, B.; Simova, I.; Donoghue, J.C., 2nd; Svenning, J.C.; McGill, B.J.; Boyle, B.; et al. Functional trait space and the latitudinal diversity gradient. *Proc. Natl. Acad. Sci. USA* **2014**, *111*, 13745–13750. [CrossRef] [PubMed]
6. Bassler, C.; Cadotte, M.W.; Beudert, B.; Heibl, C.; Blaschke, M.; Bradtka, J.H.; Langbehn, T.; Werth, S.; Müller, J. Contrasting patterns of lichen functional diversity and species richness across an elevation gradient. *Ecography* **2016**, *39*, 689–698. [CrossRef]
7. Mittelbach, G.G.; Schemske, D.W.; Cornell, H.V.; Allen, A.P.; Brown, J.M.; Bush, M.B.; Harrison, S.P.; Hurlbert, A.H.; Knowlton, N.; Lessios, H.A. Evolution and the latitudinal diversity gradient: Speciation, extinction and biogeography. *Ecol. Lett.* **2007**, *10*, 315–331. [CrossRef] [PubMed]
8. Devictor, V.; Mouillot, D.; Meynard, C.; Jiguet, F.; Thuiller, W.; Mouquet, N. Spatial mismatch and congruence between taxonomic, phylogenetic and functional diversity: The need for integrative conservation strategies in a changing world. *Ecol. Lett.* **2010**, *13*, 1030–1040. [CrossRef]
9. Stuart-Smith, R.D.; Bates, A.E.; Lefcheck, J.S.; Duffy, J.E.; Baker, S.C.; Thomson, R.J.; Stuart-Smith, J.F.; Hill, N.A.; Kininmonth, S.J.; Airoldi, L. Integrating abundance and functional traits reveals new global hotspots of fish diversity. *Nature* **2013**, *501*, 539–542. [CrossRef] [PubMed]
10. MacArthur, R.H. Patterns of species diversity. *Biol. Rev.* **1965**, *40*, 510–533. [CrossRef]
11. Violle, C.; Jiang, L. Towards a trait-based quantification of species niche. *J. Plant Ecol.* **2009**, *2*, 87–93. [CrossRef]
12. Pellissier, V.; Barnagaud, J.-Y.; Kissling, W.D.; Şekercioğlu, Ç.; Svenning, J.-C. Niche packing and expansion account for species richness-productivity relationships in global bird assemblages. *Glob. Ecol. Biogeogr.* **2018**, *27*, 604–615. [CrossRef]
13. Huang, J.; Yu, H.; Guan, X.; Wang, G.; Guo, R. Accelerated dryland expansion under climate change. *Nat. Clim. Chang.* **2016**, *6*, 166–171. [CrossRef]
14. Shiono, T.; Kusumoto, B.; Maeshiro, R.; Fujii, S.-J.; Götzenberger, L.; de Bello, F.; Kubota, Y.; Linder, P. Climatic drivers of trait assembly in woody plants in Japan. *J. Biogeogr.* **2015**, *42*, 1176–1186. [CrossRef]
15. Swenson, N. *Functional and Phylogenetic Ecology in R*; Springer: New York, NY, USA, 2014.
16. Ochoa-Ochoa, L.M.; Mejia-Dominguez, N.R.; Velasco, J.A.; Marske, K.A.; Rahbek, C.; Schrodt, F. Amphibian functional diversity is related to high annual precipitation and low precipitation seasonality in the New World. *Glob. Ecol. Biogeogr.* **2019**, *28*, 1219–1229. [CrossRef]
17. Ceulemans, T.; Merckx, R.; Hens, M.; Honnay, O. Plant species loss from European semi-natural grasslands following nutrient enrichment—Is it nitrogen or is it phosphorus? *Glob. Ecol. Biogeogr.* **2013**, *22*, 73–82. [CrossRef]
18. Mahdi, A.; Law, R.; Willis, A. Large niche overlaps among coexisting plant species in a limestone grassland community. *J. Ecol.* **1989**, *77*, 386–400. [CrossRef]
19. Pigot, A.L.; Trisos, C.H.; Tobias, J.A. Functional traits reveal the expansion and packing of ecological niche space underlying an elevational diversity gradient in passerine birds. *Proc. R. Soc. B Biol. Sci.* **2016**, *283*, 20152013. [CrossRef] [PubMed]
20. Van de Perre, F.; Willig, M.R.; Presley, S.J.; Mukinzi, I.J.-C.; Gambalemoke, M.S.; Leirs, H.; Verheyen, E. Functional volumes, niche packing and species richness: Biogeographic legacies in the Congo Basin. *R. Soc. Open Sci.* **2020**, *7*, 191582. [CrossRef]
21. Mason, N.W.H.; Mouillot, D.; Lee, W.G.; Wilson, J.B. Functional richness, functional evenness and functional divergence: The primary components of functional diversity. *Oikos* **2005**, *111*, 112–118. [CrossRef]
22. Mouchet, M.A.; Villéger, S.; Mason, N.W.H.; Mouillot, D. Functional diversity measures: An overview of their redundancy and their ability to discriminate community assembly rules. *Funct. Ecol.* **2010**, *24*, 867–876. [CrossRef]
23. Villéger, S.; Mason, N.; Mouillot, D. New multidimensional functional diversity indices for a multifaceted framework in functional ecology. *Ecology* **2008**, *89*, 2290–2301. [CrossRef] [PubMed]
24. Swenson, N.G.; Weiser, M.D. On the packing and filling of functional space in eastern North American tree assemblages. *Ecography* **2014**, *37*, 1056–1062. [CrossRef]
25. Brown, L.E.; Khamis, K.; Wilkes, M.; Blaen, P.; Brittain, J.E.; Carrivick, J.L.; Fell, S.; Friberg, N.; Füreder, L.; Gislason, G.M. Functional diversity and community assembly of river invertebrates show globally consistent responses to decreasing glacier cover. *Nat. Ecol. Evol.* **2018**, *2*, 325–333. [CrossRef] [PubMed]
26. Mouillot, D.; Stubbs, W.; Faure, M.; Dumay, O.; Tomasini, J.A.; Wilson, J.B.; Chi, T.D. Niche overlap estimates based on quantitative functional traits: A new family of non-parametric indices. *Oecologia* **2005**, *145*, 345–353. [CrossRef]
27. Kraft, N.J.; Adler, P.B.; Godoy, O.; James, E.C.; Fuller, S.; Levine, J.M. Community assembly, coexistence and the environmental filtering metaphor. *Funct. Ecol.* **2015**, *29*, 592–599. [CrossRef]
28. Hardin, G. The Competitive Exclusion Principle: An idea that took a century to be born has implications in ecology, economics, and genetics. *Science* **1960**, *131*, 1292–1297. [CrossRef] [PubMed]
29. Craven, D.; Hall, J.S.; Berlyn, G.P.; Ashton, M.S.; van Breugel, M. Environmental filtering limits functional diversity during succession in a seasonally wet tropical secondary forest. *J. Veg. Sci.* **2018**, *29*, 511–520. [CrossRef]
30. Laliberté, E.; Zemunik, G.; Turner, B.L. Environmental filtering explains variation in plant diversity along resource gradients. *Science* **2014**, *345*, 1602–1605. [CrossRef]
31. Yu, G.; Chen, Z.; Piao, S.; Peng, C.; Ciais, P.; Wang, Q.; Li, X.; Zhu, X. High carbon dioxide uptake by subtropical forest ecosystems in the East Asian monsoon region. *Proc. Natl. Acad. Sci. USA* **2014**, *111*, 4910–4915. [CrossRef] [PubMed]

32. Shi, W.; Wang, Y.Q.; Xiang, W.S.; Li, X.K.; Cao, K.F. Environmental filtering and dispersal limitation jointly shaped the taxonomic and phylogenetic beta diversity of natural forests in southern China. *Ecol. Evol.* **2021**, *11*, 8783–8794. [CrossRef]
33. Kröber, W.; Böhnke, M.; Welk, E.; Wirth, C.; Bruelheide, H. Leaf trait-environment relationships in a subtropical broadleaved forest in South-East China. *PLoS ONE* **2012**, *7*, e35742. [CrossRef] [PubMed]
34. Condit, R. *Tropical Forest Census Plots: Methods and Results from Barro Colorado Island, Panama and a Comparison with Other Plots*; Springer Science & Business Media: Berlin, Germany, 1998.
35. Perez-Harguindeguy, N.; Diaz, S.; Garnier, E.; Lavorel, S.; Poorter, H.; Jaureguiberry, P.; Bret-Harte, M.S.; Cornwell, W.K.; Craine, J.M.; Gurvich, D.E. New handbook for standardised measurement of plant functional traits worldwide. *Aust. J. Bot.* **2013**, *61*, 167–234. [CrossRef]
36. Chen, Y.; Huang, P. *Seeds of woody plants in China*; China Forestry Publishing House: Beijing, China, 2000.
37. Gardens, R.B.; Kew, U. *Seed Information Database (SID). Version 7.1*; Kew Publishing, Royal Botanic Gardens, Kew: London, UK, 2014.
38. USDA. *The PLANTS Database*; National Plant Data Team: Greensboro, NC, USA, 2018; pp. 24901–27401. Available online: http://plants.usda.gov (accessed on 2 November 2019).
39. Maitner, B.S.; Boyle, B.; Casler, N.; Condit, R.; Donoghue, J.; Durán, S.M.; Guaderrama, D.; Hinchliff, C.E.; Jørgensen, P.M.; Kraft, N.J.; et al. The bien r package: A tool to access the Botanical Information and Ecology Network (BIEN) database. *Methods Ecol. Evol.* **2018**, *9*, 373–379. [CrossRef]
40. Huang, C.; Xu, Y.; Zang, R. Variation patterns of functional trait moments along geographical gadients and their environmental determinants in the subtropical evergreen broadleaved forests. *Front. Plant Sci.* **2021**, *12*, 686965. [CrossRef] [PubMed]
41. Laliberté, E.; Legendre, P. A distance-based framework for measuring functional diversity from multiple traits. *Ecology* **2010**, *91*, 299–305. [CrossRef] [PubMed]
42. R Core Team. *R: A Language and Environment for Statistical Computing*; R Foundation for Statistical Computing: Vienna, Austria, 2013.
43. Andrew, S.C.; Mokany, K.; Falster, D.S.; Wenk, E.; Wright, I.J.; Merow, C.; Adams, V.; Gallagher, R.V. Functional diversity of the Australian flora: Strong links to species richness and climate. *J. Veg. Sci.* **2021**, *32*, e13018. [CrossRef]
44. Yang, J.; Zhang, G.; Ci, X.; Swenson, N.G.; Cao, M.; Sha, L.; Li, J.; Baskin, C.C.; Slik, J.F.; Lin, L. Functional and phylogenetic assembly in a Chinese tropical tree community across size classes, spatial scales and habitats. *Funct. Ecol.* **2014**, *28*, 520–529. [CrossRef]
45. Luo, W.; Lan, R.; Chen, D.; Zhang, B.; Xi, N.; Li, Y.; Fang, S.; Valverde-Barrantes, O.J.; Eissenstat, D.M.; Chu, C.; et al. Limiting similarity shapes the functional and phylogenetic structure of root neighborhoods in a subtropical forest. *New Phytol.* **2020**, *229*, 1078–1090. [CrossRef] [PubMed]
46. Bernard-Verdier, M.; Navas, M.L.; Vellend, M.; Violle, C.; Fayolle, A.; Garnier, E. Community assembly along a soil depth gradient: Contrasting patterns of plant trait convergence and divergence in a M editerranean rangeland. *J. Ecol.* **2012**, *100*, 1422–1433. [CrossRef]
47. Lhotsky, B.; Kovács, B.; Ónodi, G.; Csecserits, A.; Rédei, T.; Lengyel, A.; Kertész, M.; Botta-Dukat, Z. Changes in assembly rules along a stress gradient from open dry grasslands to wetlands. *J. Ecol.* **2016**, *104*, 507–517. [CrossRef]
48. Bates, D.; Mchler, M.; Bolker, B.; Walker, S. Fitting linear mixed-effects models using lme4. *J. Stat. Softw.* **2014**, *67*, 1–51.
49. Nakagawa, S.; Schielzeth, H. A general and simple method for obtaining R2 from generalized linear mixed-effects models. *Methods Ecol. Evol.* **2013**, *4*, 133–142. [CrossRef]
50. Kuznetsova, A.; Brockhoff, P.B.; Christensen, R.H.B. lmerTest package: Tests in linear mixed effects models. *J. Stat. Softw.* **2017**, *82*, 1–26. [CrossRef]
51. Ehbrecht, M.; Seidel, D.; Annighofer, P.; Kreft, H.; Kohler, M.; Zemp, D.C.; Puettmann, K.; Nilus, R.; Babweteera, F.; Willim, K.; et al. Global patterns and climatic controls of forest structural complexity. *Nat. Commun.* **2021**, *12*, 519. [CrossRef] [PubMed]
52. Asner, G.P.; Martin, R.E.; Anderson, C.B.; Kryston, K.; Vaughn, N.; Knapp, D.E.; Bentley, L.P.; Shenkin, A.; Salinas, N.; Sinca, F.; et al. Scale dependence of canopy trait distributions along a tropical forest elevation gradient. *New Phytol.* **2017**, *214*, 973–988. [CrossRef]
53. Walsh, C.; MacNally, R. *Hierarchical Partitioning*; (Part of Documentation for R: A Language and Environment for Statistical Computing); R Foundation for Statistical Computing: Vienna, Austria, 2013; pp. 1–4.
54. Nally, R.M. Multiple regression and inference in ecology and conservation biology: Further comments on identifying important predictor variables. *Biodivers. Conserv.* **2002**, *11*, 1397–1401. [CrossRef]
55. Vázquez, D.P.; Stevens, R.D. The latitudinal gradient in niche breadth: Concepts and evidence. *Am. Nat.* **2004**, *164*, E1–E19. [CrossRef] [PubMed]
56. MacArthur, R.H. *Geographical Ecology: Patterns in the Distribution of Species*; Princeton University Press: Princeton, NJ, USA, 1984.
57. Liu, S.; Yan, Z.; Chen, Y.; Zhang, M.; Chen, J.; Han, W.; Kerkhoff, A. Foliar pH, an emerging plant functional trait: Biogeography and variability across northern China. *Glob. Ecol. Biogeogr.* **2019**, *28*, 386–397. [CrossRef]
58. Gherardi, L.A.; Sala, O.E. Enhanced interannual precipitation variability increases plant functional diversity that in turn ameliorates negative impact on productivity. *Ecol. Lett.* **2015**, *18*, 1293–1300. [CrossRef] [PubMed]
59. Knapp, A.K.; Fay, P.A.; Blair, J.M.; Collins, S.L.; Smith, M.D.; Carlisle, J.D.; Harper, C.W.; Danner, B.T.; Lett, M.S.; McCarron, J.K. Rainfall variability, carbon cycling and plant species diversity in a mesic grassland. *Science* **2002**, *298*, 2202–2205. [CrossRef] [PubMed]

60. Han, W.; Chen, Y.; Zhao, F.J.; Tang, L.; Jiang, R.; Zhang, F. Floral, climatic and soil pH controls on leaf ash content in China's terrestrial plants. *Glob. Ecol. Biogeogr.* **2012**, *21*, 376–382. [CrossRef]
61. Powell, T.L.; Koven, C.D.; Johnson, D.J.; Faybishenko, B.; Fisher, R.A.; Knox, R.G.; McDowell, N.G.; Condit, R.; Hubbell, S.P.; Wright, S.J. Variation in hydroclimate sustains tropical forest biomass and promotes functional diversity. *New Phytol.* **2018**, *219*, 932–946. [CrossRef] [PubMed]
62. Foulds, W. Nutrient Concentrations of Foliage and Soil in South-Western Australia. *New Phytol.* **1993**, *125*, 529–546. [CrossRef]
63. Wang, T.; Yang, Y.; Wenhong, M.A.; Ecology, D.O. Storage, patterns and environmental controls of soil phosphorus in China. *Acta Sci. Nat. Univ. Pekin.* **2008**, *44*, 945–952.
64. Cornwell, W.K.; Schwilk, D.W.; Ackerly, D.D. A trait-based test for habitat filtering: Convex hull volume. *Ecology* **2006**, *87*, 1465–1471. [CrossRef]
65. Hoiss, B.; Krauss, J.; Potts, S.G.; Roberts, S.; Steffan-Dewenter, I. Altitude acts as an environmental filter on phylogenetic composition, traits and diversity in bee communities. *Proc. R. Soc. B Biol. Sci.* **2012**, *279*, 4447–4456. [CrossRef] [PubMed]
66. Thakur, D.; Chawla, A. Functional diversity along elevational gradients in the high altitude vegetation of the western Himalaya. *Biodivers. Conserv.* **2019**, *28*, 1977–1996. [CrossRef]
67. Enquist, B.J.; Norberg, J.; Bonser, S.P.; Violle, C.; Savage, V.M. Scaling from traits to ecosystems: Developing a general Trait Driver Theory via integrating trait-based and metabolic scaling theories. *Adv. Ecol. Res.* **2015**, *52*, 249–318.
68. Wieczynski, D.J.; Boyle, B.; Buzzard, V.; Duran, S.M.; Henderson, A.N.; Hulshof, C.M.; Kerkhoff, A.J.; McCarthy, M.C.; Michaletz, S.T.; Swenson, N.G.; et al. Climate shapes and shifts functional biodiversity in forests worldwide. *Proc. Natl. Acad. Sci. USA* **2019**, *116*, 587–592. [CrossRef] [PubMed]
69. Montaño-Centellas, F.A.; McCain, C.; Loiselle, B.A.; Grytnes, J.A. Using functional and phylogenetic diversity to infer avian community assembly along elevational gradients. *Glob. Ecol. Biogeogr.* **2019**, *29*, 232–245. [CrossRef]
70. Evans, K.L.; Warren, P.H.; Gaston, K.J. Species–energy relationships at the macroecological scale: A review of the mechanisms. *Biol. Rev.* **2005**, *80*, 1–25. [CrossRef] [PubMed]
71. Mason, N.W.H.; Irz, P.; Lanoiselee, C.; Mouillot, D.; Argillier, C. Evidence That Niche Specialization Explains Species-Energy Relationships in Lake Fish Communities. *J. Anim. Ecol.* **2008**, *77*, 285–296. [CrossRef] [PubMed]
72. Münkemüller, T.; Gallien, L.; Pollock, L.J.; Barros, C.; Carboni, M.; Chalmandrier, L.; Mazel, F.; Mokany, K.; Roquet, C.; Smyčka, J. Dos and don'ts when inferring assembly rules from diversity patterns. *Glob. Ecol. Biogeogr.* **2020**, *29*, 1212–1229. [CrossRef]
73. Mayfield, M.M.; Levine, J.M. Opposing effects of competitive exclusion on the phylogenetic structure of communities. *Ecol. Lett.* **2010**, *13*, 1085–1093. [CrossRef] [PubMed]
74. Connolly, S.R.; Keith, S.A.; Colwell, R.K.; Rahbek, C. Process, mechanism, and modeling in macroecology. *Trends Ecol. Evol.* **2017**, *32*, 835–844. [CrossRef] [PubMed]
75. Pontarp, M.; Brännström, Å.; Petchey, O.L. Inferring community assembly processes from macroscopic patterns using dynamic eco-evolutionary models and Approximate Bayesian Computation (ABC). *Methods Ecol. Evol.* **2019**, *10*, 450–460. [CrossRef]

forests

MDPI

Article

Interspecific Association and Community Stability of Tree Species in Natural Secondary Forests at Different Altitude Gradients in the Southern Taihang Mountains

Shan-Shan Jin †, Yan-Yan Zhang †, Meng-Li Zhou, Xiao-Ming Dong, Chen-Hao Chang, Ting Wang and Dong-Feng Yan *

College of Forestry, Henan Agricultural University, Zhengzhou 450002, China; jinshanshanzzz@126.com (S.-S.J.); zyylxyhenau@126.com (Y.-Y.Z.); dreamer170808@126.com (M.-L.Z.); dxm13592503501@126.com (X.-M.D.); changchenhao2022@163.com (C.-H.C.); tingwang126@126.com (T.W.)

* Correspondence: ydflx@henau.edu.cn

† These authors contributed equally to this work.

Abstract: An interspecific association represents an inter-relatedness of different species in spatial distribution and combined with the altitude factor, is key for revealing the formation and evolution of an ecological community. Therefore, we analyzed the changes in interspecific association and community stability at different altitudes in the southern Taihang Mountains using the variance ratio (*VR*), χ^2 test, association coefficient (*AC*), percentage of co-occurrence (*PC*) and Godron stability method. In total, 27 sample plots measuring 20 × 20 m were set up and were divided into lower altitude (700~1100 m), medium altitude (1100~1500 m) and higher altitude areas (1500~1900 m) into. The results showed that the overall interspecies association of communities exhibited an insignificant negative association in both the lower (*VR* = 0.79, *W* = 7.15) and higher (*VR* = 0.81, *W* = 7.36) altitude areas, while an insignificant positive association was observed in the medium (*VR* = 1.48, *W* = 13.34) altitude area. Besides, the χ^2 test showed the ratio of positively and negatively correlated species pairs decreased as altitude increased with values of 1.39, 1.22 and 0.95 in the lower, medium and higher altitude areas, respectively. Moreover, the *AC* and *PC* indices stated that most species pairs had a weaker association in the three altitude areas, but the *AC* indices also suggested the number of positive association species pairs was more than that of negative association only in medium altitude area. Meanwhile, the Godron stability method showed the distances from the intersection point to the stable point (20 and 80) were still far away, with values of 22.53, 11.92 and 21.34 in the lower, medium and higher altitude areas, respectively, which indicated an unstable succession stage, though the community appeared steadier in the medium altitude area. This study can provide some guidance for effective afforestation and vegetation restoration.

Keywords: altitude; interspecific association; Godron stability; the southern Taihang Mountains

Citation: Jin, S.-S.; Zhang, Y.-Y.; Zhou, M.-L.; Dong, X.-M.; Chang, C.-H.; Wang, T.; Yan, D.-F. Interspecific Association and Community Stability of Tree Species in Natural Secondary Forests at Different Altitude Gradients in the Southern Taihang Mountains. *Forests* **2022**, *13*, 373. https://doi.org/10.3390/f13030373

Academic Editors: Runguo Zang and Yi Ding

Received: 23 December 2021
Accepted: 21 February 2022
Published: 23 February 2022

1. Introduction

Tree species are interdependent in the forest successional process, with a greatly changed abundance and composition, which showed a certain interspecific association and then affected the forest stability [1–4]. This interspecific association represents an inter-relatedness of different species in the spatial distribution affected by different habitats of tree populations and is essential for the formation and evolution of ecological communities [5–7], so it can be generally reflected by their corresponding habitats [8,9]. Additionally, interspecific association is mainly used to explain the community composition, structure, function, succession trend and competition status [10,11]. Therefore, relevant studies on interspecific association can provide an important scientific and theoretical basis for vegetation reconstruction and biodiversity protection.

Community stability is a foundation for the continuous functioning of the forest ecosystem and has a comprehensive feature of the structure and function of the plant community [12]. Generally, the biological ecology method and the Godron stability measurement method were used to describe the community stability in previous studies. The former is mainly analyzed by using tree species compositions and community age structures, and the latter is reflected by the means of mathematical models [13]. In this study, we selected the second method, in view of the actual situation of the study area. Community stability and interspecific association are usually combined to reveal the competition status of plant populations, community structure and the trend of community succession [14–16].

Altitude greatly influenced species distribution and forest stability and was the main terrain factor [17–20]. For example, Zhang et al. [17] considered the vegetation patterns in the middle part of the Taihang Mountain Range to be significantly correlated with altitude, Mu et al. [21] found a higher continuity existed along an altitudinal gradient in the secondary forest community in Changbai Mountains and Cabrera et al. [22] showed that altitude was responsible for the division of structural and floristic groups, as the most important factor for analyzing the diversity and structure of the vanishing montane forest of southern Ecuador. Besides, Bhutia et al. [23] suggested that low-altitude forests (900–1700 m) had the highest Shannon diversity index by examining the forest structure in the eastern Himalayas. These examples demonstrate that previous studies on forest communities principally focused on the relationships between altitude and species diversity or distribution. Moreover, researches on interspecific relationships are mostly conducted under a similar habitat condition, such as at a same altitude level [17,21–23]. However, there are few studies that have linked altitude factors with interspecies associations and community stability and analyzed their interactions.

The Taihang Mountains are located at the intersection of the north and south flora, with abundant species and a complex community structure [24]. However, the forests in this area are subject to frequent human disturbances and suffered severe soil erosion since the 1950s, resulting in heavy habitat destruction and biodiversity loss, as well as poor ecosystem self-recovery capabilities [25,26], but vegetation there is gradually recovering, with a series of natural forest protection and reconstruction projects in recent years [27]. Therefore, the study of vegetation restoration in this area has become a hot topic today. For example, Yan et al. [28] analyzed the niche characteristics of tree populations at different altitudes, Zhao et al. [26] revealed the relationship between secondary forests and environment and Zhao et al. [29] explored the mechanism of plant community diversity changes. Moreover, it is of great significance to study the relationship between different altitudes and interspecies association and the community stability of the dominant tree species in this area, but relevant research is still rare.

It is necessary to explore the current state and regularity of the interspecific association and community stability at different altitudes in this local area, which is important for formulating protection policies of species diversity and forest conservation. Therefore, we aim to answer the following questions in this study: (1) As altitude changes, has the interspecific association changed significantly? (2) Has altitude change affected vegetation community stability? (3) What is the current stage of vegetation succession in this area?

2. Materials and Methods

2.1. Location Overview of Study Area

The Taihang Mountains are located on the eastern border of the Loess Plateau. They belong to a typical mountainous landform and an important ecological barrier for the North China Plain. The study area was located in the southern Taihang Mountains in northern Jiyuan City, Henan Province (Figure 1). The climate here was warm temperate and semi-humid, with the continental characteristic of a mean annual temperature of 14.5 °C. The annual sunshine duration is 20,400 h, the frost-free period is 213 days, and the annual rainfall is 696 mm, which mainly falls from July to September [30]. The altitude ranges from 150 m to 1955 m. The Taihang Mountains are highly heterogeneous in terms of topography, soil, climate and vegetation due to the wide range of altitude and typical geological features [20]. The landform in this region is mostly hilly and mountainous, and the soil type belongs to brown and cinnamon soil, which is mostly acid and neutral. In 1998, the Taihang Mountain Macaque Nature Reserve was set up in the southern Taihang Mountains to protect the local forest ecosystem. At present, most forests in this area are relatively young, and most of the areas below 600 m above sea level grow secondary forests, shrubs and cultivated land, while primary secondary forests distribute widely above 600 m above sea level. The main vegetation types there are secondary broadleaved deciduous forests (including the tree species *Quercus aliena, Quercus variabilis, Cotinus coggygria*, etc.), the shrubs are mostly *Vitex negundo* var. *heterophylla* and *Viburnum mongolicum* and the grasses are mostly *Elsholtzia ciliate* and *Carex rigescens* [28]. In this field, we found there are fewer newer seedlings under the forest and the shrubs are mostly *Viburnum dilatatum, Forsythia suspensa, Vitex negundo* var. *Heterophylla*., etc. *Carex rigescens, Clematis florida* and *Elsholtzia ciliate* distribute widely in grass layer.

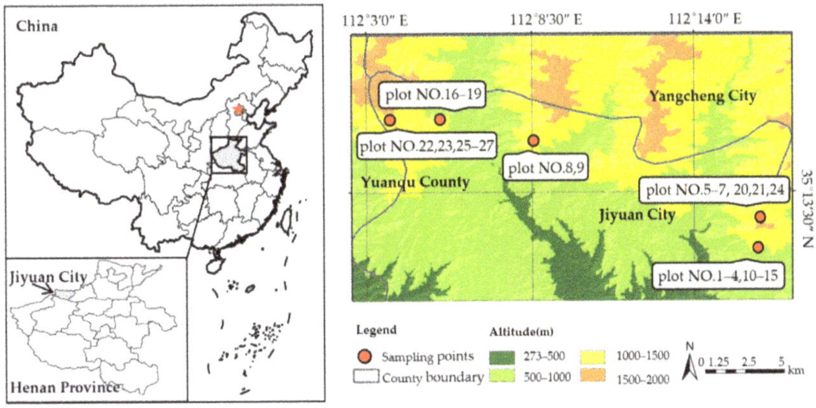

Figure 1. Location map of the sampling sites in northern Jiyuan City.

2.2. Field Measurements

A field survey was conducted in the Macaque Natural Reserve of Taihang Mountain in August 2020 in the southern Taihang Mountains. We selected the natural forest areas with typical characteristics and good growth as test objects from 700 m to 1900 m, and then divided the survey area into three altitude areas: the lower altitude area (700–1100 m), medium altitude area (1100–1500 m), and higher altitude area (1500–1900 m). In each altitude area, 9 permanent 20 × 20 m sample plots were established (27 in total). All trees with a diameter at breast height (DBH; 1.3 m) $\geq$ 1 cm were marked, and their locations in the plots were recorded using a forest locator (POSTEX). Meanwhile, we used hand-held GPS to measure the plots' locations and elevations. DBH, height, names, and growth of trees, canopy cover, slope aspect and the interference situation of the sample plots were also recorded (Table 1). Table 1 shows that the habitat conditions were similar among the nine sample plots under the same altitude area since it had a low degree of variation (CV < 15%) according to the basic indication information of trees and sample plots.

Table 1. Basic information of the permanent samples.

Altitude Zone	Sample Plot No.	Mean DBH * (cm)	Mean Tree Height (m)	Tree Density (Tree/hm^2)	Canopy Density	Slope (°)	Elevation (m)
Lower altitude area	1	10.44	5.99	3750	0.80	29	787
	2	9.99	6.90	3975	0.75	32	852
	3	10.76	6.63	4025	0.70	25	868
	4	10.94	7.44	3625	0.78	23	871
	5	12.70	7.14	3925	0.75	22	955
	6	8.92	5.10	4050	0.80	31	971
	7	11.37	6.41	3950	0.70	30	1016
	8	10.41	6.74	4050	0.65	32	1054
	9	12.32	7.58	4150	0.75	30	1066
	CV	10.03	10.85	3.89	6.41	12.91	9.86
Medium altitude area	10	14.72	6.33	3225	0.65	24	1114
	11	12.83	9.45	3375	0.65	25	1145
	12	14.34	7.71	3150	0.68	30	1193
	13	15.79	10.32	3375	0.65	30	1222
	14	12.86	7.52	3600	0.70	25	1321
	15	10.68	8.58	3900	0.78	28	1324
	16	11.31	8.46	3175	0.80	32	1415
	17	14.21	9.00	3325	0.80	26	1430
	18	9.73	7.66	3475	0.75	35	1495
	CV	14.79	13.40	6.55	8.56	12.34	9.80
Higher altitude area	19	11.94	8.97	4825	0.75	31	1510
	20	8.94	8.23	4875	0.80	25	1521
	21	11.29	7.31	4650	0.85	25	1535
	22	11.82	8.16	4750	0.70	25	1539
	23	12.97	8.12	4950	0.65	33	1554
	24	12.80	8.06	4825	0.70	36	1582
	25	10.75	8.97	4775	0.85	28	1720
	26	9.86	8.49	4850	0.85	23	1787
	27	10.31	8.29	4950	0.85	30	1813
	CV	11.36	5.73	1.87	9.64	14.46	7.03

* DBH: diameter at breast height (forestry).

2.3. Data Analysis

2.3.1. Importance Values

The importance value (IV) of a species was used to characterize the status and role of each species in the community and defined as the average of the relative abundance (RA), relative frequency (RF) and relative dominance (RD) of the species [4,6]. In this study, the IV was used as an index for selecting the dominant tree species. It was calculated with the following equations [6]:

$$\mathrm{IV} = (\mathrm{RA} + \mathrm{RD} + \mathrm{RF})/3 \tag{1}$$

$$\mathrm{RA} = a_i / \sum_{i=1}^{S} a_i \quad \mathrm{RD} = d_i / \sum_{i=1}^{S} d_i \quad \mathrm{RF} = f_i / \sum_{i=1}^{S} f_i \tag{2}$$

where a_i is the number of individuals of population i, d_i is the basal area at the height of 1.3 m of population i, f_i is the number of quadrats in which the population i appears and S is the total number of species.

2.3.2. Interspecific Association Quantification

The variance ratio (VR) test was used to gain insight into the overall association among the different species, and significance was further tested using the W statistic value. The formulas are listed below [31,32]:

$$P_i = n_i / N \tag{3}$$

$$VR = S_T^2/\delta_T^2 = [\frac{1}{N}\sum_{i=1}^{N}(T_j - t)^2]/\sum_{i=n}^{S}P_i(1-P_i) \quad (4)$$

$$W = VR \times N \quad (5)$$

where n_i is the number of quadrats containing species i, N is the total number of quadrats, S is the total number of species, T_j is the number of species occurring in quadrat j, and t is the average number of species in the quadrats.

If $VR > 1$, the species have a positive association, and if $VR < 1$, species have a negative association. $VR = 1$ indicates that species have no associations because they are assumed independent. If $\chi^2_{0.95(N)} < W < \chi^2_{0.05(N)}$, the overall interspecific association is not significant ($P > 0.05$). Conversely, the association is significant ($P < 0.05$) when $W < \chi^2_{0.95(N)}$ or $W > \chi^2_{0.05(N)}$.

The degree of association was conducted based on a 2 × 2 contingency column table that was generated by the existence or absence of the two species. For each pair of species A and B, we can obtain a contingency table such as in the example of Table 2 [6]:

Table 2. Example of a 2 × 2 contingency table.

		Species B		Sum
		Present	**Absent**	
Species A	Present	a	b	$a + b$
	Absent	c	d	$c + d$
	Sum	$a + c$	$b + d$	$N = a + b + c + d$

a, the number of quadrats in which species A and B co-occurred; b, the number of quadrats in which species A occurred, but not B; c, the number of quadrats in which species B occurred, but not A; d, the number of quadrats in which neither A nor B were found; N, the total number of quadrats.

χ^2 was corrected by the Yates continuous correction formula since the study was a discontinuous sample, and we determined the sign of the association between species pairs by the sign of the V value [33,34]. The b and d values were weighted to 1 to avoid a non-computable situation when the denominator was 0 and the frequency of occurrence of a certain species was 100% [35]. These were calculated as follows:

$$\chi^2 = N(|ad - bc| - \frac{1}{2}n)^2/[(a+b)(c+d)(a+c)(b+d)] \quad (6)$$

$$V = [(a+d) - (b+c)]/(a+b+c+d) \quad (7)$$

When $\chi^2 < 3.841$, there is an insignificant interspecific association between two species pairs ($P > 0.05$), when $3.841 \leq \chi^2 \leq 6.635$, the interspecific association between two species pairs is significant ($0.01 \leq P \leq 0.05$) and when $\chi^2 > 6.635$, the interspecific association between two species pairs is highly significant ($P < 0.01$). In addition, if $V > 0$, there is a positive association. Conversely, if $V < 0$, there is a negative association [33].

The χ^2 test results were further verified by the association coefficient (*AC*) and the percentage of co-occurrence (*PC*). The association coefficient (*AC*) was used to quantify the interspecific association of each species pair, and the percentage of co-occurrence (*PC*) can further reflect the strength of the positive association between tree species. These formulas are as follows:

$$\text{When } ad \geq bc,\ AC = (ad - bc)/[(a+b)(b+d)] \quad (8)$$

$$\text{When } ad < bc \text{ and } d \geq a,\ AC = (ad - bc)/[(a+b)(a+c)] \quad (9)$$

$$\text{When } ad < bc \text{ and } d < a,\ AC = (ad - bc)/[(b+d)(d+c)] \quad (10)$$

$$PC = a/(a+b+c) \quad (11)$$

AC index values range from 1 for complete positive associations ($b = 0$ and $c = 0$) to −1 for complete negative associations ($a = 0$ and $d = 0$). When *AC* equals zero, it shows there is no association [6]. The *PC* range is (0,1). The closer the *PC* is to 1, the more positive associations between tree species pairs; when the *PC* is equal to 0, there is no association between the tree species pairs [35].

2.3.3. Community Stability Analysis

The Godron stability index was used to determine the community stability. The 27 plots with size of 20 × 20 m were taken as a unit and arranged using the frequencies of all tree species in an ascending order. Next, the relative frequency of each tree species was calculated (the frequency of each species/the total frequency of all species), as well as the reciprocal of total species (1/the number of all species), its accumulative relative frequency (as dependent variable) and the accumulative reciprocal one by one (as an independent variable). Finally, using the smooth curve of scatter points, the binomial equation was simulated, and the coordinate of the intersection point between this simulation equation and the equation $y = -x + 100$ was calculated. According to the Godron stability judgment method, the closer a coordinate is to the community stability point coordinate (20, 80), the higher the community stability is [36].

2.4. Statistics and Analysis

In this study, R 4.0.3 (R Core Team, Vienna, Austria), Excel 2016 (Microsoft Corporation, Redmond, WA, USA) and Origin 2018 (OriginLab, Northampton, MA, USA) were used for all statistical analyses. The species association indices were conducted using the R packages "spaa" [37] and "corrplot" [38]. The calculation and drawing of community stability were derived by Excel 2016 and Origin 2018, respectively. In addition, ArcGIS 10.2 was used to generate the map of the sampling plots.

3. Results

3.1. Composition of Trees Species

A total of 68 different tree species were found during our investigation in this field, with 23 tree species in the lower altitude area, 22 tree species in the medium altitude area, and 23 tree species in higher altitude area. Only 10 (14.71%) tree species existed in all three altitude areas, and 24 (35.29%) tree species only appeared in one altitude area, which indicated that there was an obvious difference in tree species composition in the different altitude areas. In this study, 21 dominant species with more than one frequency and importance values greater than 1% were selected for interspecific association analysis among the plant communities in the 3 different altitude areas (Table 3).

Table 3. Dominant species, abbreviations, and importance values.

Species	Abbreviation	Importance Value/%		
		Lower Altitude Area	Medium Altitude Area	Higher Altitude Area
Quercus aliena	Qa	26.56	27.45	22.05
Quercus variabilis	Qv	23.98	19.73	3.87
Koelreuteria paniculata	Kp	6.17	/ [1]	/ [2]
Pinus tabuliformis	Pt	5.47	/ [2]	13.45
Acer mono	Am	4.65	7.33	4.27
Diospyros lotus	Dl	4.36	1.81	/ [1]
Carpinus cordata	Co	3.27	9.71	5.58
Cotinus coggygria	Cc	3.16	/ [1]	/ [1]
Cornus macrophylla	Cm	3.04	3.45	1.19
Crataegus pinnatifida	Cp	2.31	3.93	/ [1]
Ziziphus jujuba	Zj	1.33	/ [1]	/ [1]
Malus honanensis	Mh	1.32	/ [2]	1.51
Acer davidii	Ad	1.18	8.47	15.05
Toxicodendron vernicifluum	Tv	1.10	2.78	4.53

Table 3. *Cont.*

Species	Abbreviation	Importance Value/%		
		Lower Altitude Area	Medium Altitude Area	Higher Altitude Area
Fraxinus chinensis	Fc	/ [1]	2.66	2.43
Zelkova schneideriana	Zs	/ [1]	2.19	/1
Quercus mongolica	Qm	/ [1]	1.39	5.64
Picrasma quassioides	Pq	/ [2]	1.22	/ [1]
Rhus chinensis	Rc	/ [1]	1.84	/ [1]
Betula platyphylla	Bp	/ [1]	/ [1]	3.75
Pinus armandii	Pa	/ [1]	/ [1]	10.49

[1]: The species does not occur in the altitude area; [2]: The species occurs in the altitude area with a frequency less than 1 and has an importance value less than or equal to 1%.

3.2. Overall Interspecific Association

The overall interspecific association of the communities in three altitude areas is presented in Table 4. The results showed that the communities in the lower and higher altitude areas both exhibited a negative association for $VR < 1$ and were also insignificant when combined with W statistics and the χ^2 test. Conversely, the community in the medium altitude area indicated a positive association for $VR > 1$, and the χ^2 test result further revealed an insignificant overall interspecific association. This indicated that the tree populations in medium altitude areas maintained a relatively stable stage and appeared to exist in a mutually beneficial symbiotic relationship.

Table 4. Overall association among dominant tree species.

Altitude Zone	Variance Ratio (*VR*)	Test Statistics (*W*)	$\chi^2_{(0.95,N)}, \chi^2_{(0.05,N)}$	Test Results
Lower altitude area	0.79	7.15	3.325, 16.92	Not a significant association
Medium altitude area	1.48	13.34	3.325, 16.92	Not a significant association
Higher altitude area	0.81	7.36	3.325, 16.92	Not a significant association

3.3. Associations between Dominant Species Pairs

3.3.1. Test of Dominant Species Pair Associations

A χ^2 test determined the level of significance of dominant species pairs based on the 2×2 contingency table. The results demonstrated that among the dominant tree populations, the proportion of positive associations decreased slightly as altitude increased from the lower altitude area (58.24%) to the medium altitude area (54.95%) and the higher altitude area (51.28%) (Figure 2). In the lower altitude area, there were no significantly associated pairs. In the medium altitude area, a significantly positively associated pair was *Cornus macrophylla* and *Carpinus cordata* ($\chi^2 = 5.41$; $0.01 < P < 0.05$), and a negatively associated pair was *Zelkova schneideriana* and *Acer mono* ($\chi^2 = -4.14$; $0.01 < P < 0.05$). In the higher altitude area, there were five pairs that reached significant associations. *Quercus mongolica* and *Acer mono* ($\chi^2 = 5.06$; $0.01 < P < 0.05$) showed a significant positive association, while *Pinus armandii* and *Quercus variabilis* ($\chi^2 = -5.06$; $0.01 < P < 0.05$), *Cornus macrophylla* and *Carpinus cordata* ($\chi^2 = -4.14$; $0.01 < P < 0.05$), *Quercus mongolica* and *Pinus tabuliformis* ($\chi^2 = -5.06$; $0.01 < P < 0.05$) and *Acer mono* and *Pinus tabuliformis* ($\chi^2 = -5.06$; $0.01 < P < 0.05$) all showed negative associations. Overall, our results indicated that most of the pairs among the dominant tree populations were not significant associations. This indicated that most species pairs were weak associations for most species, and the distribution of tree species is independent.

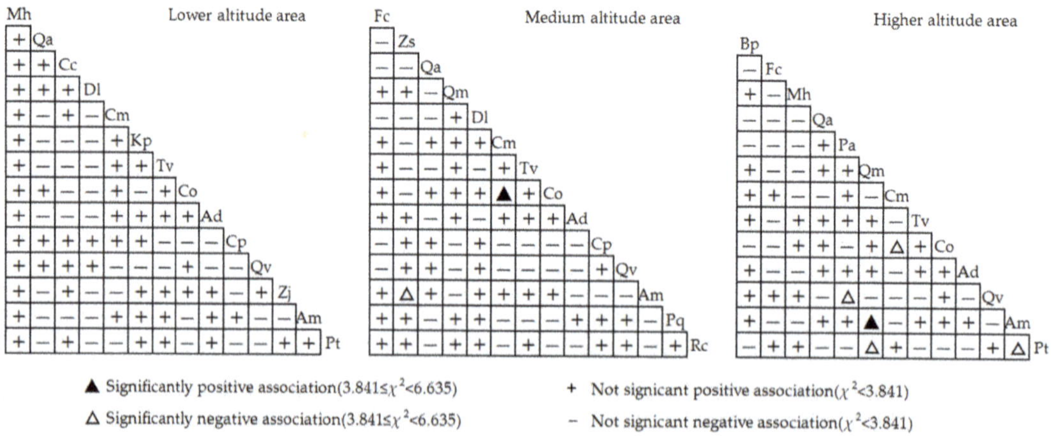

Figure 2. Half matrix graph of the interspecific association χ^2 test among the dominant tree species.

3.3.2. Measures of Dominant Species Pair Associations

The association coefficients (*AC*) and the percentage of co-occurrence (*PC*) results of the dominant tree species in the three altitude areas further distinguished the association strength between the species pairs (Figures 3–5 and Table 5). The number of positive association species pairs were 40, 47 and 34, with corresponding positive and negative species pair association ratio values of 0.78, 1.07 and 0.94 in the lower, medium and higher altitude areas, respectively, based on the *AC* results.

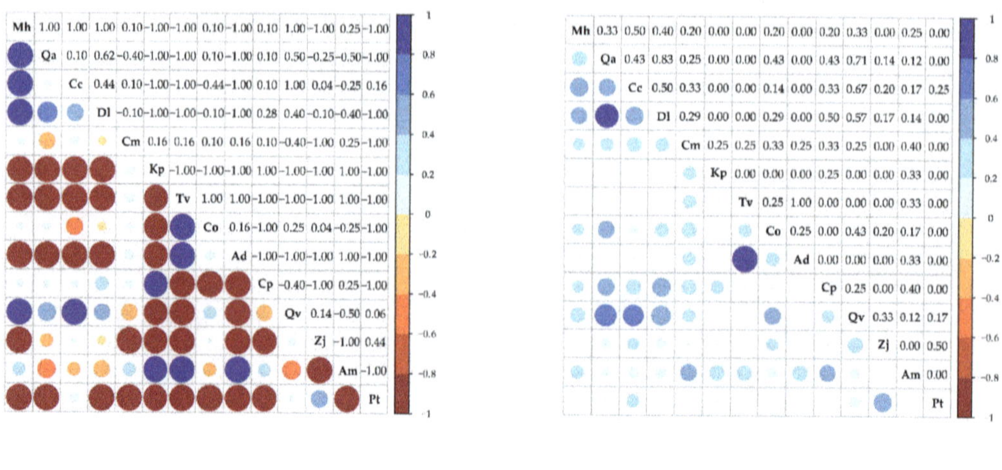

(**a**) (**b**)

Figure 3. *AC* and *PC* values in lower altitude area: (**a**) the association coefficients (*AC* value); (**b**) the percentage of co-occurrence (*PC* value). Notes: The size and color depth of the circles represent the absolute value of the interspecific association. Blue represents a positive association between the tree pairs, and red represents a negative association between the tree pairs. The same below.

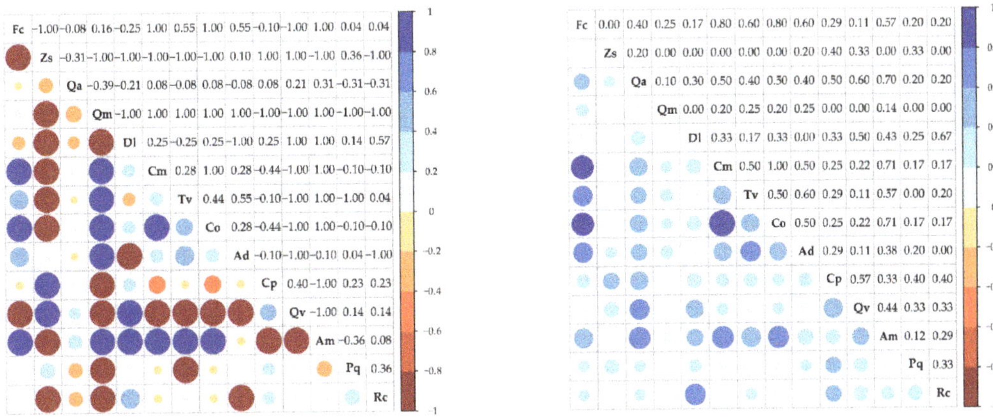

(a) (b)

Figure 4. *AC* and *PC* values in the medium altitude area: (**a**) the association coefficients (*AC* value); (**b**) the percentage of co-occurrence (*PC* value).

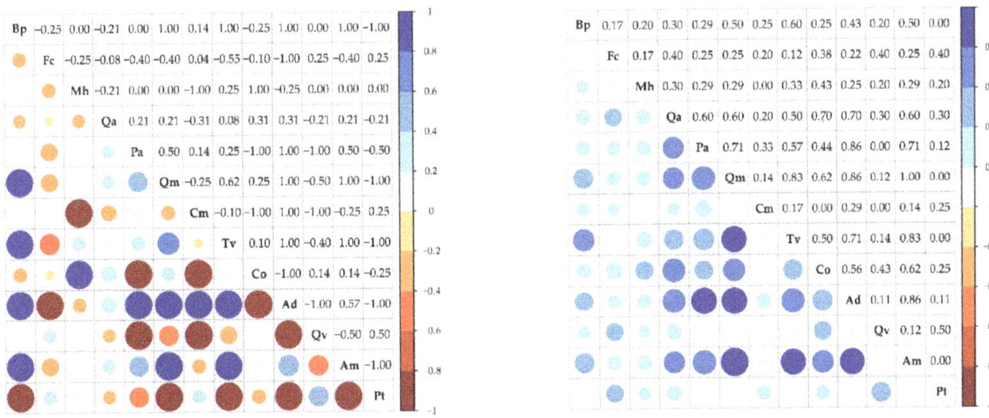

(a) (b)

Figure 5. *AC* and *PC* values in the higher altitude area: (**a**) the association coefficients (*AC* value); (**b**) the percentage of co-occurrence (*PC* value).

In the lower altitude area, 12 species pairs showed obvious significant positive associations ($AC \geq 0.6$), such as *Quercus aliena* and *Diospyros lotus* ($AC = 0.62$, $PC = 0.83$), *Malus honanensis* and *Quercus aliena* ($AC = 1.00$, $PC = 0.33$) *and Toxicodendron vernicifluum* and *Acer davidii* ($AC = 1.00$, $PC = 1.0$) ($AC = 1.00$, $PC > 0.25$), indicating that these tree species were more likely to appear in the same habitats. Besides, there was an obvious significant negative association between 38 species pairs ($AC < -0.6$), such as *Pinus tabuliformis* with other tree species ($AC = -1.00$, $PC = 0$) except for *Quercus variabilis*, *Ziziphus jujuba* and *Cotinus coggygria*, demonstrating no appearances in the same plots simultaneously. Additionally, there were 41 species pairs with weak associations ($-0.6 \leq AC < 0.6$).

Table 5. The association coefficients and the percentage of co-occurrence among the dominant tree species.

Association Type	Type	Value Range	Lower Altitude Area		Medium Altitude Area		Higher Altitude Area	
			Species Pair Number	%	Species Pair Number	%	Species Pair Number	%
Association coefficient (AC)	Positive association	$AC \geq 0.6$	12	13.19	16	17.58	12	15.38
		$0.2 \leq AC < 0.6$	9	9.89	18	19.78	15	19.23
		$0 < AC < 0.2$	19	20.88	13	14.29	7	8.97
	No association	$AC = 0$	0	0.00	0	0.00	8	10.26
	Negative association	$-0.2 \leq AC < 0$	3	3.30	11	12.09	3	3.85
		$-0.6 \leq AC < -0.2$	10	10.99	10	10.99	20	25.64
		$AC < -0.6$	38	41.76	23	25.27	13	16.67
Percentage of co-occurrence (PC)		$PC = 1$	1	1.10	1	1.10	1	1.28
		$0.5 \leq PC < 1$	8	8.79	21	23.08	23	29.49
		$0 < PC < 0.5$	44	48.35	53	58.24	46	58.97
		$PC = 0$	38	41.76	16	17.58	8	10.26
Total			91	100	91	100	78	100

In the medium altitude area, there were 16 species pairs showing obvious significant positive associations ($AC \geq 0.6$), such as *Acer mono* and *Carpinus cordata* (*Toxicodendron vernicifluum* and *Cornus macrophylla*) ($AC = 1.00$, $PC > 0.5$), suggesting a similar requirement for habitats. Apart from this, 23 species pairs showed obvious significant negative associations ($AC < -0.6$), such as *Quercus variabilis* and *Acer davidii* (*Carpinus cordata*) ($AC = -1.00$, $PC < 0.5$). Moreover, 52 species pairs had unremarkable associations ($-0.6 \leq AC < 0.6$).

In the higher altitude area, 12 species pairs exhibited strong positive associations ($AC \geq 0.6$), such as *Acer davidii and Toxicodendron vernicifluum* (*Cornus macrophylla, Quercus mongolica, Pinus armandii* and *Betula platyphylla)* ($AC = 1.00$), with corresponding larger PC values. Conversely, 13 species pairs had strong negative associations ($AC \leq -0.6$), such as *Carpinus cordata* and *Acer davidii* and *Acer davidii* and *Quercus variabilis* ($AC = -1$), which meant there were large ecological differences among them. Besides, 53 species pairs were relatively weakly associated ($-0.6 \leq AC < 0.6$), and there were 8 species pairs that showed independent relationships among them ($AC = 0$), such as *Malus honanensis* and *Pinus armandii* (*Quercus mongolica*) ($AC = 0$, $PC < 0.5$), which manifested completely independently and cannot appear in the same plots at the same time.

3.4. Analysis of Community Stability in the Different Altitude Areas

Godron scatter plots and the calculation results of communities in the three altitude areas are shown in Table 6 and Figure 6, respectively. The results showed that the distances from the intersection point of the three regression models to the stable point (20, 80) were 22.53, 11.92, 21.34 in the lower, medium and higher altitude areas, respectively. According to the Godron stability judgment method, we can conclude that the medium altitude area had a more stable community compared with the other altitude areas.

Table 6. Community stability analysis results.

Altitude Area	Curve Equation	Correlation Coefficient	Coordinates	Distance of Intersection Point and Stable Point
Lower altitude area	$y = -0.0105x^2 + 1.9003x + 9.3519$	0.982	35.93, 64.07	22.53
Medium altitude area	$y = -0.0104x^2 + 1.915x + 8.7367$	0.995	28.43, 71.57	11.92
Higher altitude area	$y = -0.012x^2 + 2.0825x + 6.6125$	0.992	35.09, 64.91	21.34

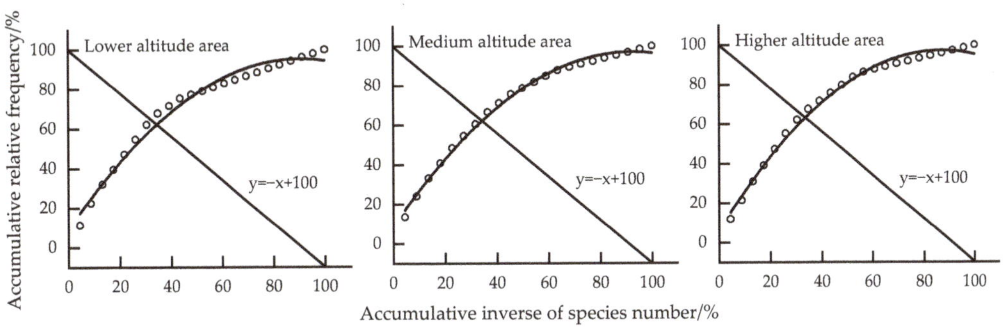

Figure 6. Godron scatter plots of communities in the three different altitude areas.

4. Discussion

4.1. Analysis of Overall Interspecific Association

The overall interspecific association reflects the stability of the community structure and species composition and better describes the community succession stage [39–41]. Generally, when a plant community reaches top-level succession, each species usually can achieve maximum utilization of the resource environment and can realize mutual promotion of species growth. Besides, the community structure tends to be complete and balanced [41–43], and, correspondingly, there is a positive overall species association. This means that when species show a negative interspecific association, the community was at an early succession or a secondary succession, with a relatively unstable community structure and composition and a low degree of interspecific association between species pairs. We calculated the overall interspecific association of the dominant species pairs at different altitude areas and found it was positive only in the medium altitude area, which suggested that, compared with other altitude areas, its community was in a stable phase, and the population of dominant trees appeared to exist in a mutually beneficial relationship.

4.2. Analysis of the Associations among the Dominant Tree Species Pairs

Interspecific associations represent a relationship among the species pairs in different habitats and show their ecological adaptability to environmental factors [44,45]. The χ^2 test results showed that the ratio of positive and negative species pair associations appeared to have a downward trend as altitude increased, which, thus, illustrated a weakened correlation and interdependence among the tree populations. In fact, several significant positive species pair associations existed in the medium altitude area, so it seemed that the medium altitude area had a stronger association between the dominant species pairs, which was consistent with the result of overall interspecific association. However, it should be clearly recognized that the interspecific association was still loose, and the distribution of each tree species was relatively independent because most species pairs were not significantly associated.

A positive correlation between species pairs means that they have a same or similar demand for environmental resources and a reciprocal symbiotic relationship, while a negative correlation reflects the adaptability of species pairs to environmental heterogeneity due to their great differences in biological characteristics, thus resulting in an exclusion and niche separation [40,44–46]. For example, *Carpinus cordata* and *Cornus macrophylla* in the medium altitude area and *Quercus Mongolica* and *Acer mono* in the higher altitude area displayed a significant positive association because they are all fond of light and resistant to poor soil and cold environment, while *Carpinus cordata* and *Cornus macrophylla* in the higher altitude area displayed a significant negative association because, unlike *Cornus macrophylla, Carpinus cordata* is not resistant to water and humidity. However, the scale of habitats also had a great impact on the result of interspecies association. If the scale of habitats was too large, they mostly tended to have positive associations; otherwise, they had negative associations [47]. Generally, the appropriate area of a community in the temperate forest was 200–500 m^2 according to the empirical value. In our study, the sample plot area was 400 m^2, which was reasonable for the study, so we inferred that this biological characteristic was the key factor in interspecies association.

Similarly, the number of positive species pair associations was more than that of the negative associations only in the medium altitude area based on the AC and PC analyses, which is consistent with the overall interspecific association and the χ^2 test. Moreover, the associations of some species pairs were not related to altitude change. For example, *Quercus variabilis* and *Acer davidii* were negatively associated throughout, while *Toxicodendron vernicifluum* and *Carpinus cordata* were positively associated from 700 m to 1900 m. However, the associations of some species pairs evolved from positive to negative as altitude increased, such as *Cornus macrophylla* and *Toxicodendron vernicifluum* and *Quercus aliena* and *Quercus variabilis*, while some evolved from negative to positive, such as *Quercus aliena* and *Acer davidii*, which clarified that the altitude factor could surely change the associations between some species pairs.

The competition theory considered that the associations between species pairs would change due to external conditions, and a species pair may show different associations in different habitats [48,49], which confirmed the results of this study. The dominant species in the lower altitude area prefer sunshine and tend to compete for the limited light resources and nutrients, thus causing a negative association. Furthermore, forests in this area suffered from heavy logging activities in the last century and human interference in recent years. Therefore, it might change tree species composition, promote or inhibit the survival of certain tree species and have a certain negative impact on the interactions among species as well [50]. Moreover, our study showed that more positive species pair associations were observed than negative species pairs in the medium altitude area, owing to its better environmental conditions and protection, so we inferred that historical and realistic disturbances could affect the ratio of positive and negative associations. This was quite the same as the research results of Gu et al. [40]. Moreover, the number of negatively associated species was greater in the higher altitude area that was restricted by harsh climatic and soil conditions.

4.3. Analysis of the Stability of Tree Communities

The Godron method is comprehensive and systematic in describing community stability and can further improve the analysis results of interspecific association [48,50]. The Godron method analysis results in this study showed that the coordinates of the stability point change from (35.93, 64.07) in the lower altitude area to (28.43, 71.57) in the medium altitude area, and then to (35.09, 64.91) in the higher altitude area. It can be seen that the coordinate of stability point in medium altitude area was closer to the Godron stable point (20, 80) compared with the other altitude areas. This result is basically consistent with the results of *VR*, *AC* and *PC*, because the lower altitude areas suffer from a certain human interference, and the habitat in higher altitude areas was inferior because of the lower temperature and poor soil fertility, which influence the status and role of the tree species in plant communities, thus affecting the community stability [21,22,44,51].

In summary, the communities in the southern foothills of the Taihang Mountains are at a relatively unstable succession stage, with fluctuating species compositions, community structures and competitions, though the community appeared to be steadier in medium altitude area, which was contrary to the results of Li et al. [52] on the interspecific association of the main trees in the tropical rainforest. The reason might be that tree species in the study areas are mainly oak trees, which are concentrated in the plot and more easily form dominant species groups and constructive tree species compared with tropical rain forests with complex community composition and high heterogeneity.

4.4. Vegetation Protection Strategy and Prospect

In this study, we found that communities were at an unstable stage of secondary succession in this area, with *Quercus aliena* and *Quercus variabilis* as the main ones in the tree layers. These tree species had strong vitality, which distributed centrally and occupied a large part in community [28] Their niches were generally negatively associated with *Fraxinus chinensis*, *Swida alba* and other tree species, with an extremely small coupling coefficient and competitions for environmental resources. Therefore, we should not only fully understand the ecological and biological characteristics of tree species in the process of community reconstruction and restoration but also consider the influence of different habitats on the relationships between tree species. The specific measures are to select tree species with strong environmental adaptability and strong positive interspecific association according to the altitude division for collocation planting to improve the community structure, prevent vicious competition among species [53] and to promote the restoration of vegetation and the stability of communities in the southern Taihang Mountains.

Altitude could affect interspecies associations and community stability. However, it is still difficult to explain the specific reasons for the formation of interspecific associations, which are generally affected by complex factors. Therefore, the forest spatial structure index, soil, topography, climate and other factors should also be considered jointly to obtain a more comprehensive analysis result in future relevant research.

5. Conclusions

The overall interspecific association, association between dominant species pairs and community stability in natural secondary forests at different altitudes were studied in this research. We concluded that the altitude factor can change the interspecific associations between tree species pairs. The communities of three altitude areas were at a relatively unstable succession stage, though it was steadier in the medium altitude area. Tree species should be selected for planting in accordance with altitude gradients, and the ratio of positive and negative species pair correlations should also be adjusted reasonably. Studying the effects of the altitude factor on community stability and the interspecific association of natural secondary forest after long-term restoration is important to understand the effectiveness of ecosystem restoration in the local area.

Author Contributions: Methodology, S.-S.J. and Y.-Y.Z.; software, Y.-Y.Z. and M.-L.Z.; investigation, Y.-Y.Z., X.-M.D., C.-H.C. and T.W.; data curation, T.W., X.-M.D. and C.-H.C.; writing—original draft preparation, S.-S.J., Y.-Y.Z. and D.-F.Y.; writing—review and editing, Y.-Y.Z., S.-S.J. and M.-L.Z.; visualization, M.-L.Z. and S.-S.J.; project administration and funding acquisition, D.-F.Y. All authors have read and agreed to the published version of the manuscript.

Funding: This study was supported by Major Science and Technology Special Projects Research in Henan Province (Sub-project of Key Technologies for Cultivating High-efficiency and Stable Plantation in the Yellow River Basin Construction Technology of Stable Plant Community in Funiu Mountain Ecological Barrier Zone) (201300111400-2), and Science and Technology Projects Research in Henan Province (Key Carbon Sink Management Technologies for the Young and Middle-aged Oak Forests Based on Close to Nature Management) (222102110418).

Institutional Review Board Statement: Not applicable.

Informed Consent Statement: Not applicable.

Data Availability Statement: The following is available online at http://www.gscloud.cn/ (25 November 2021), Figure 1: Location map of the sampling sites in the northern Jiyuan City.

Acknowledgments: We are grateful to the staff from Yugong forest farm and Huanglianshu forest farm, Wang Qunxing and Sun Yijie from Henan Agricultural University for their support during fieldwork and Yu Xiaoya from Qiannan Normal University for Nationalities for his teaching in data processing.

Conflicts of Interest: The authors declare no conflict of interest.

References

1. Liu, Y.Y.; Jin, G.Z. Spatial patterns and associations of four species in an old-growth temperate forest. *J. Plant Interact.* **2014**, *9*, 745–753. [CrossRef]
2. Luo, D.; Wang, C.S.; Dao, B.H.; Zhao, Z.G.; Guo, J.J.; Zeng, J. Species composition and diversity of *Betula alnoides* natural forests at Dehong Prefecture, Yunnan Province. *For. Res.* **2021**, *34*, 159–167.
3. Chen, Q.; Chen, J.; Zhong, J.J.; Ji, L.T.; Kang, B. Interspecific association and functional group classification of the dominant populations in shrub layer in secondary forest of *Pinus tabuliformis* in Qinling Mountain, China. *Chin. J. Appl. Ecol.* **2018**, *29*, 1736–1744.
4. Liu, Z.W.; Zhu, Y.; Wang, J.J.; Ma, W.; Meng, J.H. Species association of the dominant tree species in an old-growth forest and implications for enrichment planting for the restoration of natural degraded forest in subtropical China. *Forests* **2019**, *10*, 957. [CrossRef]
5. Maihaiti, M.; Zhang, W.J. A mini review on theories and measures of interspecific associations. *Selforganizology* **2014**, *1*, 206–210.
6. Chai, Z.Z.; Sun, C.L.; Wang, D.X.; Liu, W. Interspecific associations of dominant tree populations in a virgin old-growth oak forest in the Qinling Mountains, China. *Bot. Stud.* **2016**, *57*, 23. [CrossRef] [PubMed]
7. Greig-Smith, P. *Quantitative Plant Ecology*; Blackwell Science Publications: Oxford, PA, USA, 1983; pp. 105–112.
8. Cole, L.C. The measure of interspecific association. *Ecology* **1949**, *30*, 411–424. [CrossRef]
9. Ofomata, V.C.; Overholt, W.A.; Van, H.A.; Egwuatu, R.I.; Ngi-Song, A.J. Niche overlap and interspecific association between *Chilo partellus* and *Chilo orichalcociliellus* on the Kenya coast. *Entomol. Exp. Appl.* **1999**, *93*, 141–148. [CrossRef]
10. Wang, B.S.; Li, M.G.; Zan, Q.J. An analysis of interspecific associations in secondary succession forest communities in Heishiding natural reserve, Guangdong Province. *Chin. J. Plant Ecol.* **2000**, *24*, 332–339.
11. Zhang, Z.H.; Hu, G. Interspecific relationships of dominant species in *Cyclobalanopsis glauca* community in karst mountain area. *Ecol. Env. Sci.* **2011**, *20*, 1209–1213.
12. Lu, L.L.; Guo, Z.L.; Fan, C.N.; Zheng, J.P. Community characteristic and stability analysis of secondary deciduous broad-leaved forest in Mopan Mountains, Jilin Province, China. *Chin. J. Appl. Ecol.* **2018**, *29*, 2079–2087.
13. Yu, X.W.; Song, X.S.; Kang, F.F.; Han, H.R. Evaluation on the stability of typical forest community in source region of Liaohe river in North Hebei. *J. Arid Land Res. Environ.* **2015**, *29*, 93–98.
14. Hubalek, Z. Coeffident of association and similarity based on binary data: An evaluation. *Biol. Rev.* **1982**, *57*, 669–689. [CrossRef]
15. Zhang, Y.; Guo, L.P.; Yi, X.M.; Cao, W.; Wang, Y.X.; Wu, P.L.; Ji, L.Z. Analysis of interspecific associations among major tree species in three forest communities on the north slope of Changbai Mountain. *Acta Ecol. Sin.* **2015**, *35*, 0106–0115.
16. Jiang, H.; Zhang, H.; Long, W.X.; Fang, Y.S.; Fu, M.Q.; Zhu, K.X. Interspecific associations and niche characteristics of communities invaded by *Decalobanthus boisianus*. *Biodivers. Sci.* **2019**, *27*, 388–399.
17. Zhang, J.T.; Xi, Y.; Li, J. The relationships between environment and plant communities in the middle part of Taihang Mountain Range, North China. *Community Ecol.* **2006**, *7*, 155–163. [CrossRef]
18. Saha, S.; Rajwar, G.S.; Kumar, M. Forest structure, diversity and regeneration potential along altitudinal gradient in Dhanaulti of Garhwal Himalaya. *For. Syst.* **2016**, *25*, e058.
19. Tian, Z.P.; Wang, X.L.; Zhao, X.Y.; Zhuang, L. A unique mountains vertical distribution patterns and related environmental interpretation- a case on the northern slope of the ILI river valley. *Pak. J. Bot.* **2016**, *48*, 1877–1886.
20. Geng, S.B.; Shi, P.L.; Song, M.H.; Zong, N.; Zu, J.X.; Zhu, W.R. Diversity of vegetation composition enhances ecosystem stability along elevational gradients in the Taihang Mountains, China. *Ecol. Indic.* **2019**, *104*, 594–603. [CrossRef]
21. Mu, C.C.; Ni, Z.Y.; Li, D.; Chen, J.L. Distribution patterns of woody plant diversity in stream riparian forests along an altitudinal gradient in Changbai Mountains. *Chin. J. Appl. Ecol.* **2007**, *18*, 943–950.
22. Cabrera, O.; Benitez, A.; Cumbicus, N.; Naranjo, C.; Ramon, P.; Tinitana, F.; Escudero, A. Geomorphology and altitude effects on the diversity and structure of the vanishing montane forest of southern Ecuador. *Diversity* **2019**, *11*, 32. [CrossRef]
23. Bhutia, Y.; Gudasalamani, R.; Ganesan, R.; Saha, S. Assessing Forest structure and composition along the altitudinal gradient in the state of Sikkim, Eastern Himalayas, India. *Forests* **2019**, *10*, 633. [CrossRef]
24. Zhang, Y.B.; Meng, Q.X.; Qin, H.; Zhang, F. Flora and geographic pattern of mountain forests at community level in Taihang Mountains: Results based on plant community survey. *J. Appl. Ecol.* **2019**, *30*, 3395–3402.
25. Yang, X.L.; Xu, Q.H.; Zhao, H.P. The vegetation succession since the last glaciation in Taihang Mountains. *Geogr. Territ. Res.* **1999**, *15*, 82–89.
26. Zhao, H.; Wang, Q.R.; Fan, W.; Song, G.H. The relationship between secondary forest and environmental factors in the Southern Taihang Mountains. *Sci. Rep.* **2017**, *7*, 2474–2475. [CrossRef]
27. Wang, J.; Feng, J.W.; Niu, S.; Zhao, L.X.; Ye, Y.Z. Comparative analysis of niche characteristics of the species and interspecific correlation in Taihang Mountains. *J. Henau Agric. Univ.* **2016**, *50*, 181–188.
28. Yan, D.F.; Zhang, Y.Y.; Lv, K.T.; Zhou, M.L.; Wang, T.; Zhao, N. Niche characteristics of dominant tree species in natural forests at different altitudes in the south of Taihang Mountains. *J. Ecol. Env. Sci.* **2021**, *30*, 1571–1580.
29. Zhao, Y.; Fan, W.; Ye, Y.Z.; Yan, L.; Zhao, D. Analysis of species diversity of difference plant communities in hilly region of Taihang mountain. *Sci. Soil Water Conser.* **2007**, *5*, 64–71.
30. Wan, M.; Tian, D.L.; Wei, W. Structure characteristics of the *Quercus* variabilis forest community in the south area of Taihang Mountains. *J. Henau Agric. Univ.* **2009**, *43*, 139–144, 150.

31. Schluter, D. A variance test for detecting species associations, with some example applications. *Ecology* **1984**, *65*, 998–1005. [CrossRef]
32. Shao, L.Y.; Zhang, G.F. Niche and interspecific of dominant tree populations of *Zelkova Schneideriana* community in eastern China. *Bot. Sci.* **2021**, *99*, 823–833. [CrossRef]
33. Dai, J.; Liu, H.; Xu, C.; Qi, Y.; Zhu, X.; Zhou, M.; Liu, B.; Wu, Y. Divergent hydraulic strategies explain the interspecific associations of co-occurring trees in forest–steppe ecotone. *Forests* **2020**, *11*, 942. [CrossRef]
34. Yarranton, G.A. A plotless method of sampling vegetation. *J. Ecol.* **1966**, *54*, 229–237. [CrossRef]
35. Wang, B.S.; Peng, S.L. Studies on the measuring techniques of interspecific association of lower-subtropical evergreen-broadleaved forests. *Chin. J. Plant Ecol.* **1985**, *4*, 274–285.
36. Godron, M. Some aspecets of heterogeneity in grasslands of Cantal. *Statis. Ecol.* **1972**, *3*, 397–415.
37. Zhang, J.L. *Spaa: Species Association Analysis, R Package Version 0.2.2*; R Foundation for Statistical Computing: Vienna, Austria, 2016.
38. Wei, T.; Simko, V.R. Package "Corrplot": Visualization of a Correlation Matrix (Version 0.84). Available online: https://github.com/taiyun/corrplot (accessed on 1 November 2021).
39. Su, S.J.; Liu, J.F.; He, Z.S.; Zheng, S.Q.; Hong, W.; Xu, D.W. Ecological species groups and interspecific association of dominant tree species in Daiyun Mountain National Nature Reserve. *J. Mt. Sci.* **2015**, *12*, 637–646. [CrossRef]
40. Gu, L.; Gong, Z.W.; Li, W.Z. Niches and interspecific associations of dominant populations in three changed stages of natural secondary forests on Loess Plateau, P.R. China. *Sci. Rep.* **2017**, *7*, 6604–6616. [PubMed]
41. Ma, F.F.; Pan, G.; Li, X.Q.; Han, Y.J. Interspecific relationship and canonical correspondence analysis within woody plant communities in the karst mountains of Southwest Guangxi, southern China. *J. Beijing For. Univ.* **2017**, *39*, 32–44.
42. Li, T.T.; Rong, L.; Wang, M.J.; Ye, T.M.; Qi, W. Dynamic changes in niche and interspecific association of major species of karst secondary forest in central Guizhou. *J. Trop. Subtrop. Bot.* **2021**, *29*, 9–19.
43. Jin, S.; Wu, S.K. Niche and interspecific association of dominant species in herb layer of burned *Pinus tabuliformis* forest in the southern Taihang Mountain of northern China. *J. Beijing For. Univ.* **2021**, *43*, 35–46.
44. Liu, R.H.; Jiang, Y.; Chang, B.; Li, J.F.; Rong, C.Y.; Liang, S.C.; Yang, R.A.; Liu, X.T.; Zeng, H.F.; Su, X.L.; et al. Interspecific associations and correlations among the main woody plants in a *Pterocarya stenoptera* community in a riparian zone of Lijiang River, Guilin, Southwest China. *Acta Ecol. Sin.* **2018**, *38*, 6881–6893.
45. Wu, D.T.; Wu, C.P.; Sheng, G.; Wei, X.; Jiao, J.J.; Jiang, B.; Zhu, J.R.; Yuan, W.G. Interspecific association dynamics of Nanmu natural forest in Jiande, Zhejiang Province. *J. Zhejiang AF Univ.* **2021**, *38*, 671–681.
46. Lv, A.Q.; Li, D.H.; Yang, X.B.; Wu, L.X. Plant community diversity and inter-specific relationship of coastal rain forest, semi-deciduous monsoon forest to deciduous monsoon forest in coastal hills of Sanya City, Hainan Province. *Guihaia* **2021**, *41*, 384–395.
47. Xu, M.H.; Liu, M.; Zhai, D.T.; Liu, T. A review of contents and methods used to analyze various aspects of plant interspecific associations. *Acta Ecol. Sin.* **2016**, *36*, 8224–8233.
48. Liu, R.X.; Chen, L.Q. Effect of flooding disturbance on plant community stability and interspecific relationship in the riparian zone of reservoir. *Acta Ecol. Sin.* **2021**, *41*, 6566–6579.
49. Luo, J.F.; Zheng, J.M.; Zhou, J.X.; Zhang, X.; Cui, M. Analysis of the interspecific associations present in an alpine meadow community undergoing revegetation on the railway-construction affected land of the Qinghai-Tibet Plateau. *Acta Ecol. Sin.* **2016**, *36*, 6528–6537.
50. Xie, C.P.; Liu, D.W.; Nan, C.H.; Li, H.; Huang, C.Y.; Zhang, J. Study on community stability and interspecific correlations among dominant populations in *Pseudolarix amabilis* communities. *Ecol. Sci.* **2021**, *40*, 62–70.
51. Ye, Q.P.; Zhang, W.H.; Yu, S.C.; Xue, W.Y. Interspecific association of the main tree populations of the *Quercus acutissima* community in the Qiaoshan forest area. *Acta Ecol. Sin.* **2018**, *38*, 3165–3174.
52. Li, J.L.; Lin, Y.C.; Wang, X.; Liu, H.W.; Wen, G.S.; Wag, S.; Zhang, X.L. Comparative study on rainforest species association in different altitude in Mountain Diaoluo of Hainan. *Chin. J. Trop. Crop.* **2013**, *34*, 584–590.
53. Liu, R.H.; Chen, L.; Tu, H.R.; Liang, S.C.; Jiang, Y.; Li, Y.J.; Huang, D.L.; Nong, J.L. Niche and interspecific association of main species in shrub layer of *Cyclobalanopsis glauca* community in karst hills of Guilin, southwest China. *Acta Ecol. Sin.* **2020**, *40*, 2057–2071.

forests

MDPI

Article

Phylogenetic and Functional Structure of Wood Communities among Different Disturbance Regimes in a Temperate Mountain Forest

Peikun Li [1,2], Zihan Geng [1,2], Xueying Wang [3], Panpan Zhang [1,2], Jian Zhang [1,2], Shengyan Ding [1,2,*] and Qiang Fu [4,*]

1 Key Laboratory of Geospatial Technology for the Middle and Lower Yellow River Regions, Ministry of Education, Kaifeng 475004, China; peikunlee@126.com (P.L.); 104753190167@henu.edu.cn (Z.G.); zhangpanpan@henu.edu.cn (P.Z.); zj_0305@henu.edu.cn (J.Z.)

2 College of Environment and Planning, Henan University, Kaifeng 475004, China

3 Henan Forestry Investigation and Planning Institute, Zhengzhou 450002, China; wangsnow0909@163.com

4 College of Landscape Architecture and Art, Henan Agricultural University, Zhengzhou 450002, China

* Correspondence: syding@henu.edu.cn (S.D.); fqlandscape@126.com (Q.F.); Tel.: +86-0371-23881102 (S.D.)

Abstract: The mechanisms responsible for biodiversity formation and maintenance are central themes in biodiversity conservation. However, the relationships between community assembly, phylogeny, and functional traits remain poorly understood, especially following disturbance. In this study, we examined forest community assembly mechanisms in different disturbance regimes across spatial scales and including tree life history classes, using phylogenetic and functional trait metrics. Across disturbance regimes, phylogenetic structure tended to be over-dispersed, while functional structure tended to be clustered. The over-dispersion of phylogenetic structure also increased from small to large diameter species. Moreover, the explanation of spatial distance for the turnover of phylogenetic and functional structure was increased, while environmental distance explained less structure as disturbance intensity decreased. Our findings suggest that niche theory largely explains forest community assembly in different disturbance regimes. Furthermore, environmental filtering plays a major role in moderate to high disturbance regimes, while competitive exclusion is more important in undisturbed and slightly disturbed ecosystems.

Keywords: net relatedness index; functional trait; niche theory; habitat filtering; competitive exclusion; deciduous broad-leaved forests

Citation: Li, P.; Geng, Z.; Wang, X.; Zhang, P.; Zhang, J.; Ding, S.; Fu, Q. Phylogenetic and Functional Structure of Wood Communities among Different Disturbance Regimes in a Temperate Mountain Forest. *Forests* **2022**, *13*, 896. https://doi.org/10.3390/f13060896

Academic Editor: Brian J. Palik

Received: 16 May 2022
Accepted: 6 June 2022
Published: 8 June 2022

1. Introduction

Biodiversity formation and maintenance mechanisms, and community assembly mechanisms in particular, are central themes in biodiversity conservation [1]. Niche theory holds that the niche differentiation among coexisting species strongly affects community construction, which results from forces including habitat filtering and competitive exclusion [2,3]. In contrast, neutral theory posits that stochastic factors, such as diffusion and random effects, are the determinants of community construction [4]. Based on the phylogenetic niche conservation theory of Webb [5], the phylogenetic distance of species within communities can be used to infer the relative strengths of niche and neutral progress in community assembly. If the evolution of species functional traits is relatively slow, habitat filtering is predicted to lead to clustered community phylogenetic structures, while competitive exclusion leads to over-dispersed communities [6–9]. Random phylogenetic structures may result from diffusion and habitat filtering or a combination of random effects and competitive exclusion [10,11]. Community functional trait structure, therefore, represents a comprehensive pattern of species functional traits [12]. The existing community trait distributions result from differences in the selection of environmental and non-environmental factors by species, and thus provide important clues to understanding the relative importance of ecological processes in community construction [13,14]. With improved phylogenetic and functional ecology methods, community phylogenetic studies of plant functional traits have become common tools for assessing community construction mechanisms. The roles of niche and neutral processes in community construction, based on phylogenetic or functional traits and α− and β−diversity, have attracted much attention [15–19]. Studies have shown that the α- and β-diversity of community phylogenetic and functional traits are closely related to study scales, both in time and space [20,21]. Interspecific interactions and diffusion restrictions are more prevalent at smaller community scales, while environmental filtering is generally a feature of larger scales [22,23]. Meanwhile, the α-diversity of community phylogenetic and functional traits also show different response patterns for tree species at different life history stages, due to different environmental needs and tolerances [20]. For example, small and medium diameter tree species are commonly subject to habitat filtering, leading to clustered community phylogenetic structures [24]. Whereas, competitive exclusion is more likely to occur between large diameter trees, due to the need for more resources, resulting in over-dispersion [9]. The relative importance of diffusion and environmental filtering in community construction can be inferred from changes in the phylogenetic signals and ecological characteristics of species between communities [25,26]. Although phylogenetic and functional trait diversity is increasingly used to infer community assembly mechanisms individually, most studies do not combine them [27].

Disturbances, from human activities to natural fires and earthquakes, have profound impacts on regional community construction and species diversity [28]. With the increasing frequency of human activities, human disturbance has become the primary factor affecting the construction of regional communities [29]. The influence of human disturbance on ecosystems has long been a focus of multidisciplinary research in geography, ecology, and natural resources science [30]. Disturbance theory is a vital part of ecology, and the "intermediate-disturbance hypothesis" is currently the most studied [31,32]. This hypothesis suggests that moderate disturbances help maintain high biodiversity [33,34]. Generally, unmanaged forests after human disturbance are in the early and middle stages of succession [35]. However, some extreme disturbances can reverse the succession of secondary forests, which seriously threatens healthy forest development [36].

Deforestation is among the most common human disturbances [37] and affects forest phylogenetic and functional trait structure, and subsequently alters forest community biodiversity and ecosystem function [11]. Differences in the biotic (e.g., community structure and species composition) and abiotic (e.g., soil and light) environments [38–42] resulting from different deforestation methods, intensities, and intervals lead to different effects on the structure, function, and biodiversity of forest ecosystems [43]. Therefore, studying the effects of deforestation disturbance on regional community construction and species diversity is of great significance for the renewal and development of forest communities [44]. Most studies have focused on the impact of disturbances on forest community structure, stability, and species diversity [29,44–48]. However, there are few studies of community assembly mechanisms that examined the phylogenetic and functional trait structure of woody plants as succession progressed.

In this study, we examined the community assembly mechanisms of woody plants in forests subject to different disturbance regimes, at spatial and diameter at breast height (DBH) scales, using phylogenetic signals and the phylogenetic and functional trait structure. We hypothesized that (1) species with similar genetic relationships would have similar functional traits, as a result of significant phylogenetic signals; (2) the phylogenetic and functional trait structure at small and medium spatial and DBH class scale could have higher clustering, due to habitat filtering and competitive exclusion; and (3) environmental filtering could tend to be more important following high and moderate disturbance, due to increased resource availability and species richness, and competitive exclusion could tend to be more important for slightly disturbed and undisturbed communities, due to diffusional limitation.

2. Materials and Methods

2.1. Study Site and Sampling

The study site was located in the Baiyun Mountain National Nature Reserve (111°48′–112°16′ E, 33°33′–33°56′ N), Luoyang, south of Henan Province, China (Figure 1), which is about 168 km^2 and 1500–2216 m above sea level [29]. The slope of the mountain is mostly 40–80°. Long-term mean annual precipitation is approximately 1200 mm, with most occurring from July to September, and the mean annual relative humidity is 70–78%. Mean annual temperature was 13.1–13.9 °C, and extreme minimum and maximum temperatures were −14.4 and 42.1 °C, respectively. The soil texture is mainly light soil, with a pH of 5.5–6.5 [49].

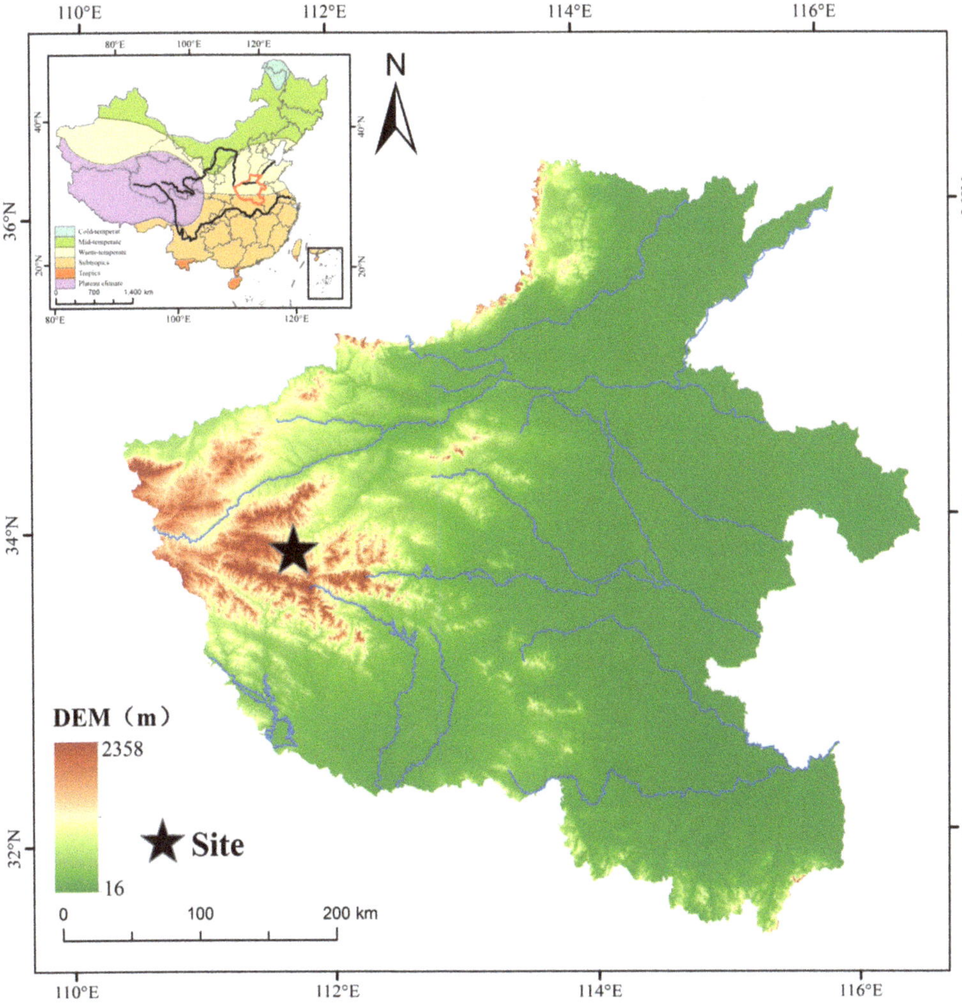

Figure 1. Location of different disturbance dynamics plots in the temperate−subtropical ecological transition zone of the Baiyun Mountain Nature Reserve. DEM represents altitude.

The Baiyun Mountain Nature Reserve is located in a temperate–subtropical ecological transition zone, with mostly deciduous broad-leaved forests. The forest coverage in the reserve reaches 98.5%, which consists of 1991 species of plants, including the following dominant species: *Quercus aliena* var. *acutiserrata*, *Carpinus turczaninowii*, *Betula platyphylla*, *Pinus armandii Franch*, and *Toxicodendron vernicifluum* [49].

Forest monitoring plots were randomly selected and stratified by disturbance regime in the Baiyun Mountain Nature Reserve, where plant growth and ecosystem functions are sensitive to climate change [50]. Four disturbance regimes of the forest were estimated, based on knowledge of local logging events and forest physiognomy. Four 1 hm^2 plots (100 m × 100 m), namely, plantation, twice-cut, once-cut, and old-growth forests, were randomly selected within each disturbance regime in the reserve (Table 1). Four 1 hm^2 plots were divided into 100 grids (10 m × 10 m). All trees with diameter at breast height (DBH) ≥1 cm in the plot were tagged, mapped, and measured [51]. Topographic variables (elevation, convexity, slope, and aspect) were measured using the methodology of Harms [52] for each 10 m × 10 m grid in the plot.

Table 1. Summaries of disturbance regime forest plots in Baiyunshan Nature Reserve.

	Plantation Forest	Twice-Cut Forest	Once-Cut Forest	Old-Growth Forest
Average elevation (m)	1647.4	1578.66	1413.15	1772.62
Mean DBH (cm)	14.23	7.21	7.76	9.5
Total basal area (m^2)	22.925	25.95	33.33	31.9
Number of species	42	46	57	52
Individual number	953	2534	3671	2318
Disturbance regimes	*Larix kaempferi* forest planted after logging and clearing.	Natural regeneration occurred after once-cutting. Twice-cutting and breeding were carried out after about 30 years natural recovery, followed again by natural recovery.	The forest was restored after comprehensive once-cutting.	The forest has been undisturbed for more than 100 years.
Age of forest (years)	20	50	50	100
Degree of disturbance	High disturbance	Moderate disturbance	Slight disturbance	Undisturbed
Dominant species	*Quercus aliena* var. *acutiserrata*; *Larix gmelinii*	*Quercus aliena* var. *acutiserrata*; *Pinus armandii Franch*; *Corylus heterophylla*	*Quercus aliena* var. *acutiserrata*; *Pinus armandii Franch*; *Forsythia suspensa*	*Quercus aliena* var. *acutiserrata*; *Sorbus hupehensis*; *Litsea tsinlingensis*

To assess the relationship between phylogenetic structure and spatial scale, we further divided each 1 hm^2 plot into 10 m × 10 m, 20 m × 20 m, and 25 m × 25 m grids, with a total of 100, 25, and 16 of each size, respectively. To investigate the relationship between time scale (as measured by tree size) and phylogenetic structure, we divided all woody plants with a DBH ≥1 cm into three different diameter classes following [53]; namely, small (1 cm ≤ DBH < 5 cm), medium (5 cm ≤ DBH < 10 cm), and large (DBH ≥ 10 cm).

2.2. Phylogenetic Tree Construction

Phylogenetic trees were constructed using the database Phylomatic [8]. All species information in each plot was imported into the community phylogenetic software Phylocom Version 3.21 (available online http://www.phylodiversity.net/phylocom, accessed on 4 February 2018) [54]. First, major relationships were taken from the Angiosperm Phylogeny Group classification (APG IV 2016). Second, the BLADJ algorithm was implemented within Phylocom to calibrate each species pool supertree, by applying known molecular and fossil dates [55] to nodes on the supertree, resulting in ultrametric phylogenetic trees of each community (Figure 2).

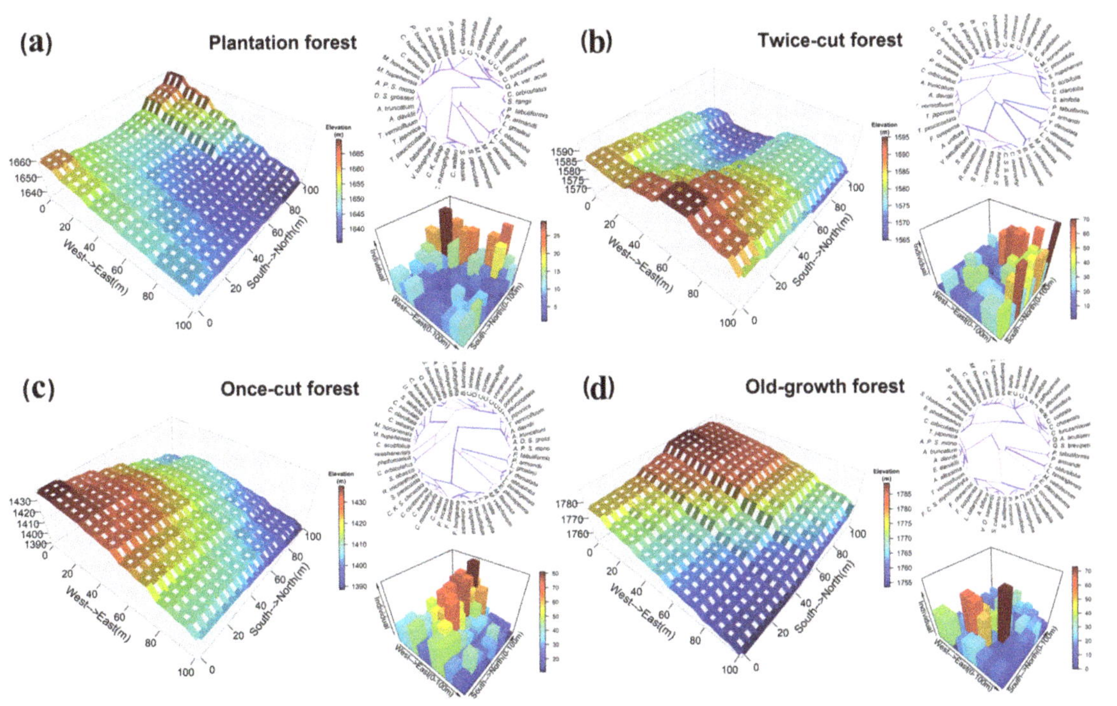

Figure 2. Topographic maps, phylogenetic trees, and spatial species abundances of woody plants in the plantation (**a**), twice-cut (**b**), once-cut (**c**), and old-growth (**d**) forests of the Baiyun Mountain Nature Reserve.

2.3. Functional Trait Clustering

We measured seven key functional traits representing the leaf (specific leaf area, stomatal conductance), stem (maximum tree height, wood density), and physiological (minimal fluorescence, non−photochemical quenching, transpiration rate) traits of the tree species in the community, according to the handbook for standardized measurement of plant functional traits by Pérez−Harguindeguy [56]. We reduced the dimensionality of these traits through principal component analysis (PCA). The first three axes were selected as comprehensive functional trait factors to transform the trait matrix into a distance matrix, and hierarchical clustering was conducted according to the trait distances between species, to generate a functional trait clustering tree [57].

2.4. Data Analysis

2.4.1. Phylogenetic Signal

The phylogenetic signal was analyzed using Blomberg's K [58], which is a measure of the observed trait variance compared to that expected under Brownian motion. The null expectation of $K = 0$ represents no phylogenetic signal, while $K = 1$ indicates a strong phylogenetic signal, and that the trait evolves according to Brownian motion. A weak phylogenetic signal is indicated by $0 < K < 1$, whereas $K > 1$ indicates a very strong signal, and that the trait values are more similar than expected under Brownian motion. The significance of the phylogenetic signal can be obtained by comparing variance observations of standardized independent differences across the phylogenetic tree for functional traits with a random test of the null model.

2.4.2. Community Phylogenetic and Functional Trait Structure

The net relatedness index (*NRI*) and standardization mean pairwise trait distance (*S.E.S PW*) were calculated to reflect the phylogenetic and functional trait character structures, respectively, of tree species in each spatial scale and diameter class, for each disturbance regime [5,59]. First, the mean phylogenetic distances (*MPDs*) and mean pairwise trait distances (*PWs*) for all species pairs in the quadrat were quantified. Then, we used a richness null modelling approach to estimate the expected subplot species richness distributions under random processes; we randomly permuted the species set of the phylogenetic tree or functional trait clustering tree 999 times to obtain the *MPD* or *PW* of each species pair in the quadrat under the random null model [60]. Finally, the observed values were normalized using the random distribution result, to obtain the values of *NRI* and *S.E.S PW*, calculated using the following formula [5]:

$$NRI = -1 \times \frac{(MPD_{obs} - \text{mean}(MPD_{rnd}))}{\text{sd}(MPD_{rnd})}, \quad (1)$$

$$S.E.S\ PW = -1 \times \frac{(PW_{obs} - \text{mean}(PW_{rnd}))}{\text{sd}(PW_{rnd})} \quad (2)$$

where MPD_{obs} and PW_{obs} represent the observed *MPD* and *PW* values; MPD_{rnd} and PW_{rnd} represent the *MPD* and *PW* values of 999 randomly generated null communities; and sd (MPD_{rnd}) and sd (PW_{rnd}) are the standard deviations of the 999 MPD_{rnd} and PW_{rnd} values, respectively. Negative values of *NRI* and *S.E.S PW* indicate higher mean phylogenetic distances and mean pairwise trait distance, respectively, than expected given the random assemblages, and are indicative of phylogenetic and functional trait over-dispersion. Whereas positive *NRI* and *S.E.S PW* values indicate lower mean distances and phylogenetic and functional trait clustering, respectively.

Previous phylogenetic studies have shown that the distributions of *NRI* and *S.E.S PW* scores from multiple equally sized quadrats are generally right-biased [9]. Therefore, we used the nonparametric Wilcoxon test to test for significant deviations between *NRI* or *S.E.S PW* and zero [61]. Moran's I was used to test the spatial autocorrelation of species *NRI* and *S.E.S PW* at different scales [62]. Spatial autoregression analyses (SAR) were used to analyze the effects of removing spatial autocorrelation on community phylogenetic structure and functional trait structure [54].

2.4.3. Beta Diversity of Phylogenetic and Functional Traits

The mean pairwise distance (D_{pw}) index was used to measure the phylogenetic or functional dissimilarity among the four disturbance regimes at different scales [63]:

$$D_{pw} = \frac{\sum_{i=1}^{n_{k1}} f_i \overline{\delta_{ik2}} + \sum_{j=1}^{n_{k2}} f_j \overline{\delta_{jk1}}}{n_{k1} + n_{k2}} \quad (3)$$

where $\overline{\delta_{ik2}}$ is the mean pairwise phylogenetic or pairwise trait distance between species *i* in community k_1 to all species in community k_2 and $\overline{\delta_{ik1}}$ is the mean pairwise phylogenetic or pairwise trait distance between species *j* in community k_2 to all species in community k_1; and f_i and f_j are the relative abundances of species *i* and species *j*, respectively.

The four disturbance regime forests were divided into 20 m × 20 m subplots, and the Euclidean distances between the centers of the 25 subplots in each plot were calculated as a spatial distance. Environmental distances were measured as the Euclidean distances between environmental factors (standardize slope, aspect, elevation, and convexity) to create a standardized environmental matrix. We calculated D_{pw} values between the 100 quadrats and used Mantel tests to measure the correlations between D_{pw} and environment matrices. Multiple regression on distance matrices (MRM) was used for the partial Mantel tests of spatial distance, environmental distance, and D_{pw}. The MRM was used to decompose the variance of the phylogenetic β-diversity value into three parts: spatial distance, environmental distance, and the interaction between the two. MRM was used to assess the effects of spatial and environmental distance on community phylogenetic and functional trait turnover [64].

Phylogenetic and functional indices, Blomberg's K, and associated p-values were estimated with the "picante" package [25]. Moran's I and spatial autoregression analyses were conducted with the "spdep" package [64]. Mantel tests and MRM were conducted with the "ecodist" package [64]. All statistical analyses were conducted in R 3.4.0 (R Development Core Team, http://www.Rproject.org, accessed on 4 February 2018) [65].

3. Results

3.1. Phylogenetic Signals

Across the four disturbance plots, we detected phylogenetic signals ($K > 0$, $p < 0.05$) for all traits, except transpiration rate (TR), in the twice-cut forest (Table 2). Blomberg's K was smallest for maximum tree height (MTH) and greatest for non-photochemical quenching (NPQ), ($K > 1$ in twice-cut and old-growth plots). Therefore, the evolutionary history explained much of the functional trait variation of the plant species in the Baiyun Mountain plots; that is, species with similar kinship had similar functional traits.

Table 2. Phylogenetic signal as measured by Blomberg's K of functional traits in four disturbance regimes. MTH = maximum tree height (m); SLA = specific leaf area ($cm^2 \cdot g^{-1}$); WD = wood density ($g \cdot cm^{-3}$); F0 = minimal fluorescence; NPQ = non-photochemical quenching; TR = transpiration rate ($mol \cdot m^{-2} \cdot s^{-1}$); SC = stomatal conductance ($mmol \cdot m^{-2} \cdot s^{-1}$). "*" and "**" represent $p < 0.05$ and $p < 0.01$, respectively.

Trait	Plantation		Twice-Cut		Once-Cut		Old-Growth	
	K	p	K	p	K	p	K	p
MHT	0.376	0.007 **	0.27	0.035 *	0.248	0.02 *	0.406	0.001 **
SLA	0.649	0.001 **	0.531	0.001 **	0.686	0.001 **	0.616	0.001 **
WD	0.462	0.037 *	0.593	0.039 *	0.425	0.049 *	0.617	0.003 **
F0	0.899	0.001 **	0.662	0.022 *	0.918	0.001 **	0.702	0.001 **
NPQ	0.968	0.004 **	1.206	0.005 **	0.915	0.003 **	1.098	0.004 **
TR	0.499	0.017 **	0.515	0.074	0.474	0.02 *	0.588	0.005 **
SC	0.75	0.001 **	0.387	0.004 *	0.394	0.001 **	0.318	0.029 *

3.2. Phylogenetic and Functional Structure at Spatial and DBH Scales

The phylogenetic structure tended to be over-dispersed across disturbance regimes, both overall and at different spatial scales and diameter classes (Table 3). Specifically, we observed over-dispersion ($NRI < 0$, $p < 0.05$) in all four disturbance regimes overall and with large DBH species, twice-cut and once-cut plots with medium DBH species, and in the twice-cut plot with small DBH species (Figure 3a, Table 3). Moreover, we observed over-dispersion in the once-cut plots at 20 × 20 m and in medium diameter DBH species at 10 × 10 m and 20 × 20 m.

Within DBH classes we found evidence of a clustered functional structure in different disturbance regimes. With the exception of the plantation plot with all DBH species and the once-cut plot with medium DBH species, we detected functional clustering ($S.E.S\ PW > 0$, $p < 0.05$) at 10 × 10 m, 20 × 20 m, and 25 × 25 m scales across disturbance regimes with overall, small, and medium DBH (Figure 3b, Table 3). However, the NRI in plantation and once-cut plots with large DBH species at 10 × 10 m and 20 × 20 m scales were functionally over-dispersed.

Table 3. Results of t−test for the hypothesis that the mean values of *NRI* and *S.E.S PW* is zero at different spatial scales and DBH classes in four disturbance regimes. D1, D2, D3, and D4 represent the plantation, twice-cut, once-cut, and old-growth forests, respectively. "*", "**", and "***" represent $p < 0.05$, $p < 0.01$, and $p < 0.001$, respectively.

Space	DBH	*NRI*				*S.E.S PW*			
		D1	D2	D3	D4	D1	D2	D3	D4
10 × 10 m	Overall	3.969 ***	7.428 ***	26.777 ***	6.159 ***	1.354	17.931 ***	6.93 ***	5.965 ***
	Small	0.415	1.589	−2.516 *	−1.186	2.429 *	14.449 ***	10.541 ***	3.507 ***
	Medium	1.102	−2.43 *	18.422 ***	−3.018 **	3.88 ***	9.53 ***	−0.117	4.419 ***
	Large	−2.635 *	6.847 ***	18.515 ***	−3.302 **	−2.542 *	11.116 ***	−3.188 **	1.453
20 × 20 m	Overall	6.677 ***	−5.24 ***	16.486 ***	5.764 ***	0.553	9.396 ***	3.1 **	5.332 ***
	Small	0.432	−0.343	−3.503 **	−2.052*	2.897 *	11.695 ***	9.0 ***	3.456 **
	Medium	−0.239	−2.64 *	15.024 ***	−2.256*	2.572 *	7.745 ***	−0.732	3.376 **
	Large	5.179 ***	3.863 ***	12.809 ***	−3.033 **	−2.19 *	5.319 ***	−3.154 **	1.437
25 × 25 m	Overall	6.499 ***	4.947 ***	10.141 ***	4.309 ***	1.535	9.661 ***	3.912 **	3.995 **
	Small	0.192	−0.352	−3.74 **	−1.449	4.341 ***	14.522 ***	7.311 ***	2.656 *
	Medium	−0.124	−2.631 *	14.267 ***	−1.973	3.155 **	6.039 ***	0.045	2.711 *
	Large	4.336 ***	6.393 ***	−8.01 ***	−2.232*	−1.845	4.5 ***	−1.943	1.294

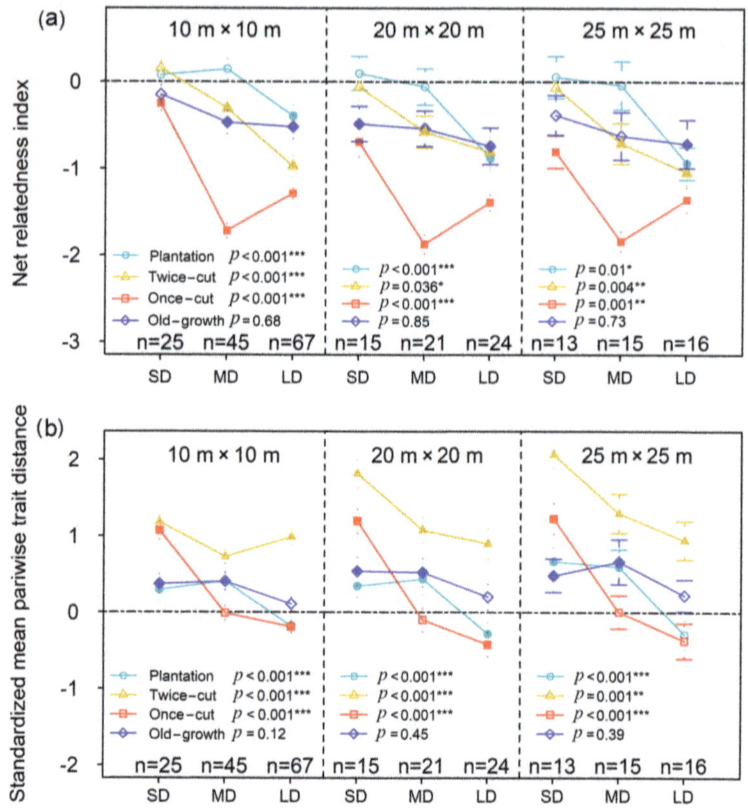

Figure 3. Phylogenetic (**a**) and functional trait structure (**b**) (mean ± SE) of different diameter classes in four disturbance plots at three spatial scales. Solid markers represent means of *NRI* or *PW* that are significantly different from 0 and the open markers represent non-significant differences based on the Wilcoxon test. "*", "**", and "***" represent $p < 0.05$, $p < 0.01$, and $p < 0.001$, respectively.

The communities tended to be more phylogenetically over-dispersed as DBH class increased ($p < 0.05$, Figure 3a), while also shifting from functional clustering to functional randomness, and even over-dispersion ($p < 0.05$, Figure 3b). With respect to spatial scale, we found that the phylogenetic and functional structure was relatively scale-independent within DBH classes, as we detected few significant differences between scales ($p > 0.05$, Figure S1). However, the phylogenetic structure decreased significantly in once-cut forests with overall and small DBH species ($p < 0.05$, Figure S1a,b), and the functional structure increased significantly in twice-cut forest in overall, small, and medium DBH species ($p < 0.05$, Figure S1c,d).

3.3. Phylogenetic and Functional Structure in Different Disturbance Regimes

All disturbance regimes exhibited over-dispersion of the overall DBH class, and the once-cut community was significantly more over-dispersed than the plantation and twice-cut communities ($p < 0.05$, Figure 4a). Within DBH classes, the once-cut community showed higher over-dispersion than more disturbed communities in the small DBH class ($p < 0.05$, Figure 4c), and the highest over-dispersion in the medium and large DBH classes ($p < 0.05$, Figure 4e,g).

In the overall DBH class, the functional structure trended to be clustered in twice-cut, once-cut, and old-growth communities, but was random in the plantation forest (Figure 4b). The twice-cut community had the strongest clustering in all DBH classes ($p < 0.05$, Figure 4b,d,f,h). Functional clustering tended to decrease with increasing disturbance in small diameter species, although the plantation (most disturbed) showed a clustering similar to the old-growth forest ($p < 0.05$, Figure 4d).

3.4. Beta Diversity of Community Phylogenetic and Functional Traits

The turnover in phylogenetic and functional traits was generally non-random in each disturbance regime and across DBH classes, as measured by *S.E.S.* D_{pw} ($p < 0.05$, Table 4). At the overall DBH level, the plantation (2.33 ± 0.62) and once-cut (2.364 ± 0.51) communities had the largest phylogenetic *S.E.S.* D_{pw} and the plantation community had the largest functional *S.E.S.* D_{pw} (1.458 ± 0.41). The phylogenetic *S.E.S.* D_{pw} of the small DBH species was consistently the smallest across DBH classes and in different disturbance regimes (Table 4). Compared with the null-model, the observed phylogenetic and functional traits varied more rapidly than expected across subplots at all scales. Both the phylogenetic and functional turnover between paired plots was greater than zero ($p < 0.05$, Table 5), and the small DBH species had the lowest turnover (Table 5, Figure 5).

Table 4. Phylogenetic and functional standardized mean pairwise distances (mean *S.E.S. Dpw* ± SE) between disturbance communities. SD, MD, and LD represent small, medium, and large diameter classes, respectively. "*", "**" and "***" represent $p < 0.05$, $p < 0.01$, and $p < 0.001$, respectively.

		Plantation	Twice-Cut	Once-Cut	Old-Growth
S.E.S. D_{pw} of phylogenetic	Overall	2.33 ± 0.62 ***	1.156 ± 0.63 ***	2.364 ± 0.51 ***	1.541 ± 0.93 ***
	SD	0.197 ± 0.97 ***	0.513 ± 0.81 **	1.521 ± 0.82 ***	1.007 ± 1.08 ***
	MD	0.576 ± 1.3 **	0.915 ± 0.91 **	2.481 ± 0.37 ***	1.037 ± 1.15 ***
	LD	1.599 ± 0.65 ***	1.285 ± 0.57 ***	2.529 ± 0.57 ***	1.073 ± 1.25 ***
S.E.S. D_{pw} of functional traits	Overall	1.458 ± 0.41 ***	0.414 ± 0.6 *	0.611 ± 0.55 **	0.878 ± 0.59 **
	SD	0.209 ± 1.03 **	−0.265 ± 0.54*	−0.407 ± 0.92 **	0.208 ± 0.56 *
	MD	0.737 ± 0.9 **	1.034 ± 0.38 ***	1.209 ± 0.48 ***	0.899 ± 0.65 **
	LD	0.234 ± 0.54 *	0.824 ± 0.44 **	0.945 ± 0.48 **	0.545 ± 0.799 **

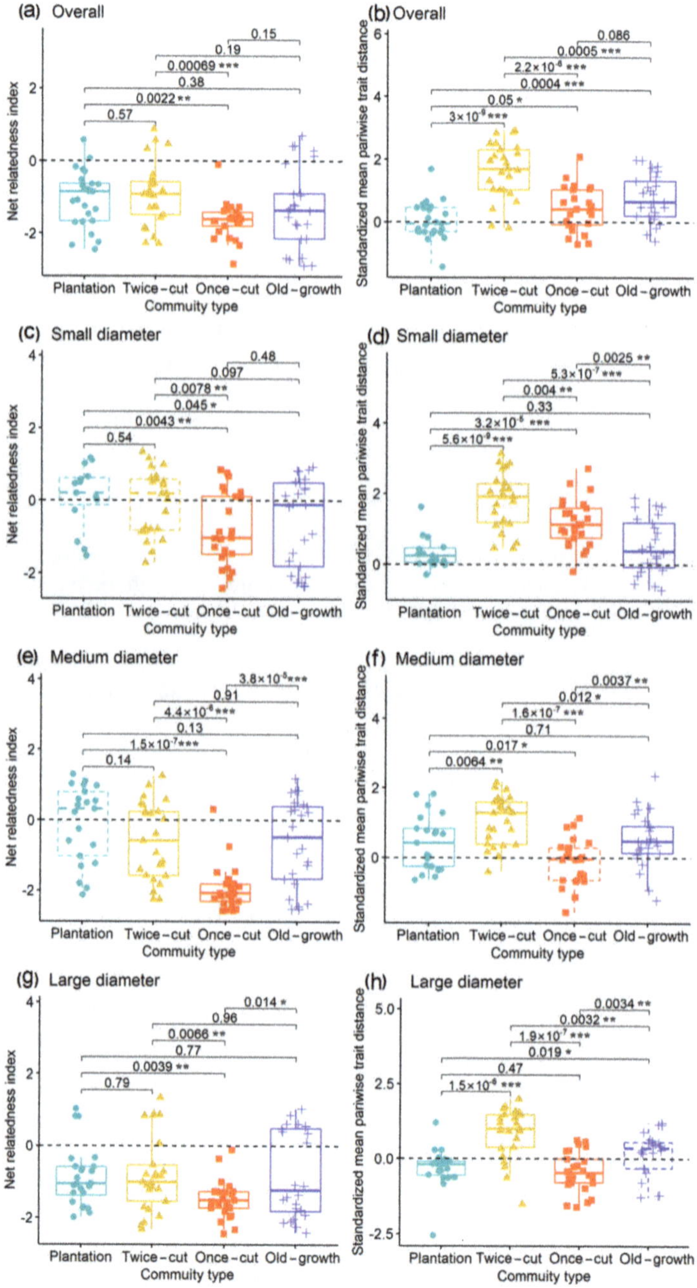

Figure 4. Phylogenetic and functional structure of woody plants in the four disturbance communities. Left, net relatedness index (*NRI*) of overall (**a**), small (**c**), medium (**e**), and large (**g**) diameter classes. Right, standardization mean pairwise distance (*S.E.S PW*) of overall (**b**), small (**d**), medium (**f**), and large (**h**) diameter classes. The black dashed lines at 0 indicate no turnover. Bold box lines represent means that are significantly different from 0, while dashed box lines represent non-significance, using t−tests. Lines joining boxes show the results of Wilcoxon tests between disturbance regimes ($p \leq 0.05$ level of significance). "*", "**", and "***" represent $p < 0.05$, $p < 0.01$, and $p < 0.001$, respectively.

Table 5. Results of *t*-tests of the hypothesis that the mean value of *NRI* or *S.E.S PW* is zero in pairs of different disturbance communities. D1, D2, D3, and D4 represent the plantation, twice-cut, once-cut, and old-growth forests, respectively. "*" and "***" represent $p < 0.05$ and $p < 0.001$, respectively.

	D1–D2		D1–D3		D1–D4		D2–D3		D2–D4		D3–D4	
	Mean	*t*	**Mean**	*t*	**Mean**	*t*	**Mean**	*t*	**Mean**	*t*	**Mean**	*t*
NRI												
Overall	1.96	47.15 ***	2.4	63.02 ***	2.11	43.08 ***	1.75	48.82 ***	1.48	27 ***	2.01	40.76 ***
SD	0.34	5.25 ***	0.69	11.22 ***	0.68	9.74 ***	1.03	19.59 ***	1.03	15.82 ***	1.47	23.27 ***
MD	0.72	9.71 ***	1.58	28.28 ***	0.83	10.52 ***	1.88	47.92 ***	1.05	15.62 ***	1.86	41.18 ***
LD	1.6	37.38 ***	2.09	54.57 ***	1.46	23.58 ***	1.92	50.09 ***	1.21	19.99 ***	1.73	30.82 ***
S.E.S PW												
Overall	1.37	41.19 ***	1.38	44.83 ***	1.52	52.26 ***	0.56	15.02 ***	0.79	21.41 ***	0.91	25.92 ***
SD	0.27	3.87 ***	0.3	3.86 ***	0.46	7.24 ***	−0.26	−5.56	0.16	4.48 ***	0.01	2.10 *
MD	0.9	21.40 ***	1.02	26.58 ***	0.89	18.78 ***	1.29	48.11 ***	1.07	33.0 ***	1.22	38.79 ***
LD	0.91	31.61 ***	0.76	25.13 ***	0.81	20.61 ***	0.95	32.61 ***	0.75	20 ***	0.9	22.58 ***

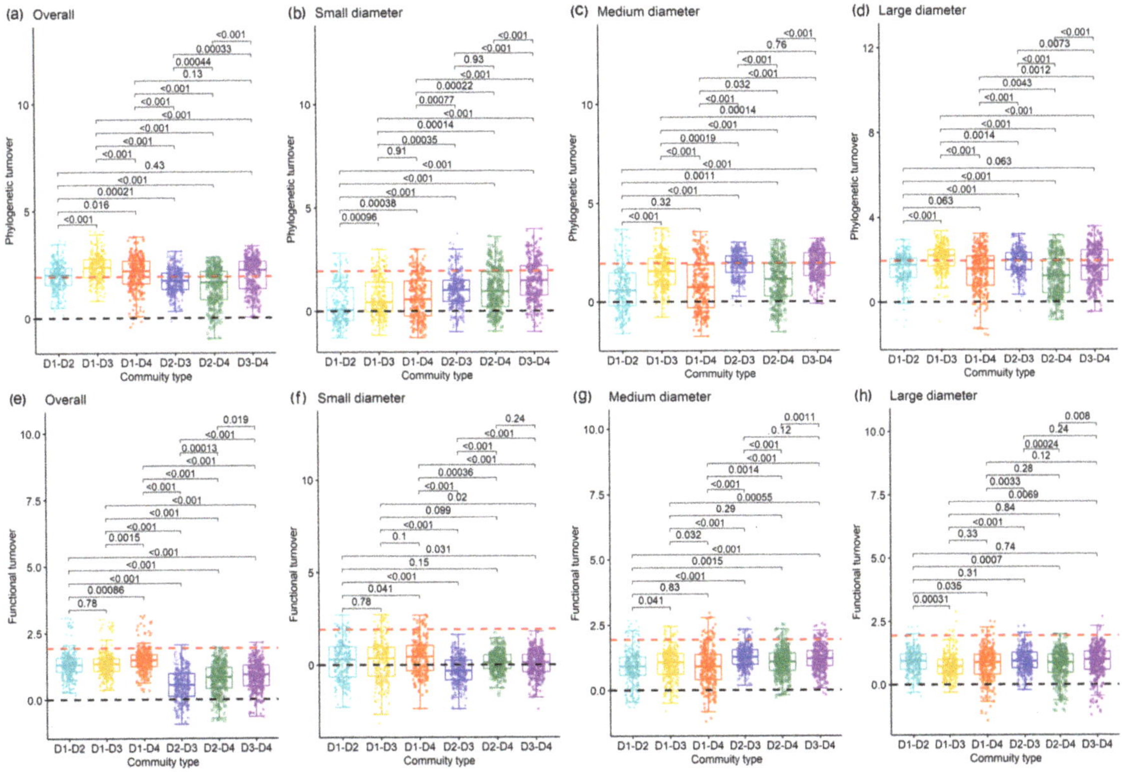

Figure 5. Phylogenetic (**a**–**d**) and functional (**e**–**h**) turnover between the four disturbance regimes; (**a**,**e**) are the overall diameter species, (**b**,**f**) are the small diameter species, (**c**,**g**) are the medium diameter species, (**d**,**h**) are the large diameter species. Black dashed lines indicate turnover = 0. Red dashed lines indicate turnover = 1.96. D1, D2, D3, and D4 represent the plantation, twice-cut, once-cut, and old-growth forests, respectively. Lines joining boxes show the results of Wilcoxon tests between disturbance regimes ($p \leq 0.05$ level of significance).

3.5. Phylogenetic and Functional—Environment Relations among Different Disturbance Plots

The Mantel tests showed that the phylogenetic and functional structures were not generally influenced by many spatial and environmental factors. However, all factors except convexity were significantly related to structure in some disturbance regimes at particular DBH scales (Table 6). Phylogenetic and functional structure were correlated with spatial distance and slope in the twice-cut and once-cut forests ($p < 0.05$, Table 6). Phylogenetic structure was also correlated with elevation in the once-cut and old-growth forests ($p < 0.05$), but functional structure was only correlated with elevation in the twice-cut forest ($p < 0.05$, Table 6). For small diameter species, only functional structure was correlated with spatial distance in the plantation forest, and only elevation in the twice-cut forest ($p < 0.05$, Table 6). For medium diameter species, phylogenetic and functional structure were correlated with slope in the once-cut forest ($p < 0.05$, Table 6). Finally, for large diameter species, phylogenetic and functional structure were correlated with slope in the twice-cut forests ($p < 0.05$, Table 6), and phylogenetic structure was correlated with spatial distance in the twice-cut forest and correlated with elevation in the once-cut and old-growth forests ($p < 0.05$, Table 6).

Table 6. Results of Mantel tests of the relationships between phylogenetic and functional structure with spatial and environment variables. D1, D2, D3, and D4 represent the plantation, twice-cut, once-cut, and old-growth forests, respectively. SD, MD, and LD represent small, medium, and large diameter classes, respectively. "*" and "**" represent $p < 0.05$ and $p < 0.01$.

Distance Matrix		Phylogenetic Index				Functional Index			
		D1	D2	D3	D4	D1	D2	D3	D4
		R	R	R	R	R	R	R	R
Overall	Spatial distance (m)	0.026	0.256 **	0.2 *	0.196 **	−0.018	0.13 *	0.157 *	−0.017
	Aspect	−0.093	−0.1	0.04	0.058	−0.01	0.003	0.015	0.13
	Slope (°)	0.024	0.22 **	0.217 **	0.069	−0.012	0.135 *	0.191*	−0.056
	Elevation (m)	0.044	0.078	0.276 *	0.254 **	0.05	0.194 *	−0.016	0.03
	Convexity (°)	0.101	−0.03	0.277	−0.058	0.019	−0.018	0.087	−0.032
SD	Spatial distance (m)	−0.088	−0.058	−0.086	0.012	0.202 *	−0.018	−0.027	0.046
	Aspect	0.073	−0.031	0.005	0.051	0.018	0.038	0.105	0.073
	Slope (°)	0.031	−0.052	−0.06	0.092	0.146	−0.017	0.007	0.063
	Elevation (m)	−0.076	−0.071	−0.122	0.067	−0.098	0.219 *	−0.068	0.049
	Convexity (°)	0.024	0.0367	−0.104	−0.035	−0.036	−0.034	−0.102	−0.052
MD	Spatial distance (m)	0.065	0.094	0.117	−0.007	−0.115	−0.007	−0.013	−0.02
	Aspect	0.015	−0.054	0.139	0.099	−0.041	0.067	0.014	0.052
	Slope (°)	0.091	0.041	0.226 **	0.083	−0.043	−0.056	0.124 *	0.054
	Elevation (m)	0.102	0.003	0.07	0.086	−0.163	0.011	−0.023	−0.053
	Convexity (°)	−0.038	−0.035	0.06	−0.028	−0.129	−0.021	−0.068	−0.088
LD	Spatial distance (m)	0.053	0.211 **	0.08	0.064	−0.073	0.121	0.108	0.116
	Aspect	−0.031	−0.082	0.013	0.061	−0.041	0.047	−0.063	0.178
	Slope (°)	0.052	0.308 **	0.039	−0.028	−0.132	0.185 *	0.115	0.013
	Elevation (m)	−0.047	0.039	0.198 *	0.23 *	−0.118	−0.021	0.095	−0.05
	Convexity (°)	0.198	0.029	0.1143	0.033	−0.105	0.04	−0.125	−0.059

The final MRM models showed that different combinations of spatial and environmental variables were correlated with the phylogenetic and functional structure of various diameter classes in different disturbance regimes (Table 7). At the overall DBH level, spatial distance better explained the phylogenetic β-diversity than environmental distance, but environmental distance better explained the functional trait β-diversity in the four disturbance regimes. Conversely, environmental distance explained more variation in phylogenetic β-diversity in small DBH species of the once- and twice-cut forests, as well as large DBH species in the plantation forest. Meanwhile, spatial distance better explained functional β-diversity than environmental distance for small DBH species in the plantation forest and large DBH species of the once- and twice-cut forests (Table 6).

Table 7. Results of multiple regression on distance matrices (MRM) of phylogenetic β-diversity, as predicted by environmental and spatial distance variables in different disturbance communities. D1, D2, D3, and D4 represent plantation, twice-cut, once-cut, and old-growth forests, respectively. M.E.S., multiple regression on distance matrices of environment and space; M.S., multiple regression on distance matrices of space; M.E., multiple regression on distance matrices of environment; M.P.S., multiple regression on distance matrices of pure space; M.P.E., multiple regression on distance matrices of pure environment.

Explanatory Variable	Phylogenetic Beta Diversity				Functional Beta Diversity			
	D1	D2	D3	D4	D1	D2	D3	D4
Overall species								
M.S.E	0.0283	0.0680	0.0336	0.0300	0.0289	0.0449	0.0260	0.0294
M.S	0.0164	0.0659	0.0191	0.0238	0.0095	0.0100	0.0164	0.0003
M.E	0.0003	0.0000	0.0000	0.0126	0.0289	0.0265	0.0001	0.0288
M.P.S	0.0279	0.0680	0.0336	0.0126	0.0296	0.0353	0.0259	0.0006
M.P.E	0.0121	0.0023	0.0147	0.0163	0.0000	0.0190	0.0098	0.0291
Small diameter								
M.S.E	0.0292	0.0093	0.0139	0.0304	0.0781	0.0884	0.0605	0.0566
M.S	0.0150	0.0034	0.0068	0.0267	0.0410	0.0243	0.0003	0.0125
M.E	0.0010	0.0040	0.0139	0.0005	0.0022	0.0466	0.0377	0.0318
M.P.S	0.0282	0.0054	0.0001	0.0300	0.0760	0.0438	0.0237	0.0256
M.P.E	0.0144	0.0059	0.0072	0.0038	0.0386	0.0657	0.0602	0.0446
Medium diameter								
M.S.E	0.0418	0.0096	0.0131	0.0067	0.0251	0.0062	0.0626	0.0340
M.S	0.0047	0.0090	0.0129	0.0064	0.0133	0.0013	0.0425	0.0143
M.E	0.0155	0.0000	0.0070	0.0000	0.0240	0.0038	0.0541	0.0270
M.P.S	0.0267	0.0096	0.0061	0.0067	0.0011	0.0025	0.0090	0.0072
M.P.E	0.0373	0.0006	0.0001	0.0003	0.0120	0.0049	0.0210	0.0200
Large diameter								
M.S.E	0.0146	0.0594	0.0022	0.0447	0.0496	0.0341	0.0430	0.0147
M.S	0.0045	0.0449	0.0022	0.0361	0.0076	0.0302	0.0426	0.0023
M.E	0.0021	0.0055	0.0010	0.0130	0.0478	0.0007	0.0171	0.0143
M.P.S	0.0126	0.0542	0.0013	0.0089	0.0019	0.0335	0.0264	0.0005
M.P.E	0.0101	0.0152	0.0000	0.0321	0.0423	0.0041	0.0004	0.0124

Generally, as the disturbance intensity decreased (and as forest age increased), the explanatory power of spatial distance for phylogenetic β-diversity structure decreased and that of environmental distance on phylogenetic structure increased. Furthermore, the spatial and environmental distances had the largest explanatory power for phylogenetic and functional β-diversity, respectively, in the moderate disturbance communities, and the smallest in the undisturbed communities.

4. Discussion

4.1. Phylogenetic Signals of Functional Traits

Determining the degree to which functional traits are evolutionarily conserved is a necessary step in the inference of species coexistence mechanisms [66]. Here, we measured the phylogenetic signals, as measured by Blomberg's *K*, in leaf, stem, and physiological traits of tree species, across disturbance regimes on Baiyun Mountain. Maximum tree height (MTH) and wood density (WD) had relatively weak phylogenetic signals, which may stem from the ubiquitous need of forest trees to grow taller and access higher light environments and as species with higher woody density can support a greater plant height [67]. Whereas non-photochemical quenching (NPQ) is a physiological trait related to chlorophyll fluorescence, which may be less affected by environmental differences and, thus, has a relatively high phylogenetic signal. All functional traits except transpiration rate (TR) in twice-cut forests showed a phylogenetic signal ($p < 0.05$, Table 3). Thus, the functional traits in the Baiyun Mountain forests tended to be evolutionarily conserved [59]; that is, species with similar genetic relationships had similar functional traits in Baiyun Mountain [68]. Our results are consistent with studies of the Changbai Mountains [69], Gutian Mountains [70], and many other forests around the world [71]. A strong phylogenetic signal may suggest environmental filtering [16,20], while over-dispersion can indicate competitive exclusion during community construction [72]. By combining patterns of community functional traits and phylogenetic structure it is possible to assess the causes of community construction [68].

4.2. Community Phylogenetic and Functional Structure

The phylogenetic and functional traits of the overall diameter class showed a non-random structure at different spatial scales, with significant β-diversity in community phylogenetic and functional traits in all disturbance regimes of the Baiyun Mountain deciduous broad-leaved forest (Table 4, Figure S1). This was not consistent with the predictions of neutral theory [27], and rather supports the notion that niche processes play an important role in community construction in this deciduous broad-leaved forest, regardless of the disturbance regime.

We found that small diameter species showed a random or slightly over-dispersed phylogenetic structure, and that over-dispersion increased significantly with diameter class. The diameter class of plants can be taken as a proxy for forest age [73]. This suggests that the growth of young trees is relatively phylogenetically clustered, perhaps due to dispersal limitations, but as individuals grow and compete, only a small number of large trees persist within communities at greater mean geographical distances [74]. This is consistent with previous findings that the phylogenetic structure of small diameter trees tends to be clustered or random, while that of large diameter trees tends to be over-dispersed [20]. We observed similar trends in functional structure, which suggests that competitive exclusion plays a major role in community construction at the large diameter size scale in the Baiyun Mountain deciduous broad-leaved forest, regardless of the disturbance regime.

4.3. Ecological Processes of Community Construction in Different Disturbance Regimes

We found significant differences in community phylogeny and functional trait structure among the different disturbance regimes in Baiyun Mountain deciduous broad-leaved forest, which indicates that the ecological processes of community construction are likely also different. Most human-disturbed forests are in the early or middle stages of succession [35], when pioneer trees play a dominant role, due to having small seeds, wide propagation, fast growth, and strong plasticity [29]. Early succession communities are often composed of closely related species, and thus moderate to highly disturbed communities tend to exhibit phylogenetic clustering [69]. However, the short life span of pioneer tree species in early succession results in their decline and replacement during forest regeneration [44]. Disturbance theory suggests that moderate disturbances increase resource availability and species richness [32]. During the later stages of succession, as dispersal is restricted and light becomes less available, competition among species for environmental resources increases, and competitive exclusion becomes a dominant process. Competitive exclusion reduces the immigration of closely related species with similar ecological niches and therefore leads to community over-dispersion [18]. Our results are consistent these aspects of disturbance theory: over-dispersion generally increased in the less disturbed plots [75,76].

The results of variance decomposition by MRM further showed that as the disturbance intensity decreased, spatial distance better explained phylogenetic and functional turnover, while the explanatory power of environmental distance decreased. That environmental distance better explained the phylogenetic and functional trait turnover in moderate to high disturbance communities suggests the importance of habitat filtering in community construction. Moreover, although competitive exclusion is often dominant in less disturbed communities [8], we found that spatial distance had a higher explanatory power of turnover in old-growth forests, consistent with diffusion limitation [27]. In conclusion, as observed in the Changbai Mountain coniferous mixed forest [69], environmental filtering plays a dominant role in community construction in the early stages of succession in high and moderate disturbance regimes, while competitive exclusion and diffusion limitation become more important in the later stages of succession [75].

Past studies of phylogenetic and functional structure across tree sizes, spatial scales, and disturbance regimes have not been entirely consistent. For example, Mo et al. found phylogenetic clustering in a young, early succession secondary forest, over-dispersion in an old secondary forest, and finally random structure in an old, late succession forest; presumably, the result of habitat filtration and competitive exclusion [77]. Whereas, Yang et al. found that medium diameter tree species showed no phylogenetic or functional structure at a small scale (5 m × 5 m), suggesting that neutral processes may play a role at small scales [68]. However, we found that community phylogenetic and functional trait structures were generally non-random, regardless of the disturbance regime or spatial scale, which is not consistent with neutral theory [4].

Finally, we observed differences in the phylogenetic and functional trait α- and β-diversity of tree species at different spatial and tree diameter scales in our Baiyun Mountain plots. The weak phylogenetic signals in functional traits ($K < 1$) may explain the inconsistent patterns in phylogenetic and functional traits. Some studies have suggested that phylogenetic distance may not be a good representation of ecological differences between species if the traits are highly differentiated [78], and studies of community assembly and species coexistence based solely on phylogenetic information may be misleading [27]. Therefore, it is necessary to combine phylogenetic and functional trait information, as we have done here, to accurately infer community assembly mechanisms [68]. It must also be said that inconsistent patterns of phylogenetic and functional traits may stem from incomplete sampling of taxa and functional traits, such that the observed data do not fully represent the actual ecological niches of species [27,79].

5. Conclusions

We examined the phylogenetic signals in leaf, stem, and physiological functional traits of tree species from different disturbance plots in Baiyun Mountain, to assess the mechanisms underlying community construction. We generally found phylogenetic signals—and thus evolutionary conservation—in functional traits, regardless of disturbance regime, diameter class, or spatial scale. Our findings suggest that niche, rather than neutral, processes played a major role in community construction in this deciduous broad-leaved forest. Furthermore, environmental filtering tended to be more important following high and moderate disturbance, and competitive exclusion was more important following slight disturbance and in undisturbed communities.

Supplementary Materials: The following supporting information can be downloaded at: https://www.mdpi.com/article/10.3390/f13060896/s1, Figure S1: phylogenetic (a,b) and functional trait (c,d) structure (mean ± SE) at different spatial scales in four disturbance regimes across tree diameter classes scales. Solid markers represent means of NRI or S. E. S. PW that are significantly different from 0 and open markers indicate that the difference was not significantly different from the 0 base in the Wilcoxon test.

Author Contributions: Conceptualization, P.L., S.D. and Q.F.; Data curation, J.Z.; Formal analysis, P.Z. and J.Z.; Funding acquisition, S.D.; Investigation, P.L. and X.W.; Methodology, P.L. and X.W.; Project administration, P.L.; Resources, S.D.; Software, P.L.; Supervision, S.D.; Validation, Z.G., P.Z. and S.D.; Visualization, P.L. and J.Z.; Writing—original draft, P.L.; Writing—review and editing, P.L., P.Z. and J.Z. All authors have read and agreed to the published version of the manuscript.

Funding: This research was funded by the National Nature Science Foundation of China (#42171091).

Acknowledgments: We would like to thank Zhendong Hong, and Pengwei Qiu for help with the research idea. We also thank Ruofan Cao, Lei Guo, and Zhiliang Yuan for help with data processing.

Conflicts of Interest: The authors declare no conflict of interest.

References

1. Thomas, J.A.; Telfer, M.G.; Roy, D.B.; Preston, C.D.; Greenwood, J.J.D.; Asher, J.; Fox, R.; Clarke, R.T.; Lawton, J.H. Comparative losses of British butterflies, birds, and plants and the global extinction crisis. *Science* **2004**, *303*, 1879–1881. [CrossRef] [PubMed]
2. Tilman, D. *Resource Competition and Community Structure*; Princeton University Press: Princeton, NJ, USA, 1982; Volume 17, pp. 1–129. ISBN 0077-0930.
3. Kunstler, G.; Lavergne, S.; Courbaud, B.; Thuiller, W.; Vieilledent, G.; Zimmermann, N.E.; Kattge, J.; Coomes, D.A. Competitive interactions between forest trees are driven by species' trait hierarchy, not phylogenetic or functional similarity: Implications for forest community assembly. *Ecol. Lett.* **2012**, *15*, 831–840. [CrossRef] [PubMed]
4. Hubbell, S.P. *The Unified Neutral Theory of Biodiversity and Biogeography*; Princeton University Press: Princeton, NJ, USA, 2001; pp. 340–348.
5. Webb, C.O.; Ackerly, D.D.; McPeek, M.A.; Donoghue, M.J. Phylogenies and community ecology. *Annu. Rev. Ecol. Syst.* **2002**, *33*, 475–505. [CrossRef]
6. Vamosi, S.M.; Heard, S.B.; Vamosi, J.C.; Webb, C.O. Emerging patterns in the comparative analysis of phylogenetic community structure. *Mol. Ecol.* **2009**, *18*, 572–592. [CrossRef]
7. Che, X.; Zhang, M.; Zhao, Y.; Zhang, Q.; Quan, Q.; Moller, A.; Zou, F. Phylogenetic and functional structure of wintering waterbird communities associated with ecological differences. *Sci. Rep.* **2018**, *8*, 1232. [CrossRef]
8. Webb, C.O.; Donoghue, M.J. Phylomatic: Tree assembly for applied phylogenetics. *Mol. Ecol. Notes* **2005**, *5*, 181–183. [CrossRef]
9. Kembel, S.W.; Hubbell, S.P. The phylogenetic structure of a neotropical forest tree community. *Ecology* **2006**, *87*, S86–S99. [CrossRef]
10. Hubbell, S.P. Neutral theory in community ecology and the hypothesis of functional equivalence. *Funct. Ecol.* **2005**, *19*, 166–172. [CrossRef]
11. Letcher, S.G. Phylogenetic structure of angiosperm communities during tropical forest succession. *Proc. R. Soc. B Biol. Sci.* **2009**, *277*, 97–104. [CrossRef]
12. Mc Gill, B.J.; Enquist, B.J.; Weiher, E.; Westoby, M. Rebuilding community ecology from functional traits. *Trends Ecol. Evol.* **2006**, *21*, 178–185. [CrossRef]
13. Kraft, N.J.; Valencia, R.; Ackerly, D.D. Functional traits and niche-based tree community assembly in an Amazonian forest. *Science* **2008**, *322*, 580–582. [CrossRef] [PubMed]
14. Lebrija-Trejos, E.; Pérez-García, E.A.; Meave, J.A.; Bongers, F.; Poorter, L. Functional traits and environmental filtering drive community assembly in a species-rich tropical system. *Ecology* **2010**, *91*, 386–398. [CrossRef] [PubMed]
15. Kembel, S.W. Disentangling niche and neutral influences on community assembly: Assessing the performance of community phylogenetic structure tests. *Ecol. Lett.* **2009**, *12*, 949–960. [CrossRef] [PubMed]
16. Kraft, N.J.; Ackerly, D.D. Functional trait and phylogenetic tests of community assembly across spatial scales in an Amazonian forest. *Ecol. Monogr.* **2010**, *80*, 401–422. [CrossRef]
17. Fine, P.V.; Kembel, S.W. Phylogenetic community structure and phylogenetic turnover across space and edaphic gradients in western Amazonian tree communities. *Ecography* **2011**, *34*, 552–565. [CrossRef]
18. González-Caro, S.; Umaña, M.N.; Álvarez, E.; Stevenson, P.R.; Swenson, N.G. Phylogenetic alpha and beta diversity in tropical tree assemblages along regional-scale environmental gradients in northwest South America. *J. Plant Ecol.* **2014**, *7*, 145–153. [CrossRef]
19. Zhao, Y.; Dunn, R.R.; Zhou, H.; Si, X.; Ding, P. Island area, not isolation, drives taxonomic, phylogenetic and functional diversity of ants on land-bridge islands. *J. Biogeogr.* **2020**, *47*, 1627–1637. [CrossRef]
20. Swenson, N.G.; Enquist, B.J.; Thompson, J.; Zimmerman, J.K. The influence of spatial and size scale on phylogenetic relatedness in tropical forest communities. *Ecology* **2007**, *88*, 1770–1780. [CrossRef]
21. Bin, Y.; Wang, Z.; Wang, Z.; Ye, W.; Cao, H.; Lian, J. The effects of dispersal limitation and topographic heterogeneity on beta diversity and phylobetadiversity in a subtropical forest. *Plant Ecol.* **2010**, *209*, 237–256. [CrossRef]
22. Weiher, E.; Keddy, P.A. Assembly rules, null models, and trait dispersion: New questions from old patterns. *Oikos* **1995**, *74*, 159–164. [CrossRef]
23. Li, P.K.; Wang, X.Y.; Wang, T.; Yao, C.L.; Yuan, Z.L.; Ye, Y.Z. Analysis on the construction and community phylogenetic structure of deciduous broad-leaved forest community in Baiyunshan nature reserve based on different models. *J. Henan Agric. Univ.* **2018**, *52*, 50–58. [CrossRef]
24. Xu, G.X.; Shi, Z.M.; Tang, J.C.; Xu, H.; Yang, H.; Liu, S.R.; Li, Y.D.; Lin, M.X. Effects of species abundance and size classes on assessing community phylogenetic structure: A case study in Jianfengling tropical montane rainforest. *Biodivers. Sci.* **2016**, *24*, 617–628. [CrossRef]
25. Rao, M.D.; Fen, G.; Zhang, J.L.; Mi, X.C.; Chen, J.H. Effects of environmental filtering and dispersal limitation on species and phylogenetic beta diversity in Gutianshan National Nature Reserve. *Chin. Sci. Bull.* **2013**, *58*, 1204–1212. [CrossRef]
26. Swenson, N.G.; Erickson, D.L.; Mi, X.; Bourg, N.A.; Forero-Montaña, J.; Ge, X.J.; Howe, R.; Lake, J.K.; Liu, X.J.; Ma, K.P.; et al. Phylogenetic and functional alpha and beta diversity in temperate and tropical tree communities. *Ecology* **2012**, *93*, S112–S125. [CrossRef]
27. Swenson, N.G. The assembly of tropical tree communities–The advances and shortcomings of phylogenetic and functional trait analyses. *Ecography* **2013**, *36*, 264–276. [CrossRef]

28. Mitchell, R.M.; Bakker, J.D.; Vincent, J.B.; Davies, G.M. Relative importance of abiotic, biotic, and disturbance drivers of plant community structure in the sagebrush steppe. *Ecol. Appl.* **2017**, *27*, 756–768. [CrossRef]
29. Xi, J.J.; Shao, Y.Z.; Li, Z.H.; Zhao, P.F.; Ye, Y.Z.; Li, W.; Chen, Y.; Yuan, Z.L. Distribution of woody plant species among different disturbance regimes of forests in a temperate deciduous broad-leaved forest. *Front. Plant Sci.* **2021**, *12*, 618524. [CrossRef]
30. Zhou, Y.; Liu, S.L.; Xie, M.M.; Sun, Y.X.; An, Y. Dynamics of regional vegetation changes under the disturbance of human activities: A case study of Xishuangbanna. *Acta Ecol. Sin.* **2021**, *41*, 565–574.
31. Fox, J.F.; Connell, J.H. Intermediate-disturbance hypothesis. *Science* **1979**, *204*, 1344–1345. [CrossRef]
32. Sven, E.J.; Brian, F. *Encyclopedia of Ecology*; Elsevier Science: Amsterdam, The Netherlands, 2008; pp. 1986–1994.
33. Roxburgh, S.H.; Shea, K.; Wilson, J.B. The intermediate disturbance hypothesis: Patch dynamics and mechanisms of species coexistence. *Ecology* **2004**, *85*, 359–371. [CrossRef]
34. Molino, J.F.; Sabatier, D. Tree diversity in tropical rain forests: A validation of the intermediate disturbance hypothesis. *Science* **2001**, *294*, 1702–1704. [CrossRef] [PubMed]
35. Feng, G.; Svenning, J.C.; Mi, X.C.; Jia, Q.; Rao, M.D.; Ren, H.B.; Bebber, D.P.; Ma, K.P. Anthropogenic disturbance shapes phylogenetic and functional tree community structure in a subtropical forest. *For. Ecol. Manag.* **2014**, *313*, 188–198. [CrossRef]
36. Xu, Y.J.; Lin, D.M.; Mi, X.C. Recovery dynamics of secondary forests with different disturbance intensity in the Gutianshan National Nature Reserve. *Biodivers. Sci.* **2014**, *22*, 358–365. [CrossRef]
37. Zhu, J.J.; Liu, Z.G. A review on disturbance ecology of forest. *Chin. J. App. Ecol.* **2004**, *15*, 1703–1710.
38. Yuan, Z.Q.; Gazol, A.; Wang, X.G.; Xing, D.L.; Lin, F.; Bai, X.J.; Bai, X.J.; Zhao, Y.Q.; Li, B.H.; Hao, Z.Q. What happens below the canopy? Direct and indirect influences of the dominant species on forest vertical layers. *Oikos* **2012**, *121*, 1145–1153. [CrossRef]
39. Song, P.; Ren, H.B.; Jia, Q.; Guo, J.X.; Zhang, N.L.; Ma, K.P. Effects of historical logging on soil microbial communities in a subtropical forest in southern China. *Plant Soil* **2015**, *397*, 115–126. [CrossRef]
40. Han, B.C.; Umaña, M.N.; Mi, X.C.; Liu, X.J.; Chen, L.; Wang, Y.Q.; Liang, Y.; Wei, W.; Ma, K.P. The role of transcriptomes linked with responses to light environment on seedling mortality in a subtropical forest. *China. J. Ecol.* **2017**, *105*, 592–601. [CrossRef]
41. Galle, A.; Czekus, Z.; Bela, K.; Horvath, E.; Ordog, A.; Csiszar, J.; Poor, P. Plant glutathione transferases and light. *Front. Plant Sci.* **2018**, *9*, 1944. [CrossRef]
42. Jia, H.R.; Chen, Y.; Wang, X.Y.; Li, P.K.; Yuan, Z.L.; Ye, Y.Z. The relationships among topographically-driven habitats, dominant species and vertical layers in temperate forest in China. *Russ. J. Ecol.* **2019**, *50*, 172–186. [CrossRef]
43. Chazdon, R.L. Tropical forest recovery: Legacies of human impact and natural disturbances. Perspect. *Plant Ecol. Evol. Syst.* **2003**, *6*, 51–71. [CrossRef]
44. Fang, Z.Y.; Li, L.Y.; Maola, A.K.E.; Zhou, L.; Lu, B. Effects of human disturbance on plant diversity of wild fruit forests in Western Tianshan Mountain. *Bull. Soil Water Conserv.* **2019**, *39*, 267–374.
45. Klopatek, J.M. Belowground carbon pools and processes in different age stands of Douglas-fir. *Tree Physiol.* **2002**, *22*, 197–204. [CrossRef] [PubMed]
46. Tang, J.; Bolstad, P.V.; Ewers, B.E.; Desai, A.R.; Davis, K.J.; Carey, E.V. Sap flux–upscaled canopy transpiration, stomatal conductance, and water use efficiency in an old growth forest in the Great Lakes region of the United States. *J. Geophys. Res. Biogeosci.* **2006**, *111*, G02009. [CrossRef]
47. Lü, G.; Wang, T.; Li, Y.X.; Wei, Z.P.; Wang, K. Herbaceous plant diversity and soil physicochemical properties on the regeneration slash of *Pinus sylvestris* var. *mongolica*. *Acta Ecol. Sin.* **2017**, *37*, 8294–8303. [CrossRef]
48. Backer, A.D.; Hoey, G.V.; Coates, D.; Vanaverbeke, J.; Hostens, K. Similar diversity-disturbance responses to different physical impacts: Three cases of small-scale biodiversity increase in the Belgian part of the North Sea. *Mar. Pollut. Bull.* **2014**, *84*, 251–262. [CrossRef]
49. Chen, Y.; Guo, L.; Yao, C.L.; Wei, B.L.; Yuan, Z.L.; Ye, Y.Z. Community characteristics of a deciduous broad-leaved forest in a temperate-subtropical ecological transition zone: Analyses of a 5-hm2 forest dynamics plot in Baiyunshan Nature Reserve, Henan Province. *Acta Ecol. Sin.* **2017**, *37*, 5602–5611.
50. Zhang, J.Y.; Dong, W.J.; Fu, C.B.; Wu, L.Y. The influence of vegetation cover on summer precipitation in China: A statistical analysis of NDVI and climate data. *Adv. Atmos. Sci.* **2003**, *20*, 1002–1006. [CrossRef]
51. Condit, R. Research in large, long-term tropical forest plots. *Trends Ecol. Evol.* **1995**, *10*, 18–22. [CrossRef]
52. Harms, K.E.; Condit, R.; Hubbell, S.P.; Foster, R.B. Habitat associations of trees and shrubs in a 50-ha Neotropical forest plot. *J. Ecol.* **2001**, *89*, 947–959. [CrossRef]
53. Kembel, S.W.; Cowan, P.D.; Helmus, M.R.; Cornwell, W.K.; Morlon, H.; Ackerly, D.D.; Blomberg, S.P.; Webb, C.O. Picante: R tools for integrating phylogenies and ecology. *Bioinformatics* **2010**, *26*, 1463–1464. [CrossRef]
54. Webb, C.O.; Ackerly, D.D.; Kembel, S.W. Phylocom: Software for the analysis of phylogenetic community structure and trait evolution. *Bioinformatics* **2008**, *24*, 2098–2100. [CrossRef] [PubMed]
55. Wikstrom, N.; Savolainen, V.; Chase, M.W. Evolution of the angiosperms: Calibrating the family tree. *Proc. R. Soc. B Biol. Sci.* **2001**, *268*, 2211–2220. [CrossRef] [PubMed]
56. Perez-Harguindeguy, N.; Diaz, S.; Garnier, E.; Lavorel, S.; Poorter, H.; Jaureguiberry, P.; Bret-Harte, M.S.; Coenwell, W.K.; Craine, J.M.; Gurvich, D.E.; et al. New handbook for standardised measurement of plant functional traits worldwide. *Aust. J. Bot.* **2013**, *61*, 167–234. [CrossRef]

57. Petchey, O.L.; Gaston, K.J. Functional diversity (FD), species richness and community composition. *Ecol. Lett.* **2002**, *5*, 402–411. [CrossRef]
58. Blomberg, S.P.; Garland, T., Jr.; Ives, A.R. Testing for phylogenetic signal in comparative data: Behavioral traits are more labile. *Evolution* **2003**, *57*, 717–745. [CrossRef]
59. Liu, X.; Swenson, N.G.; Zhang, J.L.; Ma, K.P. The environment and space, not phylogeny, determine trait dispersion in a subtropical forest. *Funct. Ecol.* **2013**, *27*, 264–272. [CrossRef]
60. Gotelli, N.J. Null model analysis of species co-occurrence patterns. *Ecology* **2000**, *81*, 2606–2621. [CrossRef]
61. Kraft, N.J.; Cornwell, W.K.; Webb, C.O.; Ackerly, D.D. Trait evolution, community assembly, and the phylogenetic structure of ecological communities. *Am. Nat.* **2007**, *170*, 271–283. [CrossRef]
62. Diniz-Filho, J.A.F.; Bini, L.M.; Hawkins, B.A. Spatial autocorrelation and red herrings in geographical ecology. *Glob. Ecol. Biogeogr.* **2003**, *12*, 53–64. [CrossRef]
63. Swenson, N.G. Phylogenetic beta diversity metrics, trait evolution and inferring the functional beta diversity of communities. *PLoS ONE* **2011**, *6*, e21264. [CrossRef]
64. Chen, Y.; Yuan, Z.L.; Bi, S.; Wang, X.Y.; Ye, Y.Z.; Svenning, J.C. Macrofungal species distributions depend on habitat partitioning of topography, light, and vegetation in a temperate mountain forest. *Sci. Rep.* **2018**, *8*, 13589. [CrossRef] [PubMed]
65. R Foundation for Statistical Computing. *R: A language and Environment for Statistical Computing*; R Foundation: Vienna, Austria, 2017.
66. Cornelissen, J.H.C.; Lavorel, S.; Garnier, E.; Diaz, S.; Buchmann, N.; Gurvich, D.E.; Reich, P.B.; ter Steege, H.; Morgan, H.D.; van der Heijden, M.G.A. A handbook of protocols for standardised and easy measurement of plant functional traits worldwide. *Aust. J. Bot.* **2003**, *51*, 335–380. [CrossRef]
67. Yang, J.; Ci, X.Q.; Lu, M.Q.; Zhang, G.C.; Cao, M.; Li, J.; Lin, L. Functional traits of tree species with phylogenetic signal co-vary with environmental niches in two large forest dynamics plots. *J. Plant Ecol.* **2014**, *7*, 115–125. [CrossRef]
68. Yang, J.; Lu, M.M.; Cao, M.; Li, J.; Lin, L.X. Phylogenetic and functional alpha and beta diversity in mid-mountain humid evergreen broad-leaved forest. *Chin. Sci. Bull.* **2014**, *59*, 2349–2358. [CrossRef]
69. Hou, M.M.; Li, X.Y.; Wang, J.W.; Liu, S.; Zhao, X.H. Phylogenetic development and functional structures during successional stages of conifer and broad-leaved mixed forest communities in Changbai Mountains, China. *Acta Ecol. Sin.* **2017**, *37*, 7503–7512.
70. Cao, K.; Rao, M.D.; Yu, J.H.; Liu, X.J.; Mi, X.C.; Chen, J.H. The phylogenetic signal of functional traits and their effects on commu-nity structure in an evergreen broad-leaved forest. *Biodivers. Sci.* **2013**, *21*, 564–571. [CrossRef]
71. Mi, X.; Swenson, N.G.; Valencia, R.; Kress, W.J.; Erickson, D.L.; Perez, A.J.; Ren, H.B.; Su, S.H.; Gnuatilleke, N.; Gunatilleke, S. The contribution of rare species to community phylogenetic diversity across a global network of forest plots. *Am. Nat.* **2012**, *180*, E17–E30. [CrossRef]
72. Mayfield, M.M.; Levine, J.M. Opposing effects of competitive exclusion on the phylogenetic structure of communities. *Ecol. Lett.* **2010**, *13*, 1085–1093. [CrossRef]
73. Niu, H.Y.; Wang, Z.F.; Lian, J.Y.; Ye, W.H.; Shen, H. New progress in community assembly: Community phylogenetic struc-ture combining evolution and ecology. *Biodivers. Sci.* **2011**, *19*, 275–283. [CrossRef]
74. Song, K.; Mi, X.C.; Jia, Q.; Ren, H.B.; Dan, B.; Ma, K.P. Variation in phylogenetic structure of forest communities along a human disturbance gradient in Gutianshan forest, China. *Biodivers. Sci.* **2011**, *19*, 190–196. [CrossRef]
75. Letcher, S.G.; Chazdon, R.L.; Andrade, A.C.; Bongers, F.; van Breugel, M.; Finegan, B.; Laurance, S.G.; Mesquita, R.C.G.; Martinez-Ramos, M.; Williamson, G.B. Phylogenetic community structure during succession: Evidence from three Neotropical forest sites. *Perspect. Plant Ecol. Evol. Syst.* **2012**, *14*, 79–87. [CrossRef]
76. Whitfeld, T.J.S.; Kress, W.J.; Erickson, D.L.; Weiblen, G.D. Change in community phylogenetic structure during tropical forest succession: Evidence from New Guinea. *Ecography* **2012**, *35*, 821–830. [CrossRef]
77. Mo, X.X.; Shi, L.L.; Zhang, Y.J.; Zhu, H.; Slik, J.W.F. Change in phylogenetic community structure during succession of traditionally managed tropical rainforest in southwest China. *PLoS ONE* **2013**, *8*, e71464. [CrossRef] [PubMed]
78. Losos, J.B. Phylogenetic niche conservatism, phylogenetic signal and the relationship between phylogenetic relatedness and ecological similarity among species. *Ecol. Lett.* **2008**, *11*, 995–1003. [CrossRef] [PubMed]
79. Cadotte, M.W.; Cardinale, B.J.; Oakley, T.H. Evolutionary history and the effect of biodiversity on plant productivity. *Proc. Natl. Acad. Sci. USA* **2008**, *105*, 17012–17017. [CrossRef]

forests

Article

Plant Diversity and Soil Nutrients in a Tropical Coastal Secondary Forest: Association Ordination and Sampling Year Differences

Muhammad Yaseen [1,2,3], Gaopan Fan [4], Xingcui Zhou [4], Wenxing Long [1,2,3,*] and Guang Feng [1,2,3]

1 College of Ecology and Environment, Hainan University, Haikou 570228, China; 20181112909153@hainanu.edu.cn (M.Y.); 995100@hainanu.edu.cn (G.F.)
2 Wuzhishan National Long-Term Forest Ecosystem Monitoring Research Station, College of Forestry, Hainan University, Haikou 570100, China
3 Administration of Tongguling National Nature Reserve, Wenchang 571300, China
4 Key Laboratory of Tropical Forest Flower Genetics and Germplasm Innovation, Ministry of Education, Haikou 570228, China; fangaopan53418368@163.com (G.F.); xingcui847876096@163.com (X.Z.)
* Correspondence: oklong@hainanu.edu.cn

Abstract: Studying the patterns of changes in species diversity and soil properties can improve our knowledge of community succession. However, there is still a gap in understanding how soil conditions are related to plant diversity in tropical coastal secondary forests. We sampled plant diversity and soil nutrients spanning two different years (2012 and 2019) to assess the patterns of species diversity and relationships of soil nutrients and species diversity on Hainan Island, southern China. Results showed that the soil pH and total nitrogen (TN) significantly decreased while the soil organic matter (OM) and total phosphorus (TP) significantly increased from 2012 to 2019. Plant species diversity was significantly higher in 2012 than in 2019, and the dominant species significantly changed in two different years. Using multiple regression analysis, we determined that soil TP and TN were significantly related to plant diversity in 2012 and 2019, respectively. Using CCA analysis, TN and OM were the strongest predictors for dominant species in 2012, whereas the soil TP and TN were the strongest predictors for dominant species in 2019. Our findings show a significant change in plant diversity and dominant species after 7 years of development in the tropical coastal secondary forest. The patterns of plant diversity and soil nutrients increase our knowledge of forest restoration in coastal areas.

Keywords: soil nutrients; plant diversity; regression dominant species; tropical monsoonal forest

Citation: Yaseen, M.; Fan, G.; Zhou, X.; Long, W.; Feng, G. Plant Diversity and Soil Nutrients in a Tropical Coastal Secondary Forest: Association Ordination and Sampling Year Differences. *Forests* **2022**, *13*, 376. https://doi.org/10.3390/f13030376

Academic Editors: Yi Ding, Runguo Zang and Timothy A. Martin

Received: 5 December 2021
Accepted: 12 February 2022
Published: 24 February 2022

1. Introduction

Tropical forests tend to have a high level of biodiversity and aboveground biomass, and they grow on strongly weathered soils. Even within a tropical region, the distribution of plant species and soils is highly diverse [1]. Plant–soil interaction is a key internal driver of ecosystem evolution. The unidirectional effects of soil nutrients on plant diversity have been described in many studies [2]. Environmental factors such as climate, soil characteristics, and herbivory have significantly influenced plant diversity [3,4], and plant community growth is heavily dependent on soil nutrient availability [5]. Plant community features such as plant biomass, vegetation cover, and species composition can be affected due to changes in soil nutrients [6], resulting in compositional dissimilarity at the local, landscape, and regional scales. However, there are still many gaps in our knowledge about how soil nutrients affect tree species diversity in tropical forests [7].

From grassland to primary forest, an increase in organic matter (OM) and total nitrogen (TN) may have a positive effect on species diversity and lead to increased species richness [8]. Despite low P availability in tropical forests, many low-P specialists have indicated that plants can maintain high growth rates [9] and exhibit muted responses due to P addition [10]. Conversely, high-P specialists have revealed that plants have much stronger responses due to P addition [10]. Soil properties influenced the existence and quantity of dominating species in a tropical forest [11]. Limited resources either increase the species richness such as the dominating species by reducing their competitive strength [12] or decrease species richness by allowing a smaller number of species to endure the environment [12]. Long et al. [13] found the greatest effects of soil nutrients OM, P, N, and potassium (K) on dominant species in two different vegetation sites, forest and shrub. However, we still lack a fundamental understanding of their causes and implications [14].

According to many recent studies on tropical forests, plant species richness is positively associated with soil nutrient concentration [15]. Plant species diversity and soil content of extractable phosphorus and potassium concentrations in the soil were discovered to have a positive relationship [16]. A high positive relationship was found between tree species richness and soil abiotic properties, i.e., nitrogen, phosphorus, and carbon concentrations in a dry deciduous forest in western India [17]. Previous studies reported that lack of soil nutrients [18] can affect the forest community structure. Contrarily, several studies have shown contradictory results [19]. Therefore, in-depth studies must be conducted, especially in tropical areas. In general, low soil nutrients and pH may be limiting factors for the loss of species diversity [20]. Soils in neotropical forests have been linked to differences in soil P limitation of plant growth [21]. It is demonstrated that differences in soil properties are mechanistically linked to plant biomass [22]. Soil N and P are major regulators of species richness, evenness, and community composition [23]. The effects of soil N and soil P on plant community structure appear to be distinct [24]. Available soil N has a positive association with the plant Simpson index [25], and soil P plays a significant role in producing and maintaining plant diversity [26].

Tropical coastal forests are typical of Hainan Island, off southern China's northeastern coast. Before 111 B.C., Hainan's forest cover reached 90% [27]. However, the forests have been severely affected by anthropogenic factors (slash and burn) and natural disasters such as typhoons [13]. Tongguling National Nature Reserve (TNNR) is located at the land–sea junction in Wenchang County of Hainan Province, surrounded by sea on three sides. This tropical coastal forest is a natural forest flora in southern China with a small number of parallel ecosystems [28]. Therefore, biodiversity conservation, natural environmental monitoring, and typhoon mitigation are critical issues in this region [13]. However, little is known about changes in forest diversity and soil conditions over different periods of time. We collected a large data set including (a) species abundance and dominance and (b) soil abiotic properties in a tropical monsoonal dwarf forest. We used plant diversity and soil nutrient data for two different years and the following questions were addressed: (1) How did species diversity, dominant species abundance, and soil abiotic properties change in 7 years? (2) Did soil nutrients influence the plant diversity and abundance of dominant species? We predicted that the richness and abundance would increase in the two different years. These variations might be associated with changes in forest ecosystem resources and the quality of soil abiotic properties [29].

2. Material and Methods

2.1. Site Conditions

The present study was carried out in the TNNR 19°36′–19°41′ N, 110°58′–111°03′ E across the northeastern part of Hainan Island in southern China. TNNR is a 44 km^2 wide natural reserve located at 338 m above sea level with a tropical monsoonal climate. The rainy season lasts from May to October, whereas the dry season lasts from November to April of the following year. The average annual temperature is 23.9 °C, with 1721.6 mm of rainfall. The tropical evergreen monsoonal forest possesses lateritic soil type [30]. Before 1980, all of these forests were deforested and converted to shrubs or secondary forests. After the establishment of TNNR in 1983, deforestation was prohibited, and forests were well recovered.

2.2. Data Collection

We randomly established nine 50 × 50 m (2500 m^2) plots in the tropical evergreen monsoonal forests in TNNR, with the distance between any two plots greater than 50 m. Each plot was divided into 64 subplots of 20 × 20 m (400 m^2) by the community size. The following criteria were used for plot selection: (a) common age of vegetation; (b) common comparison of the soil, geography, and original vegetation; and (c) without obvious additional interference. The background of each plot recovery was carried out based on official records of the forestry department or interviews with local communities. In 2012 and 2019, we conducted ground surveys to collect data on plant diversity and soil conditions. The two investigations were carried out by the same research group and kept a consistent method. Woody plants with a diameter at breast height (DBH) (i.e., 1.3 m of the stem from the ground) ≥1 cm were identified, tagged, and mapped. The DBH information of the sampled trees was collected using a measuring tape and tree height was measured using a laser rangefinder (Impulse 200, Laser Technology, Inc., Centennial, CO, USA). Flora of the Republic of China assisted in the identification of species [31].

During the rainy season (August) of 2012 and 2019, soil pH, TN, soil OM, and soil TP were measured by our research group. In each 20 × 20 m (400 m^2) plot, four soil profiles were selected at random and sampled. A topsoil sample (0–20 cm) was taken from each soil profile after the trash or grass layer above the soil was carefully removed [32] along the diagonal of each 400 m^2 plot. Each soil sample was ground and sieved to 2 mm, lightly mixed, and analyzed following [33] with minor modifications.

2.3. Soil Analysis

For the measure of OM concentration, 0.5 g of soil was mixed with 5 mL of 1N potassium dichromate ($K_2Cr_2O_7$) solution and 5 mL of 98% sulfuric acid (H_2SO_4) and then kept at room temperature for half an hour. Second, deionized water and concentrated phosphoric acid (H_3PO_4) were added followed by an amalgamation titration with 0.5 mol L^{-1} ferrous ammonium sulfate $(NH_4)_2Fe\ (SO_4)_2{\cdot}6H_2O$ solution until the color changed from purple-blue to green. Finally, the titration was repeated until the color changed to green from violet-blue.

For the measure of pH concentration, 25 g of soil was added to de-ionized water and carefully stirred. Second, pH of suspension was measured using the pH electrode (Phoenix Electrode Company, Houston, TX, USA) at temperatures ranging from 20 °C to 25 °C.

For the measure of TN concentration, 1.0 g of soil was digested in 98% sulfuric acid (H_2SO_4) with potassium sulfate, copper sulfate (K_2SO_4–$CuSO_4{\times}5$), and hydroxy selenide (H_2O–Se). Secondly, ammonium nitrogen in the digest was extracted using steam with an excess of 0.1 mol L^{-1} sodium hydroxide (NaOH) to increase the pH value. Thirdly, the distillate was obtained in 2% H^3BO^3 and titration was conducted with 0.05 H_2SO_4 to obtain a pH of 5.0. Finally, the TN concentration was calculated as per the volume change of a 0.05 M sulfuric acid (H_2SO_4) solution.

For the measure of TP concentration, 0.25 g of soil was mixed with 60% $HClO_4$ solution. Second, the mixture was diluted with a vanadium molybdate reagent, standard, and sampled at 700 nm wavelength. Third, the calibration curve was used to determine TP.

2.4. Data Analysis

The Shannon–Wiener index (*H*), Pielou's evenness index (*E*), and species richness were used to calculate the plant species diversity of each plot (50 × 50 m). The number of species was used to calculate species richness, and *H* and *E* were computed using the formula below:

Shannon–Wiener index:

$$H' = -\sum_{i=1}^{s} pi\ InPi$$

Pielou's evenness index:

$$E = \frac{H}{lnS}$$

Simpson diversity index:

$$D = 1 - \sum_{i=1}^{s} (Pi)^2$$

where S is the total number of species in the plots under investigation, and Pi denotes the relative richness of abundance of species (i) in the total number of species.

Dominant species were determined as species with an important value (IV) greater than 1.

The Importance Value Index (IVI) was calculated using the following criteria:

Important value index (IVI) = Relative density + Relative frequency + Relative dominance

Firstly, we derived a sample size formula for stratified Fisher's exact (1935) test using the values of the same input parameters as Mantel–Haenszel's test by Jung et al. [34]. Secondly, Wilcoxon's rank test was used to evaluate the variations in species diversity, evenness, and soil nutrients from 9 random plots (50 × 50 m) having 64 subplots (20 × 20 m) in the years 2012 and 2019.

Thirdly, simple regression was used to measure the relationship between plant diversity and soil conditions. The optimal model was determined through statistical model comparison approaches; R2 and the difference between the β coefficient values are examples of such metrics. The soil nutrient effects on the most common or dominant species were evaluated by using canonical correspondence analysis (CCA) at each stage of recovery. Finally, principal component analysis (PCA) was performed to identify the strongest correlation and covariation among soil variables spanning seven years. The graphical presentation was carried out by using *"ggplot2"*, *"stat"*, and *"agricole"* for Figure 1, *factominer* and *factoextra packages* for Figure 2, *factoextra package* for Figure 3, and *"stat" package* for Supplementary Table S1 in *R software* 2.9.3 [35].

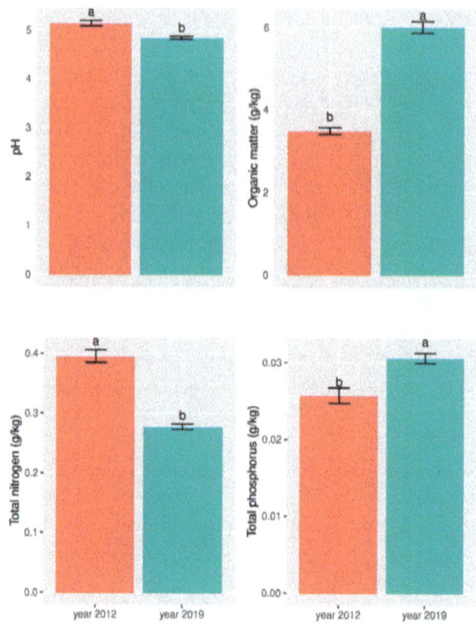

Figure 1. Bar plot showing soil pH values and organic matter, total nitrogen, and total phosphorus concentrations in 2012 and 2019. Error bars are the standard errors. Different letters indicate significant differences between the investigation years according to Wilcoxon's rank test ($p < 0.05$).

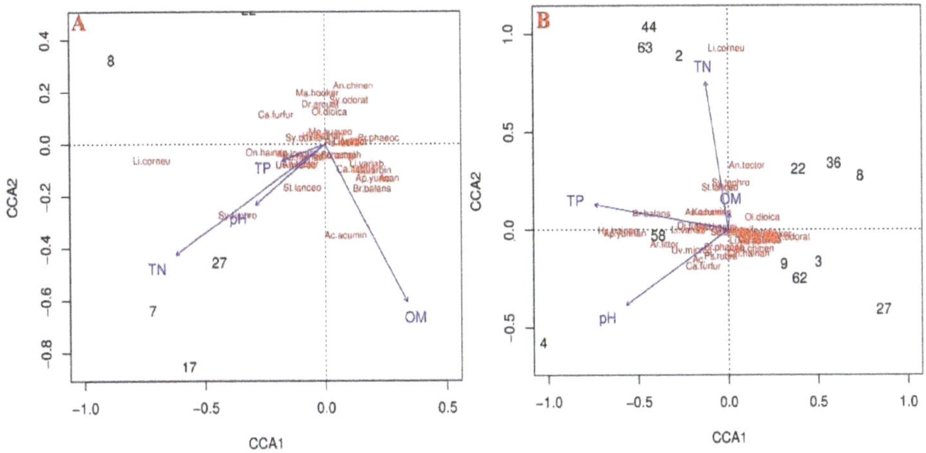

Figure 2. Canonical correspondence analysis (CCA) between dominant species richness (red species abbreviation) and soil nutrients (blue nutrient abbreviation) in the secondary tropical evergreen monsoon forest in 2012 (**A**) and 2019 (**B**). OM, organic matter; TN, total nitrogen; TP, total phosphorus.

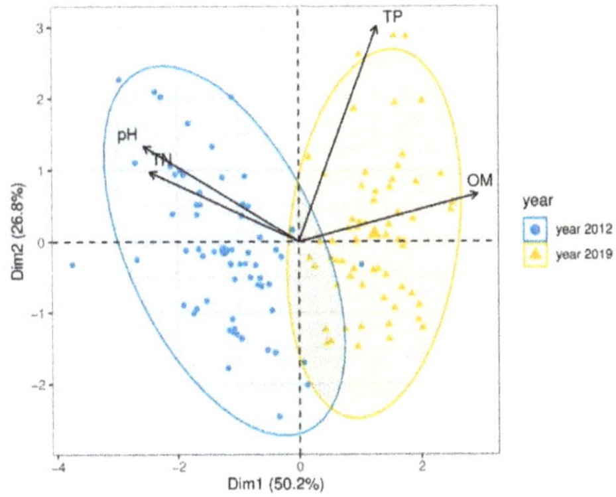

Figure 3. Principal component analysis (PCA) of soil variables in two different years (2012 and 2019). Blue and yellow dots show the sites. pH, soil pH; OM, organic matter; TN, total nitrogen; TP, total phosphorus.

3. Results

3.1. Patterns of Soil Nutrients and Effects on Plant Diversity in Two Different Years

In the forest community of TNNR, the pH (F = 22.5, p = 0.01) and TN (F = 98.54, p = 0.01) significantly decreased, while OM (F = 239.86, p = 0.01) and TP (F = 16.44, p = 0.01) significantly increased from 2012 to 2019 (Figure 1).

The Shannon–Wiener index (W = 6100.0, p = 0.01), Simpson index (W = 5907.0, p = 0.01) and species richness (W = 6462.5, p = 0.00) were significantly higher in 2012 than in 2019 (Table 1). Furthermore, Pielou's evenness (W = 4608.0, p = 0.63) remained the same in the two different periods and was highly non-significant.

Table 1. Comparison of species richness, Shannon–Wiener Index, species evenness, and Simpson index between the two years (2012 and 2019). Values are mean ± standard deviation, N = 64.

Parameters	Two Years' Difference		Wilcoxon's Rank Test	
	2012	2019	*W*	*p*-Value
Shannon–Wiener index	3.21 ± 0.02	2.84 ± 0.09	6100.0	0.01
Simpson index	0.94 ± 0.01	0.88 ± 0.02	5907.0	0.01
Species richness	45.07 ± 0.8	35.92 ± 1.56	6462.5	0.00
Pielou's evenness	0.85 ± 0.01	0.85 ± 0.01	4608.0	0.63

Based on multiple regression analysis, we found that soil TP was the strongest predictor for plant diversity in 2012, whereas soil TN was the strongest predictor for species diversity in 2019. The R2 values showed the greatest strength for the species richness in both years (Supplementary Table S1).

3.2. Patterns of Dominant Species and Effects of Soil Nutrients on Dominant Species (Based on Their Basal Area and Abundance)

We collected a total of 203 species from 57 families in 2019, and 218 total species of 71 families in 2012. In 2012, *S. octophylla*, *A. oligophlebia*, *S. buxifolium*, *A. acuminatissima*, *O. dioica*, *E. sylvestri*, *S. buxifolioideum*, *X. longispinosa*, *P. phaeocosticta*, and *L. variabilis* were dominant species based on their basal area, (Supplementary Table S2). In 2019, *H. hainanensis*, *A. acuminatissima*, *A. chinensis*, *C. furfuracea*, *S. octophylla*, *K. hainanense*, *O. dioica*, *S. buxifolium*, *L. verticillate*, and *A. oligophlebia* were dominant species (Supplementary Table S3).

Dominating species abundance was closely related to TN and OM in 2012 (Figure 2A), and to TP and TN in 2019 (Figure 2B), using canonical correlation analysis (CCA).

3.3. Variation in Soil Nutrients in Two Different Years

In Figure 3, plot numbers are represented by circular and triangular symbols representing the years 2012 and 2019, respectively, and soil nutrient indices are plotted as vectors. These vectors are pinned at the origin of PCs (Dim1 = 50.2% and Dim2 = 26.8%) (Figure 3). In the two different years, 2012 and 2019 heterogeneity levels were recorded for different sites. In 2012, pH, TN, and OM had a greater contribution to PC2 (Dim2), while TP had little influence on Dim1. Moreover, OM and TN have a negative relationship in both years. There is no significant difference in soil abiotic properties. Soil TP and OM significantly improved in 2019 and had a strong positive correlation. The soil pH was closely related to TN and had a positive association in 2012.

4. Discussion

4.1. Patterns of Soil Nutrients and Their Effects on Species Diversity in Two Different Sampling Years

The changes in soil properties affect the soil functions such as nutrient transportation and redistribution to various parts of the plant for growth and development [36]. Therefore, pH and soil nutrients, including N and P, are important regulators of plant growth (e.g., shoot and root growth) [37]. From grassland to the primary forest stage, an increase in organic matter (Figure 1) may lead to increased richness [30]. Similarly, Long et al. [30] found that the concentration of OM was increased in the tropical monsoonal forest. Our observations are consistent with these findings, as soil OM increased from 2012 to 2019, and TN and pH decreased in 2019 as compared to 2012 (Figure 1).

Soil P is the main nutrient balancing plant productivity and better development in the forestland ecosystem [38]. The species adapted to low-phosphorus soils grow fast in tropical forests [39,40]. Total phosphorus is generally a limiting factor, because the H^2PO^4-anion forms poorly soluble compounds with Al3+ and Fe3+ in tropical soils. As a result, topsoil soluble phosphorus concentrations remain low. Furthermore, the regular loss of phosphorus during long-term soil and ecosystem development may result in very low total phosphorus levels [41,42]. Our results also show that the total soil phosphorus was decreased (Figure 1) in the year 2012 with higher species diversity and not significantly correlated with diversity attributes in 2012 (Supplementary Table S1).

Soil pH has been demonstrated in several studies to have a significant effect on plant diversity, and the combination of acidic soil and low-temperature climatic conditions limits the distribution of many tropical species [43]. Furthermore, Liu et al. [20] showed the correlation between the species distribution and soil pH at the three sites. Overall, low soil nutrients and soil pH may be limiting factors for the decline in species diversity (Table 1) [20]. Our findings support the nutrient gradient hypothesis, as general mechanisms underlying the change patterns in soil properties such as pH were lower from 2012 to 2019 (Figure 1). Steege et al. [44] reported a decrease in plant diversity in tropics. A similar decreasing trend in plant diversity was observed in our study in 2019. This decrease in plant diversity may possibly be due to various natural or anthropogenic factors, including invasive species dominance, habitat fragmentation, and deforestation (Table 1) [45].

It is usually hypothesized that N availability in lowland tropical forest soils exceeds plant requirements [46]. This hypothesis was challenged by two ecologists [47] who concluded that N limitation was equally strong in temperate and tropical forests. Our findings also demonstrated TN limitation for tropical forest communities in 2019 as compared to 2012. In addition, neither total soil nitrogen nor its available mineral forms of ammonium and nitrate decreased over time. Because forest biomass was larger in older plots, the lack of a noticeable drop in soil N could be related to litter accumulation over time since abandonment. Our results also identified this, where the TN was not decreased (Figure 1) in 2012 and it was negatively correlated with species diversity (Supplementary Table S1). Organic matter contains many key elements, i.e., nitrogen. As a result, increases in OM and TN are expected to help to increase species diversity. In contrast, the limited resources of soil nutrients such as total nitrogen will limit plant growth and influence community diversity [48].

4.2. Soil Properties Differences in Two Different Years

In the assessment of the integrated soil fertility quality index ($SFQI_{PCA}$), principal component analysis (PCA) has been widely used [49]. According to PCA, such variations in the relative significance of a strategy could be mediated by changes in soil conditions [50]. Soil nutrients, such as soil TP, contributed positively to the primary axis of PCA [50]. On the other hand, Chen et al. [49] reported that first PC (PC1) was highly correlated with TN. Hence, in the current study, PCA loadings also indicated that the first PC (PC1) was highly positively correlated with TP while PC (PC1) was not correlated with TN (Figure 3).

4.3. Plant Species Diversity and Dominant Species Role in Two Different Sampling Years and Their Relationship to Soil Nutrients (2012 and 2019)

A strong positive and significant correlation was found between tree species richness with TN, TP, and OM in tropical coastal secondary forests in southern China [13]. Higher concentrations of TN and OM can maintain the loss of soil nitrogen and increase the diversity of species and the richness of dominant species [51]. We concluded that availability of soil nutrients is a key element inducing species distribution. Hence, our findings also show that the abundance of dominant species was greatly influenced by soil nutrients in the two different years (soil TN, TP, and OM) (Figure 2). In 2012, soil nutrients such as OM and TN contributed more with dominant species (Figure 2A). In 2019, the soil nutrients TN and TP contributed more with dominant species (Figure 2B). The same results were found in that diversity and richness of dominant species decreased at later stage of succession due to severe human activities such as slash and burn, as well as changing natural environmental conditions, i.e., typhoon disturbances [13], on Hainan Island in southern China. The identities of dominant species and their influence on their environments are key to linking diversity patterns to ecosystem function, as well as predicting impacts of species loss and other aspects of global change on ecosystems [52]. Likewise, the important value index (IVI) of dominated species was found in recent studies, proving that each species has its own role in promoting the balance of the ecosystem [53], and important values were used to classify major and constructive species to vegetation sites in different years. Furthermore, in 2012, we also determined the important value index (IVI) of the following dominant species based on their basal area: *S. octophylla*, *A. oligophlebia*, *S. buxifolium*, *A. acuminatissima*, *O. dioica*, *E. sylvestri*, *S. buxifolioideum*, *X. longispinosa*, *P. phaeocosticta*, and *L. variabilis* (Supplementary Table S2). In 2019, we estimated the important value index (IVI) of the following dominant species: *H. hainanensis*, *A. acuminatissima*, *A. chinensis*, *C. furfuracea*, *S. octophylla*, *K. hainanense*, *O. dioica*, *S. buxifolium*, *L. verticillate*, and *A. oligophlebia* (Supplementary Table S3).

Species richness is usually controlled by the same period of climate within the global biosphere [54], which helps to stabilize the hydraulic energy system in the ecological period [55]. In addition, our results showed that plant diversity and abundance of species were dominant in 2012. Individual dominating species were found in the tropical monsoonal forest during the two different periods of time, which could be due to changes in predominant species responses to soil nutrients (Figure 2) [56]. The decline in the dominant species in the two different years changed; the species with greater decline ratios may be removed in the future for forest restoration. Moreover, it is reported that the species with lower decline ratios possibly conserve for longer periods of time [57]. Nonetheless, these species possibly create enough space for dominant species during their maturity. However, the forest has a very complicated ecosystem, and the ecological drivers affecting species co-occurrence and replacement during the restoration period may be highly diverse [58]. Therefore, further research is needed to gain insight into the species co-occurrence techniques during second-growth species forest restoration processes.

5. Significance for Conservation

The tropical secondary forest in TNNR is well protected and had high species richness in 2012 (over 70 species in 2500 m^2). Since TNNR was established in 1983, forest have helped understand the usefulness of protection. The plant species diversity in the second-growth forest suggests that restoration was not advanced in 2012 and 2019 in this area on a large scale because of poor management practices and natural disturbances. Furthermore, evidence showed that soil nutrients (OM and TN) are closely related to the dominant species in 2012, which suggests that soil nutrient management or restoration would be a top driver in the management methods of these protected areas. Lastly, the richness of the most common or dominant species with time progression can help us identify the most suitable tree species for restoring coastal windbreaks in the area.

Supplementary Materials: The following supporting information can be downloaded at: https://www.mdpi.com/article/10.3390/f13030376/s1, Table S1: Regression analysis between plant diversity and environmental variables (2012&2019); Table S2: Species IVIs for year 2012; Table S3: Species IVIs for year 2019.

Author Contributions: M.Y. and W.L. conceived the research ideas; G.F. (Gaopan Fan) and X.Z. classified the vegetation data sets and wrote the paper with contributions; M.Y. prepared and analyzed the environmental variables; W.L. and G.F. (Guang Feng) rigorously checked and revised the manuscript; all authors discussed the results and commented on the manuscript. All authors have read and agreed to the published version of the manuscript.

Funding: This research was funded by the National Key Research and Development Program of China (2021YFD2200403).

Data Availability Statement: All the data supporting the results of this work can be found in the Supplements.

Acknowledgments: We are thankful to our lab fellows for their kind cooperation, and especially to Farhan Akhter (The Islamia University of Bahawalpur, Punjab, Pakistan) and Umair Ahmed (Huazhong Agricultural University, Wuhan, China), who supported us throughout this project.

Conflicts of Interest: The authors declare no conflict of interest.

References

1. Fujii, K.; Shibata, M.; Kitajima, K.; Ichie, T.; Kitayama, K.; Turner, B.L. Plant–soil interactions maintain biodiversity and functions of tropical forest ecosystems. *Ecol. Res.* **2018**, *33*, 149–160. [CrossRef]
2. Liu, X.; Tan, N.; Zhou, G.; Zhang, D.; Zhang, Q.; Liu, S.; Chu, G.; Liu, J. Plant diversity and species turnover co-regulate soil nitrogen and phosphorus availability in Dinghushan forests, southern China. *Plant Soil* **2021**, *464*, 257–272. [CrossRef]
3. Trindade, C.R.T.; Landeiro, V.L.; Schneck, F. Macrophyte functional groups elucidate the relative role of environmental and spatial factors on species richness and assemblage structure. *Hydrobiologia* **2018**, *823 Pt 2*, 217–230. [CrossRef]
4. Dong, S.; Sha, W.; Su, X.; Zhang, Y.; LI, S.; Gao, X.; Liu, S.; Shi, J.; Liu, Q.; Hao, Y. The impacts of geographic, soil and climatic factors on plant diversity, biomass and their relationships of the alpine dry ecosystems: Cases from the Aerjin Mountain Nature Reserve, China. *Ecol. Eng.* **2019**, *127*, 170–177. [CrossRef]
5. Deyn, D.E.; Raaijmakers, G.B.; Van, C.E.; Der Putten, W.H. Plant community development is affected by nutrients and soil biota. *J. Ecol.* **2004**, *92*, 824–834. [CrossRef]
6. Perroni-Ventura, Y.; Montaña, C.; García-Oliva, F. Relationship between soil nutrient availability and plant species richness in a tropical semi-arid environment. *J. Veg. Sci.* **2006**, *17*, 719–728. [CrossRef]
7. Bulenga, G.B.; Maliondo, S.M.S.; Katani, J.Z. Relationships between tree species diversity with soil chemical properties in semi-dry Miombo Woodland ecosystems. *Tanzan. J. For. Nat. Conserv.* **2021**, *90*, 1–17.
8. Odum, E.P. The Strategy of Ecosystem Development. *Science* **1969**, *164*, 262–270. [CrossRef]
9. Turner, B.L.; Brenes-Arguedas, T.; Condit, R. Pervasive phosphorus limitation of tree species but not communities in tropical forests. *Nature* **2018**, *555*, 367–370. [CrossRef]
10. Zalamea, P.; Turner, B.L.; Winter, K.; Jones, F.A.; Sarmiento, C.; Dalling, J.W. Seedling growth responses to phosphorus reflect adult distribution patterns of tropical trees. *New Phytol.* **2016**, *212*, 400–408. [CrossRef]
11. Jakovac, C.C.; Bongers, F.; Kuyper, T.W.; Mesquita, R.C.G.; Pena-Claros, M. Land use as a filter for species composition in Amazonian secondary forests. *J. Veg. Sci.* **2016**, *27*, 1104–1116. [CrossRef]
12. Grime, J.P. *Plant Strategies and Vegetation Processes*, 1st ed.; John Wiley & Sons: Hoboken, NJ, USA, 1979; ISBN 9780471996927.
13. Long, C.; Yang, X.; Long, W.; Li, D.; Zhou, W.; Zhang, H. Soil nutrients influence plant community assembly in two tropical coastal secondary forests. *Trop. Conserv. Sci.* **2018**, *11*, 1–9. [CrossRef]
14. Gaston, K.J. Common ecology. *Bioscience* **2011**, *61*, 354–362. [CrossRef]
15. Neri, A.V.; Schaefer, C.E.G.R.; Silva, A.F.; Souza, A.L.; Ferreira-Junior, W.G.; Meira-Neto, J.A.A. The influence of soils on the floristic composition and community structure of an area of Brazilian cerrado vegetation. *Edinb. J. Bot.* **2012**, *69*, 1–27. [CrossRef]
16. Janssen, F.; Peters, A.; Tallowin, J.R.B.; Bakker, J.P.; Bekker, R.M.; Fillat, F.; Oomes, M.J.M. Relationship between soil chemical factors and grassland diversity. *Plant Soil* **1998**, *202*, 69–78. [CrossRef]
17. Kumar, J.I.; Kumar, R.N.; Bhoi, R.K.; Sajish, P.R. Tree species diversity and soil nutrient status in three sites of tropical dry deciduous forest of western India. *Trop. Ecol.* **2010**, *51*, 273–279.
18. Nadeau, M.B.; Sullivan, T.P. Relationships between plant biodiversity and soil fertility in a mature tropical forest, Costa Rica. *Int. J. For. Res.* **2015**, *2015*, 732946. [CrossRef]
19. Cárate-Tandalla, D.; Camenzind, T.; Leuschner, C.; Homeier, J. Contrasting species responses to continued nitrogen and phosphorus addition in tropical montane forests tree seedlings. *Biotropica* **2018**, *50*, 234–245. [CrossRef]
20. Liu, H.; Chen, Q.; Liu, X.; Xu, Z.; Dai, Y.; Liu, Y.; Chen, Y. Variation patterns of plant composition/diversity in Dacrydium pectinatum communities and their driving factors in a biodiversity hotspot on Hainan Island, China. *Glob. Ecol. Conserv.* **2020**, *22*, e01034. [CrossRef]
21. John, R.; Dalling, J.W.; Harms, K.E.; Yavitt, J.B.; Stallard, R.F.; Mirabello, M.; Hubbell, S.P.; Valencia, R.; Navarrete, H.; Vallejo, M.; et al. Soil nutrients influence spatial distributions of tropical tree species. *Proc. Natl. Acad. Sci. USA* **2007**, *104*, 864–869. [CrossRef]
22. Fayolle, A.; Engelbrecht, B.; Freycon, V.; Mortier, F.; Swaine, M.; Rejou Mechain, M.; Doucet, J.-L.; Fauvet, N.; Cornu, G.; Gourlet-Fleury, S. Geological substrates shape tree species and trait distributions in African moist forests. *PLoS ONE* **2012**, *7*, e42381. [CrossRef]
23. Figueiredo, F.O.; Zuquim, G.; Tuomisto, H.; Moulatlet, G.M.; Balslev, H.; Costa, F.R. Beyond climate control on species range: The importance of soil data to predict distribution of Amazonian plant species. *J. Biogeogr.* **2018**, *45*, 190–200. [CrossRef]
24. Werden, L.K.; Becknell, J.M.; Powers, J.S. Edaphic factors, successional status and functional traits drive habitat associations of trees in naturally regenerating tropical dry forests. *Funct. Ecol.* **2018**, *32*, 2766–2776. [CrossRef]
25. Augusto, L.; Achat, D.L.; Jonard, M.; Vidal, D.; Ringeval, B. Soil parent material—A major driver of plant nutrient limitations in terrestrial ecosystems. *Glob. Change Biol.* **2017**, *23*, 3808–3824. [CrossRef] [PubMed]
26. Medvigy, D.; Wang, G.; Zhu, Q.; Riley, W.J.; Trierweiler, A.M.; Waring, B.G.; Xu, X.; Powers, J.S. Observed variation in soil properties can drive large variation in modelled forest functioning and composition during tropical forest secondary succession. *New Phytol.* **2019**, *223*, 1820–1833. [CrossRef]
27. Situ, S.J. *Study on the Historical Land Exploitation of Hainan Island*; Hainan Press: Haikou, China, 1987.
28. Waring, B.G.; Becknell, J.M.; Powers, J.S. Nitrogen, phosphorus, and cation use efficiency in stands of regenerating tropical dry forest. *Oecologia* **2015**, *178*, 887–897. [CrossRef]
29. Huang, J.; Wang, C.; Qi, L.; Zhang, X.; Tang, G.; Li, L.; Guo, J.; Jia, Y.; Dou, X.; Lu, M. Phosphorus is more effective than nitrogen in restoring plant communities of heavy metals polluted soils. *Environ. Pollut.* **2020**, *266*, 115259. [CrossRef]

30. Long, W.; Yang, X.; Li, D. Patterns of species diversity and soil nutrients along a chrono sequence of vegetation recovery in Hainan Island, South China. *Ecol. Res.* **2012**, *27*, 561–568. [CrossRef]
31. Wu, Z.Y. *Vegetation in China*; Science Press: Beijing, China, 1995.
32. Jobba'gy, E.G.; Jackson, R.B. The distribution of soil nutrients with depth: Global patterns and the imprint of plants. *Biogeochemistry* **2001**, *53*, 51–77. [CrossRef]
33. Anderson, J.M.; Ingram, J.S.I. *Tropical Soil Biology and Fertility: A Handbook of Methods*; C.A.B. International: Aberystwyth, UK, 1989; Volume 157, p. 265.
34. Jung, S.-H. Stratified Fisher's exact test and its sample size calculation. *Biom. J.* **2014**, *56*, 129–140. [CrossRef]
35. R Core Development Team (Ed.) *R: A Language and Environment for Statistical Computing*; R Foundation for Statistical Computing: Vienna, Austria, 2016; Available online: www.R-project.org/ (accessed on 10 February 2015).
36. Agbu, P.A.; Olson, K.R. Spatial variability of soil properties in selected Illinois Mollisols. *Soil Sci.* **1990**, *150*, 777–786. [CrossRef]
37. Li, X.; Zhang, X.; Wu, J.; Shen, Z.; Zhang, Y.; Xu, X.; Fan, Y.; Zhao, Y.; Yan, W. Root biomass distribution in alpine ecosystems of the northern Tibetan Plateau. *Environ. Earth Sci.* **2011**, *64*, 1911–1919. [CrossRef]
38. Chen, C.R.; Condron, L.M.; Davis, M.R.; Sherlock, R.R. Effects of afforestation on phosphorus dynamics and biological properties in a New Zealand grassland soil. *Plant Soil* **2000**, *220*, 151–163. [CrossRef]
39. Kooyman, R.M.; Laffan, S.W.; Westoby, M. The incidence of low phosphorus soils in Australia. *Plant Soil* **2017**, *412*, 143–150. [CrossRef]
40. Smithsonian Tropical Research Institute. Diverse Tropical Forests Grow Fast Despite Widespread Phosphorus Limitation. *ScienceDaily*. 7 March 2018. Available online: www.sciencedaily.com/releases/2018/03/180307153944.htm (accessed on 7 March 2018).
41. Bohn, H.; Mcneal, B.; O'Connor, G. *Soil Chemistry*, 3rd ed.; Wiley: New York, NY, USA, 2001.
42. Vitousek, P.M.; Porder, S.; Houlton, B.Z.; Chadwick, O.A. Terrestrial phosphorus limitation: Mechanisms, implications, and nitrogen–phosphorus interactions. *Ecol. Appl.* **2010**, *20*, 5–15. [CrossRef]
43. Jiang, Y.; Zang, R.; Letcher, S.G.; Ding, Y.; Huang, Y.; Lu, X.; Huang, J.; Liu, W.; Zhang, Z. Associations between plant composition/diversity and the abiotic environment across six vegetation types in a biodiversity hotspot of Hainan Island, China. *Plant Soil* **2016**, *403*, 21–35. [CrossRef]
44. ter Steege, H.; Pitman, N.C.A.; Killeen, T.J.; Laurance, W.F.; Peres, C.A.; Guevara, J.E.; Salomão, R.P.; Castilho, C.V.; Amaral, I.L.; de Almeida Matos, F.D.; et al. Estimating the global conservation status of more than 15,000 Amazonian tree species. *Sci. Adv.* **2015**, *1*. [CrossRef]
45. Kettle, C.J.; Koh, L.P. (Eds.) *Global Forest Fragmentation*; CABI: Wallingford, UK, 2014; ISBN 9781780642031.
46. Hedin, L.O.; Brookshire, E.N.J.; Menge, D.N.L.; Barron, A.R. The Nitrogen Paradox in Tropical Forest Ecosystems. *Annu. Rev. Ecol. Evol. Syst.* **2009**, *40*, 613–635. [CrossRef]
47. LeBauer, D.S.; Treseder, K.K. Nitrogen limitation of net primary productivity in terrestrial ecosystems is globally distributed. *Ecology* **2008**, *89*, 371–379. [CrossRef]
48. Avolio, M.L.; Forrestel, E.J.; Chang, C.C.; La Pierre, K.J.; Burghardt, K.T.; Smith, M.D. Demystifying dominant species. *New Phytol.* **2019**, *223*, 1106–1126. [CrossRef]
49. Chen, J.; Qu, M.; Zhang, J.; Xie, E.; Huang, B.; Zhao, Y. Soil fertility quality assessment based on geographically weighted principal component analysis (GWPCA) in large-scale areas. *CATENA* **2021**, *201*, 105197. [CrossRef]
50. Hernández-Vargas, G.; Sánchez-Velásquez, L.R.; López-Acosta, J.C.; Noa-Carrazana, J.C.; Perroni, Y. Relationship between soil properties and leaf functional traits in early secondary succession of tropical montane cloud forest. *Ecol. Res.* **2019**, *34*, 213–224. [CrossRef]
51. Whittaker, R.J.; Bush, M.B.; Richards, K. Plant Recolonization and Vegetation Succession on the Krakatau Islands, Indonesia. *Ecol. Monogr.* **1989**, *59*, 59–123. [CrossRef]
52. Schmidt, M.; Veldkamp, E.; Corre, M.D. Tree species diversity effects on productivity, soil nutrient availability and nutrient response efficiency in a temperate deciduous forest. *For. Ecol. Manag.* **2015**, *338*, 114–123. [CrossRef]
53. Chala, D.; Brochmann, C.; Psomas, A.; Ehrich, D.; Gizaw, A.; Masao, C.A.; Zimmermann, N.E. Good-bye to tropical alpine plant giants under warmer climates? Loss of range and genetic diversity in Lobelia rhynchopetalum. *Ecol. Evol.* **2016**, *6*, 8931–8941. [CrossRef] [PubMed]
54. Vázquez-Rivera, H.; Currie, D.J. Contemporaneous climate directly controls broad-scale patterns of woody plant diversity: A test by a natural experiment over 14,000 years. *Glob. Ecol. Biogeogr.* **2015**, *24*, 97–106. [CrossRef]
55. Kreft, H.; Jetz, W. Global patterns and determinants of vascular plant diversity. *Proc. Natl. Acad. Sci. USA* **2007**, *104*, 5925–5930. [CrossRef] [PubMed]
56. Whitfeld, T.J.; Lasky, J.R.; Damas, K.; Sosanika, G.; Molem, K.; Montgomery, R.A. Species richness, forest structure, and functional diversity during succession in the New Guinea lowlands. *Biotropica* **2014**, *46*, 538–548. [CrossRef]
57. Lu, X.; Zang, R.; Ding, Y.; Huang, J. Partitioning the functional variation of tree seedlings during secondary succession in a tropical lowland rainforest. *Ecosphere* **2018**, *9*, e02305. [CrossRef]
58. Li, C.; Zhao, L.; Sun, P.; Zhao, F.; Kang, D.; Yang, G.; Han, X.; Feng, Y.; Ren, G. Deep Soil C, N, and P Stocks and Stoichiometry in Response to Land Use Patterns in the Loess Hilly Region of China. *PLoS ONE* **2016**, *11*, e0159075. [CrossRef]

forests

MDPI

Article

Beta Diversity Patterns Unlock the Community Assembly of Woody Plant Communities in the Riparian Zone

Yan He [1,†], Shichu Liang [1,†], Runhong Liu [1,2] and Yong Jiang [1,*]

1 Key Laboratory of Ecology of Rare and Endangered Species and Environmental Protection, Guangxi Normal University, Ministry of Education, Guilin 541006, China; hy18278380128@163.com (Y.H.); gxlsc1965@163.com (S.L.); liurh18@lzu.edu.cn (R.L.)
2 College of Forestry, Guangxi University, Nanning 530004, China
* Correspondence: yongjiang226@126.com
† These authors contributed equally to this work.

Abstract: Beta diversity refers to changes in community composition across time and space, including species richness and replacement. Few studies have examined beta diversity patterns of riparian vegetation communities in terms of taxonomic, phylogenetic and functional attributes. In this study, we conducted a field survey of woody plant communities in the riparian zone of the Lijiang River Basin in China. We analyze variations in taxonomic, phylogenetic and functional beta diversity, the relative contributions of species richness and replacement to beta diversity and the relationships between beta diversity and environmental distance and geographical distance. The results show that: (1) replacement was the dominant component of taxonomic beta diversity and richness was the dominant component of functional and phylogenetic beta diversity; and (2) dispersal limitation and habitat filtering jointly drive the community assembly of woody plant communities in the riparian zone of the Lijiang River Basin. Therefore, when formulating conservation strategies for woody plants along the Lijiang River riparian zone, improving ecological communities and enhancing species dispersal between communities should be given equal attention. From a taxonomic perspective, it is more suitable to establish several small nature reserves, whereas from phylogenetic and functional perspectives, protection should focus on larger nature reserves.

Keywords: community assembly; beta diversity; replacement; richness

Citation: He, Y.; Liang, S.; Liu, R.; Jiang, Y. Beta Diversity Patterns Unlock the Community Assembly of Woody Plant Communities in the Riparian Zone. *Forests* **2022**, *13*, 673. https://doi.org/10.3390/f13050673

Academic Editor: Bogdan Jaroszewicz

Received: 1 March 2022
Accepted: 25 April 2022
Published: 27 April 2022

1. Introduction

Understanding the mechanisms driving the formation, maintenance and loss of biodiversity represents the focus of community ecology and contributes to sustainable development and the effective protection of biodiversity [1,2]. Beta diversity, as an important part of biodiversity, reflects turnover in species composition between communities along a predefined spatial, temporal or environmental gradient [3]. Studies have found that the loss of beta diversity could lead to biological homogenization and even reduce ecosystem functions [4]. Thus, analyzing ecological community construction processes that drive patterns of beta diversity has become a hot topic in recent years [5,6].

Beta diversity has decomposed two processes: replacement and richness [7,8]. The replacement component of beta diversity represents species replacement between locations. Mechanisms contributing to species replacement (or turnover) include habitat filtering, competition and geographic barriers [9,10]. For example, natural selection along environmental gradients can contribute to different species appearing in habitats suitable for survival [11]. Geographical isolation caused by the uplift of mountains can fragment populations and promote allopatric speciation [12]. The richness component of beta diversity represents the difference in species composition between communities caused by the loss or increase of species along the sampling axis or throughout the study area. Mechanisms that can influence richness include diversity of niches available and species thinning causing nestedness or other ecological processes (e.g., historical processes: extinction and colonization) [13,14]. Disentangling the relative roles of replacement and richness in contributing to beta diversity is necessary to understand the processes driving beta diversity and to design protected area networks.

As traditionally conceived, beta diversity quantifies the diversity of communities based on the taxonomy and abundance of species—the change in species between ecological communities [15,16]. However, communities may contain species that have redundant evolutionary relationships and functional attributes. Changes in species often lack a one-to-one relationship with phylogenetic and functional attributes. Phylogenetic beta diversity can partially remedy this gap by reflecting differences in evolutionary relatedness between communities; it also reflects the impact of historical processes on community construction [17]. Functional beta diversity refers to the value and range of functional traits in ecosystems or communities, taking into account the redundancy and complementarity of coexisting species and reflecting differences in the functional attributes of species between communities [18]. Therefore, understanding beta diversity from the perspectives of taxonomy, phylogeny and functional traits is necessary to accurately grasp the ecological processes and mechanisms driving beta diversity.

The riparian zone of the Lijiang is heterogeneous along its 168-km length. It is an important ecological transition zone for material, energy and information exchange between river ecosystems and terrestrial ecosystems [19]. It provides an excellent natural experimental platform for exploring beta diversity distribution patterns and drivers [20]. In recent years, tourism and excessive development in China's Lijiang River Basin have seriously damaged vegetation in the basin's riparian zone, and the ecology of the basin has changed significantly [21]. Riparian vegetation, particularly woody plants, contributes to most riparian ecosystem functions. Therefore, research into the restoration and protection of woody plants diversity, including beta diversity, is crucial.

Here, we investigated the taxonomic, phylogenetic and functional beta diversity patterns of woody plant communities in the riparian zone of the Lijiang River basin in China. We aim to understand and compare the processes responsible for shaping the beta diversity patterns from these three perspectives. We addressed the following questions: (1) Is the formation of the distribution of woody plant communities in the riparian zone of the Lijiang River Basin driven by replacement or richness? (2) Considering community assembly in terms of taxonomic, phylogenetic and functional beta diversity, what is the importance of environmental filtering and dispersal limitation? The answers to these questions will provide scientific guidance for the conservation of woody plant biodiversity in the region, including the selection of priority conservation sites and the formulation of conservation plans (e.g., improving the community environment or enhancing species dispersal between communities).

2. Materials and Methods

2.1. Study Area

Our study area is located in the Lijiang River Basin in the northeast of Guangxi Zhuang Autonomous Region, China. It is in the upper reaches of the Guijiang River, covering a length of approximately 164 km, with an average elevation of 158 m, as a primary tributary of the Xijiang River in the Pearl River Basin. The geographic coordinates are 24°18′–25°41′ N, 109°45′–110°40′ E. It has been listed as a World Natural Heritage Site since 2014 and is a popular tourist destination deriving from its unique karstic landscape. The karst landform is a rare geomorphological type with fragile ecosystems that are slow to recover after disturbance [22]. In recent years, excessive tourism-related development and other anthropogenic disturbances have interacted to influence the highly variable and complex Lijiang River Basin system. The area is in the humid monsoon climate zone within the mid-subtropical zone, with a mild climate and four distinct seasons. Annual rainfall is abundant but extremely uneven during the year. The flood season occurs from March to August, and the dry season occurs from September to February. The annual average temperature is 18–19 °C, the coldest January average temperature is about 8–9 °C and the hottest August average temperature is about 28 °C. Rainfall is concentrated in the warm season. The annual rainfall is 1814–1941 mm, the annual evaporation is 1377–1857 mm and the annual average relative humidity is 73%–79%. The year-round sunshine is sufficient, and the annual average sunshine time is 1670 h. Soil type is dominated by red loam and has the features of high coarse texture, gravel-type substrates, shallow soil layer and unevenly distributed soil layer thickness [19,21] for years of continuous erosion and deposition by river water. Common woody plant in the riparian zone of the Lijiang River Basin includes *Pterocarya stenoptera* C. DC., *Triadica sebifera* (L.) *Small, Celtis sinensis Pers., Cinnamomum burmannii* (Nees & T. Nees) *Blume, Cinnamomum camphora* (L.) *J. Presl, Ficus abeli Miq., Vitex negundo var. cannabifolia* (Sieb.et Zucc.) *Hand.-Mazz., Flueggea virosa* (Roxb. ex Willd.) *Royle, Rauvolfia verticillat* (Lour.) *Baill., Adina rubella Hance* and so on.

2.2. Data Collection

2.2.1. Field Survey

This study is based on a field survey of woody plant communities. A total of 12 different sites were randomly selected and they were distributed in geographically separated regions, spanning a wide spatial range, ranging from 110°19′ to 110°34′ E in longitude and from 24°44′ to 25°54′ N in latitude (Figure 1). Each site was set up with 4–8 20 m × 20 m plots—a total of 65 plots. Community surveys of all plots were performed according to the standard ForestGeo protocol (https://forestgeo.si.edu/node/145665/, accessed on 7 May 2018). The census of 65 plots was completed in 2018, recording 15,100 free-standing individuals and belonging to 78 species, representing 64 genera and 32 families. The plots covered various community types. They can be divided into nine associations on the basis of the species importance values calculation, including ASS. *P. stenoptera-F. abelii*, ASS. *C. sinensis-B. formosana + F. abelii*, ASS. *P. stenoptera-A. chinense*, ASS. *C. sinensis-F. abelii*, ASS. *C. burmannii-R. verticillate*, ASS. *R. verticillate*, ASS. *T. sebiferum-A. rubella*, ASS. *C.sinensis-V. negundo var. cannabifolia* and ASS. *T. sebiferum-B. formosana* (please see Table A1 in Appendix A).

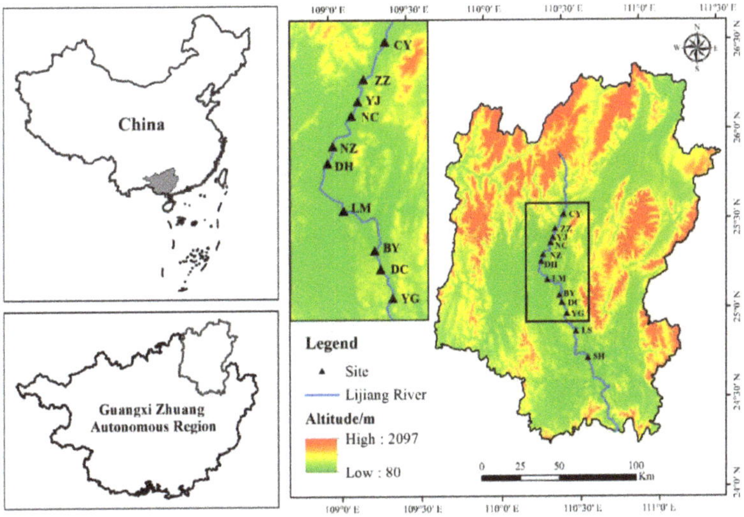

Figure 1. The distribution map of the 12 studied plots in the Lijiang River Basin. The colored background represents the elevation distribution. CY is for Caiyuan, ZZ is for Zhuzhi, YJ is for Yangjia, NC is for Nanchang, NZ is for Nanzhou, DH is for Dahe, LM is for Longmen, BY is for Biyan, DC is for Duchuan, YG is for Yueguang, LS is for Luoshi, SH is for Sanhe.

2.2.2. Functional Traits Measurement

We measured 6 functional traits for the 78 species based on standardized protocols proposed by Pérez-Harguindeguy et al. [23], including leaf chlorophyll content (LCC), leaf thickness (LT), leaf area (LA), leaf dry matter content (LDMC), specific leaf area (SLA) and twig tissue density (TTD). These traits were selected as they are known to represent the different dimensions of the functional niche, such as carbon economy, nutrient acquisition, water economy in leaves and stems among woody plants. Specifically, SLA captures species strategies for acquiring, using and conserving resources, including light, nutrients and water [24]; TTD reflects twig water and nutrient transport, structure and defense [25]. All tree individuals with a diameter at breast height (DBH, 1.3 m) $\geq$ 1 cm and shrub and woody vine individuals with a basal diameter (BD) $\geq$ 1 cm were sampled to measure functional traits data during the summer season (from June to August) in 2018. For every individual, we sampled at least three newly matured leaves and three branches at different positions. Among them, LCC was estimated by measuring red/infrared absorbance with a SPAD-502 chlorophyll meter (Spectrum Technologies, Plainfield, IL, USA). SPAD measurements were converted into chlorophyll concentrations using the homographic calibration model from Coste et al. [26]. LT was measured with a caliper (precision: 0.05 mm). LA was measured using a YMJ-C scanner with a corresponding self-developed measurement software system (Shandong, China). The leaves were put on the instrument and took real-time photos to process the data information. The leaves were then oven-dried at 70 °C for 48 h and weighed to calculate the SLA and LDMC. TTD was calculated as the oven-dried mass of a twig divided by its fresh volume.

2.2.3. Environmental Variables Measurement

We divided each 20 m × 20 m plot into four 10 m × 10 m subplots and then collected soil nutrients on 10 m × 10 m subplots for fine-scale analysis of environment factors. We measured environmental data in a total of 200 10 m × 10 m subplots. In each 10 m × 10 m subplot, a soil sampler was used to collect soil samples at five random points. At each sampling point, after removing the possible organic layer, a soil sample with a depth of 10–20 cm was collected. The soil samples at five points in each subplot were mixed as a test sample. We analyzed the physical and chemical properties of each test sample using methods that have been described by Bao Shidan [27], including soil pH (pH), water content (SWC), organic matter (SOM), total nitrogen (TN), available nitrogen (AN), total phosphorus (TP), available phosphorus (AP), total potassium (TK) and available potassium (AK). Among them, pH was determined in a 1:2.5 soil-to-water suspension ratio. SWC was measured as the fresh soil weight minus oven-dried weight divided by the fresh weight. SOM was determined by wet oxidation with KCr_2O_7 + H_2SO_4 and titrated with $FeSO_4$. TN was determined using automatic Kjeldahl analysis (KJELTECTM 8400, FOSS Quality Assurance Co., Ltd., Hillerød, Denmark). AN was quantified using the alkaline hydrolysis diffusion method. TP was examined by acid digestion with a H_2SO_4 + $HClO_4$ solution. AP was extracted with 0.5 M Na_2CO_3, and it was measured using the molybdenum blue colorimetric method. TK was digested by the HF-$HClO_4$-HNO_3 acid mixture and determined by the flame photometric method. AK was determined by flame photometry after extraction with ammonium acetate. The elevation data were recorded with a handheld GPS. Finally, we averaged the four 10 m × 10 m subplots as the environment factors of each of the 20 m × 20 m plots for subsequent analyses.

2.3. *Statistical Analyses*

2.3.1. Spatial Variables

We recorded latitude and longitude of each plot by GPS. Then, we calculated the spatial distance based on the distm function in the "geosphere" package of R software to transform the coordinates to distance [28].

2.3.2. Phylogenetic Tree Construction

For subsequent phylogenetic beta diversity analyses, we created a phylogenetic supertree that included all of our species (a data set of 78 total species sampled across 12 sites (65 20 m × 20 m)) using scenario 1 in the "V. Phylomaker" R package (see Appendix B for more details) [29].

2.3.3. Beta Diversity Calculation

Taxonomic, phylogenetic and functional beta diversity and their components were all calculated by applying pairwise-site and multiple-site dissimilarity methods [30]. Based on the taxonomic data (abundance of species at each site), phylogenetic data (phylogenetic tree across all of the sites) and functional attribute data (six traits' Euclidean distances), we calculated the Jaccard dissimilarity index to calculate taxonomic (Tβ), phylogenetic (Pβ) and functional (Fβ) beta diversity indexes. We further decomposed total beta diversity (β_{total}) into the sum of replacement (β_{repl}) and richness (β_{rich}) [17]. β_{repl} refers to the differences between the communities caused by one-to-one replacement of species composition or system evolution (functional attributes). β_{rich} emphasizes differences between communities due to differences in species composition or system evolution (functional attributes) in richness. The calculation formula is as follows:

$$\beta_{total} = \beta_{repl} + \beta_{rich} \tag{1}$$

$$\beta_{total} = \frac{b+c}{a+b+c} \tag{2}$$

$$\beta_{rich} = \frac{|b - c|}{a+b+c} \tag{3}$$

$$\beta_{repl} = 2 \times \frac{\min(b,c)}{a+b+c} \tag{4}$$

where ***a*** represents the number of species or phylogenetic (functional attribute) branch length shared by the two communities, and ***b*** and ***c*** respectively represent the number of species or phylogenetic (functional attribute) branch length unique to the two communities.

We then sought to better understand the relative contribution of environmental filtering and dispersal limitation on the three dimensions of beta diversity. First, we computed spearman correlations between variables to assess the collinearity of environmental variables, where elevations were excluded for spearman $\rho 2 > 0.5$. Then, we calculated the Euclidean distance between filtered environmental variables to the obtained environmental distance matrix. On this basis, we applied $\ln(x + 1)$ transformations on the geographic and environmental distance matrix to meet the assumptions of normality and homogeneity of variance [31,32]. We also performed a partial Mantel correlation analysis between geographic distance or environmental distance and the replacement components to test the relative importance of environmental filtering and dispersal limitation. If environmental distance was significantly correlated with beta diversity, we took a further step to test which environmental variables were more important in determining taxonomic, phylogenetic and functional beta diversity by conducting permutational MANOVA analyses [33].

All statistical analyses were conducted using R version 4.1.0 [34]. Among them, the taxonomic, phylogenetic and functional beta diversity calculations were based on the "BAT" package [35], the partial Mantel analysis was based on the "ecodist" package [36] and the Euclidean distances and permutational MANOVA analysis was performed using the package "vegan" [37].

3. Results

3.1. Patterns of Taxonomic, Functional and Phylogenetic Beta Diversity with Their Two Components

The mean values of taxonomic, functional and phylogenetic beta diversity were 0.90, 0.84 and 0.76, respectively. The relative proportions of replacement and richness components on taxonomic beta diversity were 54.17% and 45.83%, 48.51% and 51.49% on functional beta diversity and 40.53% and 59.47% on phylogenetic beta diversity. All in all, replacement was the dominant component of taxonomic beta diversity and richness was the dominant component of functional and phylogenetic beta diversity along the riparian zone of the Lijiang (Figure 2).

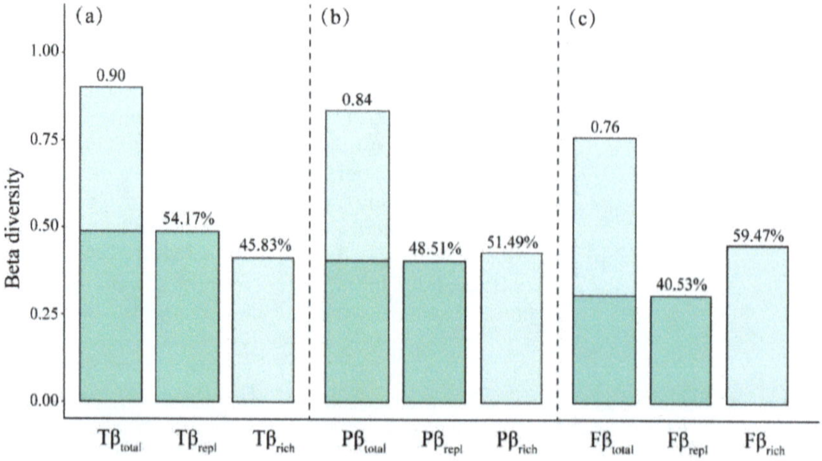

Figure 2. Beta diversity with its two components along the riparian zone of the Lijiang. (**a**) Mean taxonomic beta diversity, (**b**) mean phylogenetic beta diversity, (**c**) mean functional beta diversity. The percentages in the figure represent relative proportion of replacement and richness components, respectively. $T\beta_{total}$ represents total taxonomic beta diversity; $T\beta_{repel}$ represents the sum of replacement components on taxonomic beta diversity; $T\beta_{rich}$ represents the sum of richness components on taxonomic beta diversity. $P\beta_{total}$ represents total phylogenetic beta diversity; $P\beta_{repel}$ represents the sum of replacement components on phylogenetic beta diversity; $P\beta_{rich}$ represents the sum of richness components on phylogenetic beta diversity. $F\beta_{total}$ represents total functional beta diversity; $F\beta_{repel}$ represents the sum of replacement components on functional beta diversity; $F\beta_{rich}$ represents the sum of richness components on functional beta diversity.

3.2. *The Influence of Environmental Filtering and Dispersal Limitation on Beta Diversity Patterns*

With the increases in environmental distance and geographic distance, the replacement components of beta diversity (taxonomic, phylogenetic and functional) of the riparian plant communities in the Lijiang River Basin increased significantly (Figure 3). Partial Mantel correlation analysis revealed that the replacement components of beta diversity are subject to a combination of diffusional limitations and environmental filtering. Specifically, for taxonomic beta diversity, dispersal limitation alone explained 13.58% of the variation, and environmental filtering alone explained 9.31% of the variation. For phylogenetic diversity, dispersal limitation alone explained 14.48% of the variation, and environmental filtering alone explained 8.07% of the variation. For functional beta diversity, dispersal limitation alone explained 12.83% of the variation, and environmental filtering alone explained 9.32% of the variation.

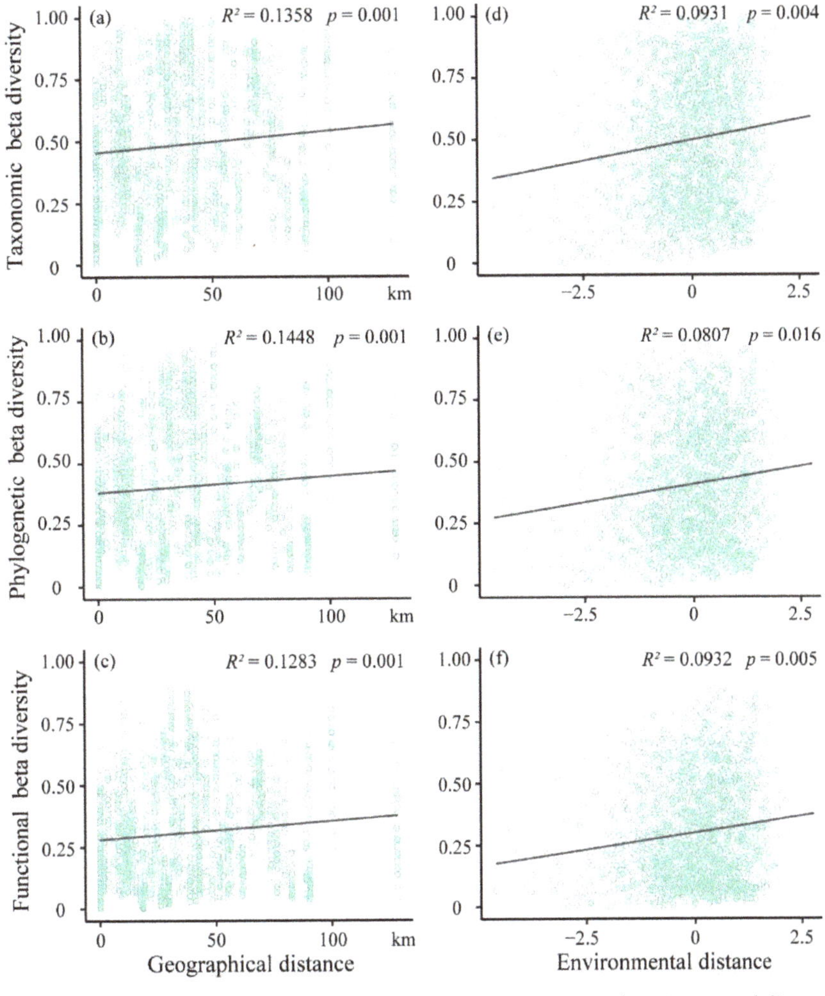

Figure 3. Variance in beta diversity along the geographic distance and environmental distances. (**a**) Variance in taxonomic beta diversity along the geographic distance, (**b**) variance in phylogenetic beta diversity along the geographic distance, (**c**) variance in functional beta diversity along the geographic distance, (**d**) variance in taxonomic beta diversity along the environmental distances, (**e**) variance in phylogenetic beta diversity along the environmental distances, (**f**) variance in functional beta diversity along the environmental distances. R^2—13.58% of variability is explained by geographic distance of taxonomic diversity, 9.31% of variability is explained by environmental distance of taxonomic diversity, 14.48% of variability is explained by geographic distance of phylogenetic diversity, 8.07% of variability is explained by environmental distance of phylogenetic diversity, 12.83% of variability is explained by geographic distance of functional diversity, 9.32% of variability is explained by environmental distance of functional diversity. p-values represent the significance of the statistical results.

3.3. The Influence of Environmental Factors on Beta Diversity

According to the results of permutational MANOVA analysis of variance (Table 1), soil pH, SOM, TN, AN, TP, AP, TK and AK had significant effects on taxonomic and phylogenetic diversity, except SWC had no significant effect. For functional beta diversity, soil pH, SOM, TN, AN, TP, AP, TK and AK had significant effects, while SWC and SOM had no significant effects. Overall, environmental factors explained a considerable portion of the variation, accounting for 50.3%–59.4%.

Table 1. Results of permutational MANOVA (adonis).

Explanatory Variables	Taxonomic Beta Diversity		Phylogenetic Beta Diversity		Functional Beta Diversity	
	R^2	*Pr (>F)*	R^2	*Pr (>F)*	R^2	*Pr (>F)*
Soil pH	0.075	***	0.063	***	0.070	***
Soil water content	0.001	ns	0.001	ns	0.001	ns
Soil organic matter	0.028	*	0.027	*	0.015	ns
Total nitrogen	0.057	***	0.057	***	0.047	**
Available nitrogen	0.040	**	0.030	*	0.064	***
Total phosphorus	0.082	***	0.082	***	0.074	***
Available phosphorus	0.056	***	0.062	**	0.083	***
Total potassium	0.091	***	0.134	***	0.120	***
Available potassium	0.073	***	0.091	***	0.120	***
Residuals	0.497	-	0.453	-	0.406	-

Note: * $p < 0.05$, ** $p < 0.01$, *** $p < 0.001$.

4. Discussion

4.1. The Composition of Taxonomic, Functional and Phylogenetic Beta Diversity with Their Two Components

Historically, taxonomic-based analyses of community assembly have been the most common approach used to characterize regional biodiversity and biogeographical processes [38]. However, recent efforts to incorporate functional and phylogenetic dimensions into the community assembly framework have provided further insights [6,39]. The findings showed the same results as those from previous studies: taxonomic > phylogenetic > functional beta diversity. One possible explanation for this pattern is that there exist redundant species tending to share the similar phylogenetic relations and functional traits in our study sites [40,41]. For example, if two different species within a community have very different functional traits, turnover in those species will increase the value of phylogenetic beta diversity and functional beta diversity. In contrast, if the similar species with redundant traits turnover, the value of phylogenetic beta diversity and functional beta diversity will often be smaller than taxonomic beta diversity. Consistent with a growing body of studies, we found that replacement was the dominant component of taxonomic beta diversity [42–44]. That suggests that the difference in taxonomic composition among sites was mainly caused by high species substitution. A recent broad study in a nearby Atlantic forest found that replacement patterns are associated with environmental heterogeneity and variability in species attributes [45]. The Lijiang River Basin, which stretches more than 160 km from the upstream to downstream, shows different environmental heterogeneity and geographic spatial gradient [46]. Many seeds of riparian species can drift along the river, but most appeared to colonize in nearby areas. It might be that habitat conditions vary with geographical distance, hindering the spread of the seeds along the riverbank so that sites further away are not suited for poorly dispersed species, leading to high species change in species composition [47]. Another result showed that richness was the dominant component of functional and phylogenetic beta diversity. Dobrovolski et al. [48] found that the nestedness component had a more important role in generating beta diversity patterns in high latitude areas affected by glaciation. However, the Lijiang River Basin without high latitude was not covered by an ice-sheet during the Last Glacial Maximum; the glaciation hypothesis proposed by previous authors may not explain the pattern of dominant nestedness component of functional and phylogenetic beta diversity. The only acceptable explanation for this general pattern is under special karst environmental stress; many of the surviving species at fine scale share large elements of their evolutionary history, which results in species convergent functional traits and phylogenetic clustering across sites [16].

4.2. The Relative Importance of Dispersal Limitation and Environmental Filtering

We found that dispersal limitation and environmental filtering jointly shaped beta diversity in taxonomic, functional and phylogenetic dimensions, and dispersal limitation effects are slightly higher than environmental filtering effects in community assembly. Some researchers theorize high-diversity forests (e.g., tropical forests, subtropical forests) reflect strong spatial correlations with the diversity, indicating the important role of dispersal limitation [49]. Our findings support this general hypothesis; many species have not reached places with favorable conditions because of dispersal limitation. A possible explanation is as follows: (1) the unique geological characteristics and hydrological structure resulting in special Lijiang River Basin karst landforms act as a geographical barrier to hinder the spread of the seeds along the riverbank; (2) excessive tourism-related development on the Lijiang River Basin midstream section act as an anthropogenic barrier to affect the species surviving rate successfully; (3) species dispersal capacity. Generally, species with larger propagules were more strongly affected by dispersal limitation than species with smaller species propagules [50]. Compared with some other factors, complex topographical features, anthropogenic disturbance and species dispersal capacity are likely to be more important drivers of the beta diversity distribution.

Additionally, we found that environmental filtering also had a significant impact on beta diversity distribution. In further testing of environmental variables, we found that environmental factors, such as soil pH, TN, AN, TP, AP, TK and AK, were significantly correlated with the three measures of beta diversities (Table 1). It is well known that nitrogen, phosphorus and potassium in soil play important roles in maintaining forest productivity [51], and soil pH may modify the nutrient availability of P and K by controlling their chemical forms [52]. These results imply that soil nutrients are important factors driving woody plant community assembly in the riparian zone of the Lijiang River Basin. Taken together, the driving mechanisms of taxonomic, phylogenetic and functional beta diversity are generally consistent in the Lijiang River riparian woody plant community, dominated by dispersal limitation and environmental filtering, but with slightly different emphases in each dimension. For example, for environmental filtering processes, taxonomic and functional beta diversity have higher convergent roles due to the complex and harsh karst habitats. However, it is also possible that this is due to the fact that we measure a limited number of local environment variables. Previous studies have shown that phylogenetic diversity was influenced by the combined effect of paleoclimate and modern climate [53], so it is necessary to incorporate more environmental variables, such as climatic factors, into the environmental analysis in future research.

5. Conclusions

Examining multiple facets and components of beta diversity is becoming an important topic in biodiversity research because it can provide information about the mechanisms driving community assembly [54,55]. In the present study, we found that (1) taxonomic beta diversity is dominated by replacement components, while phylogenetic and functional beta diversity are dominated by richness components. To conserve riparian biodiversity in the Lijiang River Basin from a taxonomic perspective, it is more suitable to establish several small nature reserves, whereas from phylogenetic and functional perspectives, the protection focus should be put on larger nature reserves. (2) Dispersal limitation and environmental filtering jointly drive the composition of woody plant communities in the riparian zone of the Lijiang River Basin. Thus, when formulating conservation strategies for woody plants along the Lijiang River riparian zone, both improving ecological community environmental conditions and enhancing species dispersal between communities should be given equal attention. Given the complexity of community assembly mechanisms, we recommend that future studies should integrate other components of biodiversity such as the intraspecific variation in functional and phylogenetic diversity [56]. These can bring further insights into species coexistence, resource use, niche overlap, plasticity and local adaptations, as well as the development of tools to assess the scales and biodiversity components for which a given management measure is the most cost-efficient (prioritization).

Author Contributions: Conceptualization, Y.J. and S.L.; methodology, Y.J.; validation, Y.J., S.L. and Y.H.; formal analysis, Y.H.; investigation, R.L.; resources, S.L.; data curation, R.L.; writing—original draft preparation, Y.H. and S.L.; writing—review and editing, R.L. and Y.J.; visualization, Y.H.; funding acquisition, Y.J. All authors have read and agreed to the published version of the manuscript.

Funding: This research was funded by the National Natural Science Foundation of China (31860124), Guangxi Natural Science Foundation (2022GXNSFAA035600), Innovation Project of Guangxi Graduate Education (JGY2021024), Key Laboratory of Ecology of Rare and Endangered Species and Environmental Protection (Guangxi Normal University), Ministry of Education, China (ERESEP2021Z08) and Guangxi Normal University Scientific Research and Education Special Project (Natural Science) (RZ2100000150).

Institutional Review Board Statement: Not applicable.

Informed Consent Statement: Not applicable.

Data Availability Statement: The data presented in this study are available on request from the corresponding author.

Acknowledgments: We are grateful to numerous students from the Chinese Guangxi Normal University for their support in conducting the tough fieldwork.

Conflicts of Interest: The authors declare no conflict of interest.

Appendix A

Table A1. Site information.

Site	Plot	Longitude	Latitude	Community Type
Caiyuan	1–7	110°27′41.00″ E	25°33′41.37″ N	ASS. *P. stenoptera-F. abelii*
Zhuzhi	8–12	110°24′20.68″ E	25°28′47.21″ N	ASS. *C. sinensis-B. formosana + F. abelii*
Yangjia	13–17	110°22′30.08″ E	25°24′48.87″ N	ASS. *P. stenoptera-F. abelii*
Nanchang	18–22	110°22′33.14″ E	25°24′46.64″ N	ASS. *P. stenoptera-A. chinense*
Nanzhou	23–26	110°19′49.48″ E	25°20′23.83″ N	ASS. *C. sinensis-F. abelii*
Dahe	27–32	110°19′22.99″ E	25°19′31.87″ N	ASS. *C. sinensis-F. abelii*
Biyan	33–37	110°25′09.45″ E	25°06′30.65″ N	ASS. *C. burmannii-R. verticillate*
Duchuan	38–41	110°25′29.78″ E	25°05′47.63″ N	ASS. *R. verticillate*
Longmen	42–46	110°20′58.79″ E	25°12′12.54″ N	ASS. *T. sebiferum-A. rubella*
Yueguang	47–52	110°27′17.94″ E	25°00′12.81″ N	ASS. *C. sinensis-V. negundo var. cannabifolia*
Luoshi	53–57	110°30′05.50″ E	24°54′22.97″ N	ASS. *T. sebiferum-B. formosana*
Sanhe	58–65	110°34′49.15″ E	24°44′58.25″ N	ASS. *T. sebiferum-B. formosana*

Appendix B

Phylogenetic Tree Construction

We constructed a phylogenetic tree that was divided into three steps. First, we assembled a species list that included all of our species sampled (a data set of 78 total species) in this study. Second, all names in the present species list were proofread through the TPL website (http://www.theplantlist.org/, accessed on 12 October 2021). Our present species list was included in the TPL. Third, a phylogenetic tree was generated based on the accepted 78 species list using the mega-tree function (scenario 1) in the R package "V. PhyloMaker" [29].

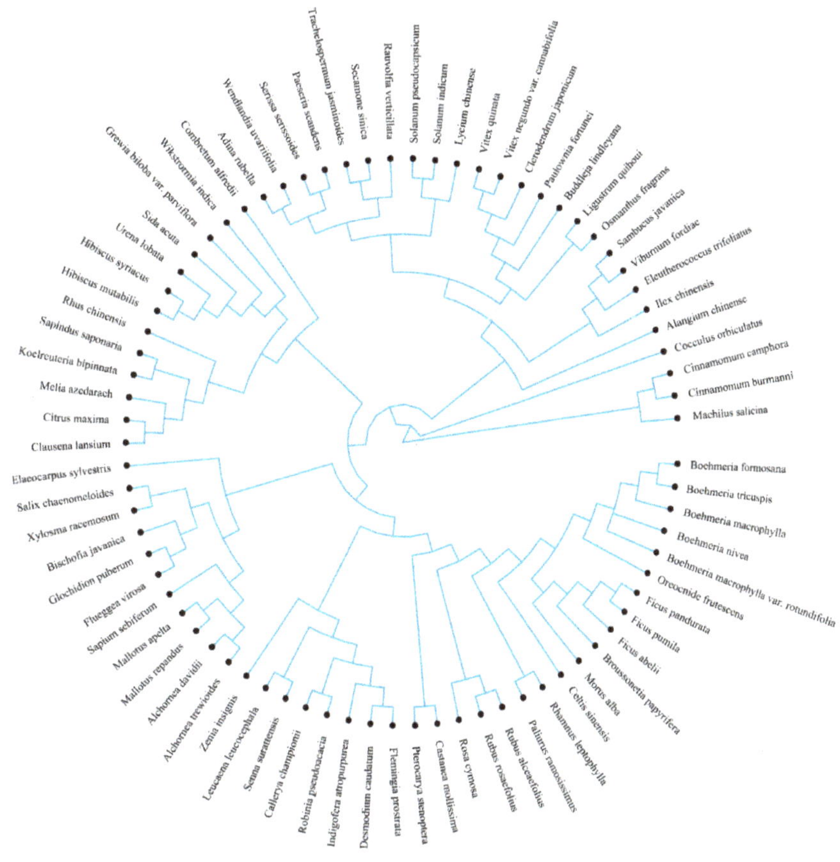

Figure A1. A super-tree representing the 78 species in the Lijiang River Basin in the northeast of Guangxi Zhuang Autonomous Region, China.

References

1. Münkemüller, T.; De, B.F.; Meynard, C.N.; Gravel, D.; Lavergne, S.; Mouillot, D.; Mouquet, N.; Thuiller, W. From diversity indices to community assembly processes: A test with simulated data. *Ecography* **2012**, *35*, 468–480. [CrossRef]
2. Jiang, L.M.; Lv, G.H.; Gong, Y.M.; Li, Y.; Wang, H.F.; Wu, D.Y. Characteristics and driving mechanisms of species beta diversity in desert plant communities. *PLoS ONE* **2021**, *16*, e0245249. [CrossRef]
3. Whittaker, R.H. Vegetation of the Siskiyou Mountains, Oregon and California. *Ecol. Monogr.* **1960**, *30*, 279–338. [CrossRef]
4. Socolar, J.B.; Gilroy, Z.J.J.; Kunin, W.E.; Edwards, D.P. How Should Beta-diversity Inform Biodiversity Conservation? *Trends Ecol. Evol.* **2016**, *31*, 67–80. [CrossRef]
5. Hu, D.; Jiang, L.M.; Hou, Z.F.; Zhang, J.; Wang, H.F.; Lv, G.H. Environmental filtration and dispersal limitation explain different aspects of beta diversity in desert plant communities. *Glob. Ecol. Conserv.* **2022**, *33*, e01956. [CrossRef]
6. García-Giróna, J.; Fernández-Aláeza, C.; Fernández-Aláeza, M.; Lahuhta, J.A. Untangling the assembly of macrophyte metacommunities by means of taxonomic, functional and phylogenetic beta diversity patterns. *Sci. Total Environ.* **2019**, *693*, 133616. [CrossRef] [PubMed]
7. Baselga, A. Partitioning the turnover and nestedness components of beta diversity. *Glob. Ecol. Biogeogr.* **2010**, *19*, 134–143. [CrossRef]
8. Baselga, A. Partitioning abundance-based multiple-site dissimilarity into components: Balanced variation in abundance and abundance gradients. *Methods Ecol. Evol.* **2017**, *8*, 799–808. [CrossRef]
9. Angeler, D.G. Revealing a conservation challenge through partitioned long-term beta diversity: Increasing turnover and decreasing nestedness of boreal lake metacommunities. *Divers. Distrib.* **2013**, *19*, 772–781. [CrossRef]
10. Legendre, P. Interpreting the replacement and richness difference components of beta diversity. *Glob. Ecol. Biogeogr.* **2014**, *23*, 1324–1334. [CrossRef]

11. Gutiérrez-Cánovas, C.; Millán, A.; Velasco, J.; Vaughan, I.P.; Ormerod, S.J. Contrasting Effects of Natural and Anthropogenic Stressors on Beta Diversity in River Organisms. *Glob. Ecol. Biogeogr.* **2013**, *22*, 796–805. [CrossRef]
12. Leprieur, F.; Tedesco, P.A.; Hugueny, B.; Beauchard, O.; Dürr, H.H.; Brosse, S.; Oberdorff, T. Partitioning global patterns of freshwater fish beta diversity reveals contrasting signatures of past climate changes. *Ecol. Lett.* **2011**, *14*, 325–334. [CrossRef] [PubMed]
13. Bergamin, R.S.; Bastazini, V.A.G.; Velez-Martin, E.; Debastiani, V.; Zanini, K.J.; Loyola, R.; Muller, S.C. Linking beta diversity patterns to protected areas: Lessons from the Brazilian Atlantic Rainforest. *Biodivers. Conserv.* **2017**, *26*, 1557–1568. [CrossRef]
14. Fu, H.; Yuan, G.; Jeppesen, E.; Ge, D.; Li, W.; Zou, D.; Huang, Z.; Wu, A.; Liu, Q. Local and regional drivers of turnover and nestedness components of species and functional beta diversity in lake macrophyte communities in China. *Sci. Total Environ.* **2019**, *687*, 206–217. [CrossRef] [PubMed]
15. Garcia-Giron, J.; Heino, J.; Baastrup-Spohr, L.; Bove, C.P.; Clayton, J.; de Winton, M.; Feldmann, T.; Fernandez-Alaez, M.; Ecke, F.; Grillas, P.; et al. Global patterns and determinants of lake macrophyte taxonomic, functional and phylogenetic beta diversity. *Sci. Total Environ.* **2020**, *723*, 138021. [CrossRef]
16. Li, F.S.; Yan, Y.Z.; Zhang, J.N.; Zhang, Q.; Niu, J.M. Taxonomic, functional, and phylogenetic beta diversity in the Inner Mongolia grassland. *Glob. Ecol. Conserv.* **2021**, *28*, e01634. [CrossRef]
17. Cardoso, P.; Rigal, F.; Carvalho, J.C.; Fortelius, M.; Borges, P.A.V.; Podani, J.; Schmera, D. Partitioning taxon, phylogenetic and functional beta diversity into replacement and richness difference components. *J. Biogeogr.* **2014**, *41*, 749–761. [CrossRef]
18. Siefert, A.; Ravenscroft, C.; Weiser, M.D.; Swenson, N.G. Functional beta-diversity patterns reveal deterministic community assembly processes in eastern North American trees. *Glob. Ecol. Biogeogr.* **2012**, *22*, 682–691. [CrossRef]
19. Jia, Z.Q.; Liu, X.J.; Cai, X.F.; Xing, L.X. Quantitative Remote Sensing Analysis of the Geomorphological Development of the Lijiang River Basin, Southern China. *J. Indian Soc. Remote Sens.* **2019**, *47*, 737–747. [CrossRef]
20. Zhang, Q.; Buyantuev, A.; Fang, X.N.; Han, P.; Li, A.; Li, F.Y.; Liang, C.Z.; Liu, Q.F.; Ma, Q.; Niu, J.M.; et al. Ecology and sustainability of the Inner Mongolian Grassland: Looking back and moving forward. *Landsc. Ecol.* **2020**, *35*, 2413–2432. [CrossRef]
21. Liu, G.; Jin, Q.W.; Li, J.Y.; Li, L.; He, C.X.; Huang, Y.Q.; Yao, Y.F. Policy factors impact analysis based on remote sensing data and the CLUE-S model in the Lijiang River Basin, China. *CATENA* **2017**, *158*, 286–297. [CrossRef]
22. Huang, W.; Chak, H.; Peng, Y.; Li, L. Qualitative risk assessment of soil erosion for karst landforms in Chahe town, Southwest China: A hazard index approach. *CATENA* **2016**, *144*, 184–193. [CrossRef]
23. Perez-Harguindeguy, N.; Diaz, S.; Garnier, E.; Lavorel, S.; Poorter, H.; Jaureguiberry, P.; Bret-Harte, M.S.; Cornwell, W.K.; Craine, J.M.; Gurvich, D.E.; et al. New Handbook for Standardised Measurement of Plant Functional Traits Worldwide. *Aust. J. Bot.* **2013**, *61*, 167–234. [CrossRef]
24. Wright, I.; Reich, P.; Westoby, M.; Ackerly, D.D.; Baruch, Z.; Bongers, F.; Cavender-Bares, J.; Chapin, T.; Cornelissen, J.H.C.; Diemer, M.; et al. The worldwide leaf economics spectrum. *Nature* **2004**, *428*, 821–827. [CrossRef]
25. Kang, M.; Chang, S.C.; Yan, E.R.; Wang, X.H. Trait variability differs between leaf and wood tissues across ecological scales in subtropical forests. *J. Veg. Sci.* **2014**, *25*, 703–714. [CrossRef]
26. Coste, S.; Baraloto, C.; Leroy, C.; Marcon, E.; Renaud, A.; Richardson, A.D.; Roggy, J.C.; Schimann, H.; Uddling, J.; Hérault, B. Assessing foliar chlorophyll contents with the SPAD-502 chlorophyll meter: A calibration test with thirteen tree species of tropical rainforest in French Guiana. *Ann. For. Sci.* **2010**, *67*, 607. [CrossRef]
27. Bao, S.D. *Soil Agrochemical Analysis*; China Agriculture Press: China, Beijing, 2000. (In Chinese)
28. Hijmans, R.J. Geosphere: Spherical Trigonometry. R Package Version 1.5-14. 2021. Available online: https://CRAN.R-project.org/package=geosphere (accessed on 12 October 2021).
29. Jin, Y.; Qian, H.V. PhyloMaker: An R package that can generate very large phylogenies for vascular plants. *Ecography* **2019**, *42*, 1353–1359. [CrossRef]
30. Baselga, A. Separating the two components of abundance-based dissimilarity: Balanced changes in abundance vs. abundance gradients. *Methods Ecol. Evol.* **2013**, *4*, 552–557. [CrossRef]
31. Talbot, J.M.; Bruns, T.D.; Taylor, J.W.; Smith, D.P.; Branco, S.; Glassman, S.E.S.; Vilgalys, R.; Liao, H.L.; Smith, M.E.; Peay, K.G. Endemism and functional convergence across the North American soil mycobiome. *Proc. Natl. Acad. Sci. USA* **2014**, *111*, 6341–6346. [CrossRef]
32. Wang, X.; Lu, X.; Yao, J.; Wang, Z.W.; Deng, Y.; Cheng, W.X.; Zhou, J.Z.; Han, X.G. Habitat-specific patterns and drivers of bacterial β-diversity in China's drylands. *ISME J.* **2017**, *11*, 1345–1358. [CrossRef]
33. Anderson, M.J. A new method for non-parametric multivariate analysis of variance. *Austral Ecol.* **2001**, *26*, 32–46. [CrossRef]
34. R Core Team. *R: A Language and Environment for Statistical Computing*; R Foundation for Statistical Computing: Vienna, Austria, 2021; Available online: https://www.R-project.org/ (accessed on 10 October 2021).
35. Cardoso, P.; Mammola, S.; Rigal, F.; Carvalho, J. BAT: Biodiversity Assessment Tools. R Package Version 2.7.0. 2021. Available online: https://CRAN.R-project.org/package=BAT (accessed on 12 October 2021).
36. Goslee, S.C.; Urban, D.L. The ecodist package for dissimilarity-based analysis of ecological data. *J. Stat. Softw.* **2007**, *22*, 1–19. [CrossRef]
37. Oksanen, J.; Blanchet, F.G.; Friendly, M.; Kindt, R.; Legendre, P.; McGlinn, D.; Minchin, P.R.; O'Hara, R.B.; Simpson, G.L.; Solymos, P.; et al. Vegan: Community Ecology Package. R Package Version 2.5-7. 2020. Available online: https://CRAN.R-project.org/package=vegan (accessed on 12 October 2021).

38. Heino, J.; Melo, A.S.; Siqueira, T.; Soininen, J.; Valanko, S.; Bini, L.M. Metacommunity organisation, spatial extent and dispersal in aquatic systems: Patterns, processes and prospects. *Freshw. Biol.* **2015**, *60*, 845–869. [CrossRef]
39. Gianuca, A.T.; Engelen, J.; Brans, K.I.; Hanashiro, F.T.T.; Vanhamel, M.; van den Berg, E.M.; Caroline, S.; De Meester, L. Taxonomic, functional and phylogenetic metacommunity ecology of cladoceran zooplankton along urbanization gradients. *Ecography* **2018**, *41*, 183–194. [CrossRef]
40. De Paula, L.F.A.; Colmenares-Trejos, S.L.; Negreiros, D.; Rosado, B.H.P.; De Mattos, E.A.; De Bello, F.; Porembski, S.; Silveira, F.A.O. High plant taxonomic beta diversity and functional and phylogenetic convergence between two Neotropical inselbergs. *Plant Ecol. Divers.* **2020**, *13*, 61–73. [CrossRef]
41. González-Trujillo, J.D.; Saito, V.S.; Petsch, D.K.; Muñoz, I.; Sabater, S. Historical legacies and contemporary processes shape beta diversity in Neotropical montane streams. *J. Biogeogr.* **2021**, *48*, 101–177. [CrossRef]
42. Soininen, J.; Heino, J.; Wang, J. A meta-analysis of nestedness and turnover components of beta diversity across organisms and ecosystems. *Glob. Ecol. Biogeogr.* **2018**, *27*, 96–109. [CrossRef]
43. Viana, D.S.; Figuerola, J.; Schwenk, K.; Manca, M.; Hobaek, A.; Mjelde, M.; Preston, C.D.; Gornall, R.J.; Croft, J.M.; King, R.A.; et al. Assembly mechanisms determining high species turnover in aquatic communities over regional and continental scales. *Ecography* **2016**, *39*, 281–288. [CrossRef]
44. Wang, X.; Wiegand, T.; Anderson-Teixeira, K.J.; Bourg, N.A.; Hao, Z.; Howe, R.; Jin, G.; Orwig, D.A.; Spasojevic, M.J.; Wang, S.; et al. Ecological drivers of spatial community dissimilarity, species replacement and species nestedness across temperate forests. *Glob. Ecol. Biogeogr.* **2018**, *27*, 581–592. [CrossRef]
45. Batista, B.C.; De Lima, I.P.; Lima, M.R. Beta diversity patterns of bats in the Atlantic Forest: How does the scale of analysis affect the importance of spatial and environmental factors? *J. Biogeogr.* **2020**, *48*, 1–10. [CrossRef]
46. Zhu, H.Y.; Li, Y.S.; Wu, L.J.; Yu, S.; Xin, C.L.; Sun, P.G.; Xiao, Q.; Zhao, H.J.; Zhang, Y.; Qin, T. Impact of the atmospheric deposition of major acid rain components, especially NH4, on carbonate weathering during recharge in typical karst areas of the Lijiang River basin, southwest China. *Appl. Geochem.* **2020**, *114*, 104518. [CrossRef]
47. Zhang, M.M.; Qin, H.; Wang, Y.; Zhang, F. Beta diversity of wetland vegetation in the middle and upper reaches of the Fenhe River watershed. *Acta Ecol. Sin.* **2016**, *36*, 3292–3299. [CrossRef]
48. Dobrovolski, R.R.; Melo, A.S.; Cassemiro, F.A.S.; Diniz-Filho, J.A.F. Climatic history and dispersal ability explain the relative importance of turnover and nestedness components of beta diversity. *Glob. Ecol. Biogeogr.* **2012**, *21*, 191–197. [CrossRef]
49. Shi, W.; Wang, Y.Q.; Xiang, W.S.; Li, X.K.; Cao, K.F. Environmental filtering and dispersal limitation jointly shaped the taxonomic and phylogenetic beta diversity of natural forests in southern China. *Ecol. Evol.* **2021**, *11*, 8783–8794. [CrossRef]
50. Qian, H.; Guo, Q. Linking biotic homogenization to habitat type, invasiveness, and growth form of naturalized alien plants in North America. *Divers. Distrib.* **2010**, *16*, 119–125. [CrossRef]
51. Kong, J.J.; Yang, J.; Bai, E. Long-term effects of wildfire on available soil nutrient composition and stoichiometry in a Chinese boreal forest. *Sci. Total Environ.* **2018**, *642*, 1353–1361. [CrossRef]
52. John, R.; Dalling, J.W.; Harms, K.E.; Yavitt, J.B.; Stallard, R.F.; Mirabello, M.; Foster, R.B. Soil nutrients influence spatial distributions of tropical tree species. *Proc. Natl. Acad. Sci. USA* **2007**, *104*, 864–869. [CrossRef]
53. Zheng, Y.; Dong, L.; Li, Z.; Zhang, J.; Li, Z.; Miao, B.; Jia, C.; Liang, C.; Wang, L.; Li, F.Y. Phylogenetic structure and formation mechanism of shrub communities in arid and semiarid areas of the Mongolian Plateau. *Ecol. Evol.* **2019**, *9*, 13320–13331. [CrossRef]
54. Sobral, F.L.; Lees, A.C.; Cianciaruso, M.V. Introductions do not compensate for functional and phylogenetic losses following extinctions in insular bird assemblages. *Ecol. Lett.* **2016**, *19*, 1091–1100. [CrossRef]
55. Jiang, X.M.; Pan, B.Z.; Jiang, W.X.; Hou, Y.M.; Yang, H.Q.; Zhu, P.H.; Heino, J. The role of environmental conditions, climatic factors and spatial processes in driving multiple facets of stream macroinvertebrate beta diversity in a climatically heterogeneous mountain region. *Ecol. Indic.* **2021**, *124*, 107407. [CrossRef]
56. Fournier, B.; Mouly, A.; Moretti, M.; Gillet, F. Contrasting processes drive alpha and beta taxonomic, functional and phylogenetic diversity of orthopteran communities in grasslands. *Agric. Ecosyst. Environ.* **2017**, *242*, 43–52. [CrossRef]

forests

MDPI

Article

Effects of Edaphic Factors at Different Depths on β-Diversity Patterns for Subtropical Plant Communities Based on MS-GDM in Southern China

Wei Xu, Miguel Ángel González-Rodríguez, Zehua Li, Zhaowei Tan, Ping Yan and Ping Zhou *

Guangzhou Institute of Geography, Guangdong Academy of Sciences, Guangzhou 510070, China
* Correspondence: pzhou@gdas.ac.cn

Abstract: Previous research on the relationship between edaphic factors and species diversity patterns has mostly focused on topsoil between 0 and 30 cm, with less attention paid to deeper layers where many plant root systems are concentrated. Since considering deeper edaphic layers might help to unravel the maintenance mechanisms of plant diversity, in the present study we explored the relationship between vegetation β-diversity and a comprehensive set of soil chemical attributes at different depths. Based on vegetation and soil data from subtropical broad-leaved forest plots in the Nanling Mountains, China, we analyzed the driving factors of β-diversity patterns of trees, shrubs, and herbs using multi-site generalized dissimilarity modeling (MS-GDM). We found that the species composition dissimilarity of trees, shrubs, and herbs layers in the study area was highly diversified and dominated by species turnover components. Topsoil chemical properties were the best explainers for the β-diversity of trees (52.5%), followed by herbs (40.3%) and shrubs (21.8%). With the increase of soil depth, especially for depth >60 cm, soil chemical elements gradually lost explanatory power. Regarding the β-diversity of trees, it was mainly affected by altitude and available nitrogen (AN), total iron (Fe), and nickel (Ni) content in the soil of 0–60 cm depth. Concerning shrubs, the best β-diversity explainers were altitude, geographical distance, and nutrient elements of the soil above 40 cm. The main factors driving the β-diversity of herbs were altitude, total boron (B), total cadmium (Cd), and total nickel (Ni) of 0–40 cm soil. Overall, our results suggest that the environmental filtration process driven by altitude and soil factors, and dispersal limitations represented by geographical distance, affected the β-diversity patterns of Nanling forest communities.

Keywords: soil depth layer; β-diversity; soil chemical elements; multi-site generalized dissimilarity modeling

Citation: Xu, W.; González-Rodríguez, M.Á.; Li, Z.; Tan, Z.; Yan, P.; Zhou, P. Effects of Edaphic Factors at Different Depths on β-Diversity Patterns for Subtropical Plant Communities Based on MS-GDM in Southern China. *Forests* **2022**, *13*, 2184. https://doi.org/10.3390/f13122184

Academic Editors: Runguo Zang and Yi Ding

Received: 26 October 2022
Accepted: 12 December 2022
Published: 19 December 2022

1. Introduction

Biodiversity patterns and changes driven by influencing factors are the core contents of biodiversity research, and studies about biodiversity patterns and mechanisms are conducive to the conservation and sustainable use of biodiversity [1,2]. As a link between local biodiversity (α-diversity) and regional biodiversity (γ-diversity), β-diversity quantifies the changes in species composition and diversity along time or space, which has been widely used in the study of biodiversity in terrestrial and aquatic ecosystems [3–5]. Admittedly, the spatial patterns of β-diversity are mainly affected by geographic distance and environmental heterogeneity between sites, reflecting the underlying processes of dispersal limitation and environmental filtering, respectively [6–8]. Since β-diversity is key to analyzing species-environment relationships, understanding the geographic patterns of β-diversity and their driving factors is essential to unravelling the mechanisms of species diversity change [9].

β-diversity is generally measured with indices representing pairwise dissimilarity, such as the Jaccard and Sørensen dissimilarity indices, which are divided into two components: species turnover and nestedness [10]. Species turnover describes the degree of species replacement between communities, while nestedness indicates that communities with low richness result from species loss in communities with high richness. Measuring the rate of change in species composition is a common method for testing species turnover. In general, as the spatial distance increases, the similarity of species composition between plots decreases (or the difference increases) [11]. Explanations for this distance-decay relationship include deterministic responses of species to biotic and abiotic conditions (e.g., niche differentiation and competitive asymmetry), as well as spatial processes by which species find suitable environments (e.g., dispersal capability) [5,7,12]. For example, Zhang et al. [13] found that environmental heterogeneity mainly affected the species richness of the Gutianshan forest plot, while spatial processes dominated the species diversity in the Barro Colorado Island forest plot. Environmental differences between communities cause species to exhibit complementary effects on resource use, occupy additional niche space at different locations, and affect β-diversity [5].

Environmental factors such as soil and physiography usually dominate species distribution patterns at regional scales [14–16]. Among them, the soil, which contains the nutrient elements required for plant growth and often exhibits strong spatial heterogeneity, is widely believed to promote species coexistence and diversity patterns by increasing niche availability and providing shelters and refuges [17,18]. Early studies provided evidence for edaphic heterogeneity accounting for most of the variation in species richness and species distributions of plant communities [19–21]. Zhang et al. [22] found that soil heterogeneity could explain 88.2% of the tree species distribution in the Gutianshan subtropical forest. Hall et al. [23] found that the distribution patterns of four species of *Entandrophragma* Genus were closely related to soil nutrients (Ca, Mg, and P) in a 100-ha forest plot in Africa. In addition, the diversity in the vertical distribution of plant roots corresponds to different plant strategies for soil resource absorption, which is tightly related to species composition [24–26]. Previous studies on the relationship between species diversity patterns and environmental factors mainly focus on 0–10 cm topsoil or 0–30 cm soil layers and pay less attention to the middle and deeper layers where 50% of tree roots are distributed [27]. Therefore, analyzing the effects of soil physicochemical properties on species diversity patterns at different depths will help to better understand the driving mechanisms of biodiversity [13].

In the present study, we simultaneously investigated (i) the relationship between turnover and nestedness components of β-diversity and their environmental correlates of different plant forms (trees, shrubs and herbs) and (ii) the effects of soil physicochemical properties at different depths on β-diversity patterns. Additionally, the 'dispersal limitation' and 'environmental filtering' hypotheses were examined to explain the influencing mechanism of β-diversity in the well-preserved natural subtropical broadleaved forests of the Nanling Mountains, in the south of China, We used multi-site generalized dissimilarity modeling to analyze the effects of geographic distances, altitude and soil chemical properties of four depth layers on β-diversity patterns of different plant forms, and quantified the effects of environmental factors on β-diversity components (turnover and nestedness). Research results could provide scientific basis for the conservation of biodiversity and forest ecosystems in the Nanling Mountains.

2. Material and Methods

2.1. Study Area

The study area is located in the Nanling National Nature Reserve in the north of Guangdong Province, China, in the middle of the Nanling Mountains. The area spans three counties (cities) of Shaoguan and Qingyuan city, namely Ruyuan, Yangshan and Lianzhou, with 24°37′ N–24°57′ N, 112°30′ E–113°04′ E, and covers an area of 58,400 ha. The highest peak in the area is Shikeng Kong, which is 1902 m above sea level and has a relative elevation difference of 1489 m. The vegetation types of the Nanling Mountains range from subtropical broad-leaved forest, mixed coniferous and broad-leaved forest, and mountain broad-leaved brushwood. The study area shows transitional monsoon climate between

Central Asia and South Asia. The annual average temperature is 17.7 °C, with an extreme minimum temperature of −4.2 °C and the extreme maximum temperature of 34.4 °C, with the annual average frost-free period 276 days. The annual average precipitation is 1705 mm, with the annual average relative humidity 84%. Additionally, the precipitation is mostly concentrated from March to October, accounting for about 82% of the annual rainfall. The soils of reserve are mainly red soil, yellow soil and mountain scrubby-meadow soil.

In 2017, 21 forest dynamic plots (40 m × 40 m) and three mountain scrubby-meadow dynamic plots (20 m × 20 m) following the Center for Tropical Forest Science of China (CTFS) standard census protocols were established in Nanling National Nature Reserve (Figure 1). The vegetation types of the plots included evergreen broad-leaved forest, evergreen coniferous broad-leaved mixed forest, mountain evergreen broad-leaved dwarf forest and mountain scrubby meadow (see Table S1 for details on the specific composition of these communities). Each plot was divided into four quadrats as vegetation survey units (20 × 20 m quadrats for forest plots and 10 × 10 cm for the mountain scrubby-meadow plots). Each forest plot included three 2 m × 2 m shrub plots and three 1 m × 1 m herb plots. All the woody stems with ≥1 cm diameter at breast height 1.3 m (DBH) in the plot were mapped, identified and their DBH, total height and crown width measured. The shrub species and saplings with DBH < 1 cm and height > 50 cm in the shrub plots were considered as shrub layer, and their species name, basal diameter and plant height were recorded. Regarding the herb plots, all the herbaceous plants, lianas, and seedlings of trees and shrubs with height < 50 cm were considered as herb layers, and their species name, plant number and height were also recorded.

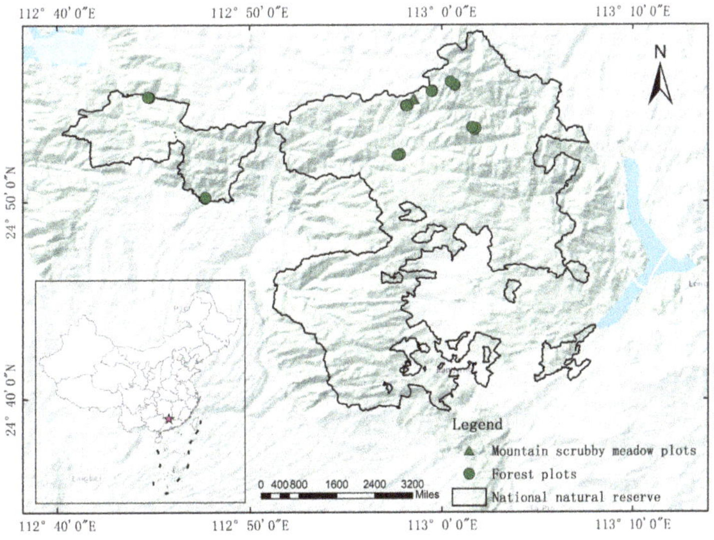

Figure 1. Geographical locations of the study area.

2.2. Soil Sampling and Analysis

In 2017, a 12 m × 15 m soil fixed observation plot was set near each fixed forest plot, which was divided into six 5 × 6 m subplots for stratification (0–20, 20–40, 40–60, 60–100 cm) sampling using soil drill. Each soil sample is a mixed sample composed of eight-ten samples from the six subplots according to the depth of the same depth layer. Consequently, every soil sample of a specific depth layer has six measurement repetitions. From each sample, 22 chemical soil properties were measured. The results included Total Nitrogen (TN), Total Phosphorus (TP), Total Potassium (TK) and Available Nitrogen (AN), Available Phosphorus (AP), Available Potassium (AK), Boron (B), Molybdenum (Mo), Manganese (Mn), Zinc (Zn), Cuprum (Cu), Ferrum (Fe), Selenium (Se), Cadmium (Cd),

Plumbum (Pb), Chromium (Cr), Nickel (Ni), Hydrargyrum (Hg) and Arsenic (As) were determined according to the Chinese standards for forest soil determination [28]. Among them, the Kjeldahl nitrogen method was used to determine TN contents. Molybdenum-antimony anti-colorimetric method was used to measure TP. The alkaline melting method was used to determine TK. The alkaline hydrolysis diffusion method was used to determine AN. The colorimetric method was used to measure AP, and the ammonium acetate leaching method was used to determine AK. The contents of B, Mo, Mn, Zn, Cu, Fe, Se, Cd, Pb, Cr, Ni, Hg and As were determined by acid digestion-spectrometer.

To consider the potential random effects of the plots, a linear mixed effects model was used to test the differences of soil chemical element contents at different soil depths. The normality of the model residuals was evaluated by the Shapiro–Wilk test, executed in R language [29]. Soil elements with significant differences at different soil depths (p-value < 0.05) were used as potential explainers in the subsequent generalized dissimilarity model. The averages of element contents and F-test values in different soil layers are shown in Table 1.

Table 1. Elemental contents in different soil depths.

Element Contents	Soil Depths (cm)				F-Value	p-Value
	0–20	20–40	40–60	60–100		
B(mg/kg)	23.72(±1.8)	27.40(±2.0)	27.92(±2.0)	27.35(±2.1)	9.19	<0.0001 ***
Mo(mg/kg)	3.85(±1.1)	4.24(±1.2)	4.20(±1.0)	4.43(±1.2)	2.23	0.0919
Mn(mg/kg)	139.15(±13.1)	155.46(±17.0)	196.49(±23.2)	244.63(±44.2)	7.34	0.0002
Zn(mg/kg)	61.92(±2.9)	80.71(±10.4)	72.79(±2.4)	81.19(±4.4)	2.41	0.0743
Cu(mg/kg)	6.76(±0.8)	5.87(±0.7)	6.43(±0.7)	6.71(±0.9)	4.92	0.0037 **
Fe(mg/kg)	23.03(±1.6)	26.53(±1.8)	26.43(±1.6)	26.06(±1.8)	9.64	<0.0001 ***
Se(mg/kg)	1.45(±0.1)	1.39(±0.1)	1.16(±0.1)	0.93(±0.1)	27.55	<0.0001 ***
Cd(mg/kg)	0.19(±0.01)	0.09(±0.01)	0.07(±0.01)	0.07(±0.01)	70.40	<0.0001 ***
Pb(mg/kg)	78.25(±4.3)	94.88(±10.0)	90.55(±7.8)	106.05(±16.5)	2.16	0.1008
Cr(mg/kg)	26.86(±1.8)	45.21(±2.9)	52.88(±3.5)	51.85(±4.8)	16.03	<0.0001 ***
Ni(mg/kg)	10.43(±0.7)	21.56(±2.1)	25.47(±2.1)	27.81(±2.8)	24.34	<0.0001 ***
Hg(mg/kg)	0.35(±0.02)	0.33(±0.02)	0.31(±0.02)	0.27(±0.01)	10.70	<0.0001 ***
As(mg/kg)	20.11(±1.2)	20.87(±1.6)	20.25(±1.4)	20.56(±1.7)	0.59	0.6237
TN(g/kg)	3.35(±0.2)	2.42(±0.3)	1.87(±0.3)	1.38(±0.2)	41.26	<0.0001 ***
TP(g/kg)	0.24(±0.02)	0.25(±0.02)	0.23(±0.02)	0.19(±0.02)	11.17	<0.0001 ***
TK(g/kg)	22.27(±1.1)	29.78(±1.3)	31.28(±1.1)	32.55(±1.4)	73.28	<0.0001 ***
AN(mg/kg)	246.39(±15.8)	264.39(±27.1)	205.83(±28.7)	150.79(±21.2)	17.77	<0.0001 ***
AP(mg/kg)	1.98(±0.2)	0.84(±0.2)	0.58(±0.2)	0.31(±0.1)	24.10	<0.0001 ***
AK(mg/kg)	80.04(±4.8)	109.88(±7.0)	101.95(±7.2)	110.56(±10.3)	10.59	<0.0001 ***

Notes: p-value < 0.001, ***; p-value < 0.01, **.

2.3. β-Diversity Index

The Sørensen dissimilarity index (β_{SOR}) was expressed by Baselga to measure β-diversity and we partitioned β_{SOR} into two components, the Simpson dissimilarity index (β_{sim}) and the nestedness component (β_{NES}) [10]. Among them, β_{SOR} is the most used β-diversity index, which is based on the proportion of common species in two communities, combined with spatial turnover and richness difference information. β_{SIM} quantifies the spatial turnover of species richness gradients, which describes species replacement. β_{NES} quantifies species loss or gain, describing lower diversity sites as a subset of higher diversity sites [4]. These indices are defined as follows:

$$\beta_{sor} = \frac{[\sum_{i<j}\min(b_{ij}, b_{ji})] + [\sum_{i<j}\max(b_{ij}, b_{ji})]}{2[\sum_i S_i - S_T] + [\sum_{i<j}\min(b_{ij}, b_{ji})] + [\sum_{i<j}\max(b_{ij}, b_{ji})]} \quad (1)$$

$$\beta_{sim} = \frac{[\sum_{i<j}\min(b_{ij}, b_{ji})]}{[\sum_i S_i - S_T] + [\sum_{i<j}\min(b_{ij}, b_{ji})]} \quad (2)$$

$$\beta_{nes} = \beta_{SOR} - \beta_{SIM} \quad (3)$$

where b_{ij} and b_{ji} are the number of species unique to plot i and plot j, S_i is the number of species in plot i, and S_T is the number of species in all plots.

2.4. Multi-Site Generalized Dissimilarity Modeling (MS-GDM)

The MS-GDM model is an extension of the generalized linear model proposed by Ferrier et al. [30]. It can effectively solve the problems of curvilinear relationship between environmental dissimilarity at extreme values (0 or 1) and community composition dissimilarity. The model uses exponential functions to connect environmental data and Bray–Curtis dissimilarity data [30]. The MS-GDM model is defined as follows:

$$-\ln(1-d_{ij}) = b + \sum_{k=1}^{p}\left|f_k(x_{ki}) - f_k\left(x_{kj}\right)\right| \quad (4)$$

where i and j indicate different plots, d_{ij} is Bray–Curtis index, and b is fitted regression coefficient. k is environmental variables (k = 1, 2, . . . , p). x_{ki} and x_{kj} are the observed values of environmental variable k in plots i and j. $f_k(x_{ki}) - f_k(x_{kj})$ provides the best possible fit between predicted and observed compositional dissimilarity. $f_k(x_k) = \sum_{k=1}^{p} a_{pk} I_{pk}(x_k)$, where a_{pk} is the fitted coefficient for I_{pk}, which is the k_{th} I-spline for variable x_k, and $a_{pk} \geq 0$. In GDM, I-spline fits the hypothesis that composition differences between communities only increase with distance along environmental gradients

To avoid model overfitting, the length of fitted nodes (Knots) should be equal to the sum of spline vectors. The default three I-spline splines of minimum (0), median (50%), and maximum (100%) were used as spline piecewise fit nodes. The GDM approach implements automatic variable selection by using matrix permutation to verify the significance of the current model, adding or deleting environmental variables to subsequently evaluate the performance gains [30]. Then, GDM can output a Partial response graph, in which the maximum height of the curve represents the relative contribution of each environmental variable to β-diversity, and the slope of the curve represents the change rate of β-diversity following environmental gradients [31–33].

In addition to the use of GDM, the relationships between β-diversity components of trees, shrubs and herbs and each soil chemical properties in different soil depth layers were estimated using the Person correlation and Mantel test. The statistics, calculations and geographic mapping in this study were carried out using R software (version 4.1.0), involving packages such as "vegan", "betapart" and "gdm" [30,33–35].

3. Results

3.1. β-Diversity Distribution Pattern of Different Plant Forms

A total of 233 species belonging to 132 genera and 70 families were found in plots, including 200 species of trees, 46 species of shrubs, and 68 species of herbs. By calculating the β-diversity index of different plant forms, we found that the β_{SOR} of trees, shrubs, and herbs were 0.87, 0.95 and 0.94, respectively. This indicated that plant species composition differed greatly among plots in the study area, and the species compositional dissimilarity of shrubs and herbs was higher. After decomposition of β-diversity into its components (i.e., species turnover and nestedness), the species turnover was predominant in the three plant forms, while the nestedness accounted for less than 10%. This implied that the species turnover component is more important in species composition in the Nanling Mountains (Figure 2).

We found that species compositional dissimilarity of trees, shrubs, and herbs in the study area increased with distance, indicating that there was an obvious species turnover of plant community in the Nanling Mountains. Additionally, this was confirmed by the use of the GDM approach, since the predicted species compositional dissimilarity matched the observed species compositional dissimilarity of trees, shrubs, and herbs (Figure 3).

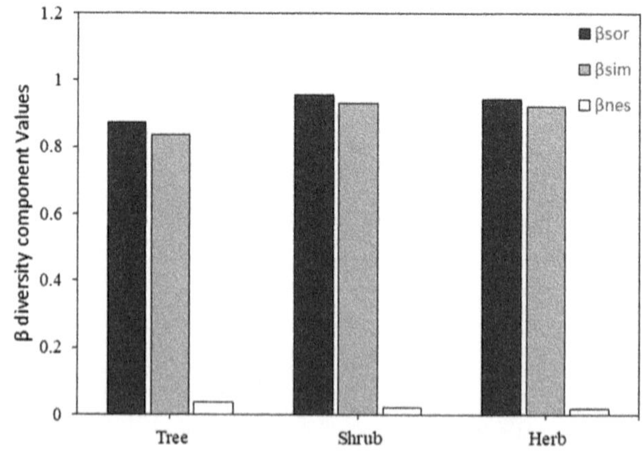

Figure 2. β-diversity and components of different plant forms of the Nanling Mountains.

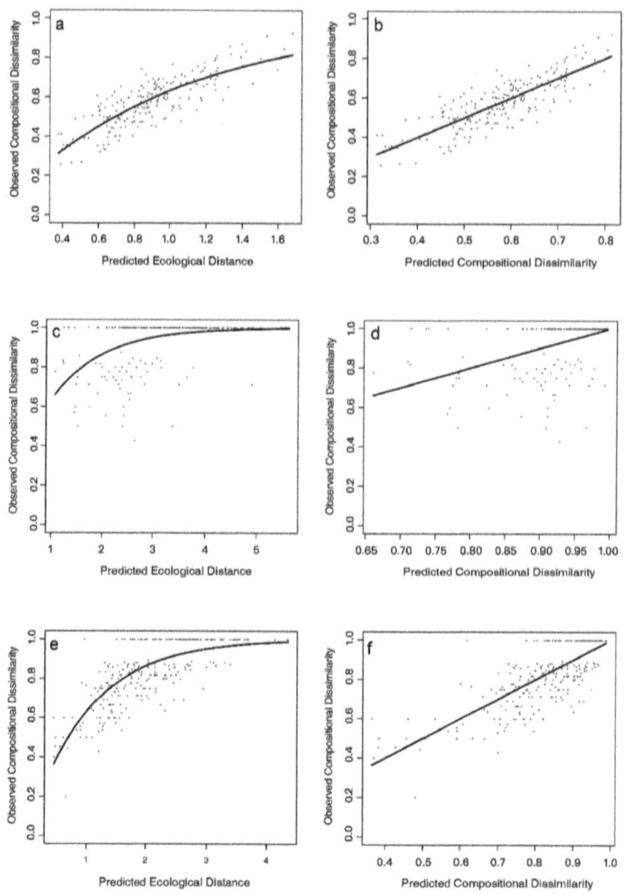

Figure 3. Relationships between compositional dissimilarity and geographical distance and relationships between predicted species compositional dissimilarity and the observed species compositional dissimilarity of different life form plants ((**a**,**b**) indicate trees, (**c**,**d**) indicate shrubs, (**e**,**f**) indicate herbs).

3.2. The Contribution of Environmental Factors to Plant β-Diversity by GDM

Soil chemical properties were the best explainers of tree β-diversity, with a mean explanation rate of 52.5%, followed by herbs (40.3%) and shrubs (21.8%) (Figure 4). With the increase in soil depth, the explanation rate of β-diversity of trees and herbs explained by soil chemical elements gradually decreased, since the explanation rate of chemical elements in the 0–20 cm soil layer remained consistently higher than the rates of deeper layers. Soil elements in the 60–100 cm layer showed the lowest explanatory power for β-diversity of trees, shrub and herbs.

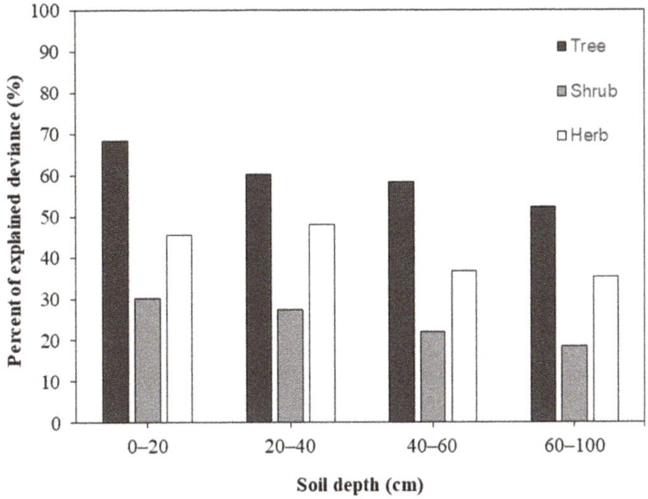

Figure 4. Contribution of environmental factors to plant β-diversity at different soil depths.

Figure 5 shows the relationship between the species composition matrix and environment matrix fitted by I-spline function. The parameters of the spline function indicate the effects of altitude, geographic distance and soil chemical elements at different soil depths on the β-diversity of trees, shrubs and herbs. The β-diversity of trees was mainly affected by altitude and the element contents in the 0–60 cm soil layer, including AN (available nitrogen), Fe and Ni. In addition, the contents of Fe and Se in the 60–100 cm soil layer also showed important effects on tree β-diversity (Figures S1–S4). The β-diversity of shrubs was mainly determined by altitude, geographical distance and soil chemical elements in the 0–60 cm layer, including AN, AP, Cu and Hg in in the 0–20 cm layer, AN, Ni and Cd in the 20–40 cm layer, and An, As and Hg in the 40–60 cm layer. The chemical elements in the 60–100 cm soil layer showed little effect on the species composition of shrubs (Figures S5–S8). The β-diversity of herbs was mainly affected by altitude, B, Cd and Pb elements, with a stronger effect of the 0–40 cm layer on herb composition in comparison to deeper soil layers (Figures S9–S12).

3.3. Relationship between Soil Chemical Properties and β-Diversity of Different Plant Forms

The species turnover components of trees were positively correlated with TN, TP, AN, Cu and Hg at the 20–60 cm soil layer, and Cu and Hg at the 60–100 cm layer. The nestedness of trees was significantly correlated with the AP, Fe, Cr and Ni contents in the 0–60 cm soil layers. The species compositional dissimilarity of shrubs was significantly correlated with TN, TP, AN and Cu at the 0–60 cm soil layer, and the turnover component was significantly correlated with TN and AN. In the 0–20 cm topsoil layer, the turnover component of herbs was positively correlated with the contents of TP, Cd and Ni, while the nestedness component was negatively correlated with AP, Cd, and Ni of the 0–20 cm layer (Tables 2–4).

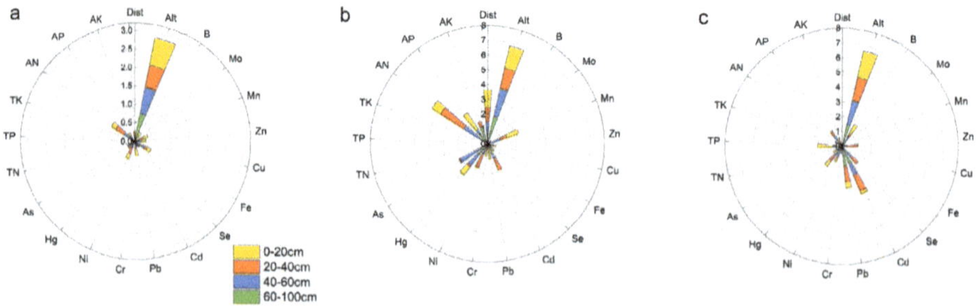

Figure 5. Sum of I-spline coefficients for environmental factors on β-diversity of trees (**a**), shrubs (**b**) and herbs (**c**). environmental factors included geographical distance (Dist), altitude (Alt), Boron (B), Molybdenum (Mo), Manganese (Mn), Zinc (Zn), Cuprum (Cu), Ferrum (Fe), Selenium (Se), Cadmium (Cd), Plumbum (Pb), Chromium (Cr), Nickel (Ni), Hydrargyrum (Hg) and Arsenic (As), Total nitrogen (TN), Total phosphorus (TP), Total kalium (TK) and Available nitrogen (AN), Available phosphorus (AP), Available kalium (AK).

Table 2. The coefficients of Pearson correlation between β-diversity components of trees and soil factors of different depths.

Soil Properties	β-sor				β-sim				β-nes			
	0–20 cm	20–40 cm	40–60 cm	60–100 cm	0–20 cm	20–40 cm	40–60 cm	60–100 cm	0–20 cm	20–40 cm	40–60 cm	60–100 cm
TN	0.01	0.09 **	0.11 *	0.12	−0.0002	0.09	0.09 **	0.00	0.001	−0.08	−0.23	−0.01
TP	0.06	0.04 *	0.03 *	0.07 *	0.01	0.03 *	0.02	0.02	0.00	0.00	0.002	0.00
TK	0.02	0.02	−0.001	0.02	0.01	0.01	−0.004	0.01	0.001	0.00	0.01	−0.01
AN	0.01	0.12 **	0.13 **	0.003	0.01	0.11 **	0.1 **	0.01	−0.001	0.00	−0.0002	−0.01
AP	−0.004	0.04	0.06	0.002	−0.02	0.01	0.05	0.00	0.05 *	0.02	0.001	−0.01
AK	0.031 *	−0.0001	0.03	0.02	0.05 *	0.003	0.05 *	0.05	−0.02	−0.03	−0.03	−0.04
B	0.05 *	0.02	−0.002	0.00	0.01	0.01	−0.01	0.00	0.03	0.01	0.03	0.01
Cu	0.07 **	0.08 **	0.09 **	0.07 *	0.06 *	0.06 *	0.06 *	0.04 *	0.00	0.00	0.00	0.01
Fe	0.05 *	0.03 *	0.00	0.01	0.01	0.01	−0.01	−0.001	0.05 *	0.02	0.08 **	0.07
Se	0.03 *	−0.0004	0.002	−0.01	0.05 *	0.001	0.001	−0.003	−0.02	−0.01	0.00	−0.001
Cd	0.004	−0.004	−0.04	0.00	0.01	−0.001	−0.01	0.01	−0.02	−0.01	−0.01	−0.01
Cr	0.001	−0.002	0.0002	−0.02	−0.01	−0.01	−0.01	−0.01	0.06 *	0.01	0.03	−0.001
Ni	0.003	0.01	0.004	−0.01	−0.01	0.001	−0.003	−0.01	0.1 **	0.04	0.08 *	0.00
Hg	0.01	0.08 ***	0.10 **	0.04 *	0.02	0.09 ***	0.13 **	0.04 *	−0.03	−0.01	−0.03	−0.004

Notes: p-value < 0.001, ***; p-value < 0.01, **; p-value < 0.05, *.

Table 3. The coefficients of Pearson correlation and Mantel test between β-diversity components of shrubs and soil factors of different depths.

Soil Properties	β-sor				β-sim				β-nes			
	0–20 cm	20–40 cm	40–60 cm	60–100 cm	0–20 cm	20–40 cm	40–60 cm	60–100 cm	0–20 cm	20–40 cm	40–60 cm	60–100 cm
TN	0.005	0.02 *	0.02 *	0.0004	0.01	0.02 *	0.02 *	0.15	−0.55	−0.01	−0.01	−0.002
TP	0.002	0.02 *	0.01	0.001	0.001	0.01	0.01	0.008	0.002	−0.003	−0.01	−0.0003
TK	0.002	−0.002	−0.01	0-.01	0.001	−0.003	−0.028	−0.02	0.0001	0.004	0.02	0.01
AN	0.02 *	0.04 **	0.03	0.001	0.02	0.03 *	0.03 *	0.001	−0.01	−0.01	−0.01	−0.001
AP	0.01	0.004	0.02	0.002	0.01	0.01	0.02	0.002	−0.003	−0.005	−0.01	−0.001
AK	−0.13	−0.002	0.01	0.001	0.001	0.002	0.004	0.001	−0.003	−0.002	−0.002	0.001
B	−0.002	−0.0003	−0.01	−0.01	−0.001	−0.002	−0.02	−0.02	0.0001	0.004	0.01	0.02
Cu	0.02	0.03 *	0.03	0.03	0.01	0.01	0.01	0.01	0.001	0.0006	−0.0001	−0.0004
Fe	0.0001	0.0001	−0.004	−0.01	0.001	−0.001	−0.01	−0.01	−0.002	0.0006	0.01	0.01
Se	0.01	−0.03	0.002	−0.01	0.0003	−0.002	0.001	−0.02	0.003	0.01	−0.03	0.01
Cd	0.003	0.01	−0.002	0.004	0.0001	0.004	−0.001	−0.01	0.002	−0.002	−0.002	0.004
Cr	0.001	0.02	−0.004	−0.003	−0.001	0.01	0.001	−0.001	0.004	−0.006	−0.01	−0.002
Ni	0.01	0.03	0.001	0.0001	0.002	0.02	0.003	0.002	0.0001	−0.001	−0.01	−0.01
Hg	−0.01	0.003	0.02	0.01	−0.001	0.0002	0.01	0.02	0.001	0.001	0.001	−0.02

Notes: p-value < 0.01, **; p-value < 0.05, *.

Table 4. The coefficients of Pearson correlation and Mantel test between β-diversity components of herbs and soil factors of different depths.

Soil Properties	β-sor				β-sim				β-nes			
	0–20 cm	20–40 cm	40–60 cm	60–100 cm	0–20 cm	20–40 cm	40–60 cm	60–100 cm	0–20 cm	20–40 cm	40–60 cm	60–100 cm
TN	0.69	0.04 *	0.01	−0.13	0.01	0.013	0.002	0.002	−0.28	0.65	0.01	0.003
TP	0.14 **	0.02	0.001	0.05	0.06 **	0.01	0	0.02 *	−0.12	−0.03	0.002	−0.01
TK	0.16	−0.003	−0.02	0.10	−0.04	−0.001	−0.01	−0.0001	0.0002	−0.001	0	−0.001
AN	0.03 *	0.04 *	0.02	0.07	0.01	0.01	0.002	0	0.001	0.01	0.01	0
AP	−0.34	0.002	0.001	−0.14	−0.01	0	0.02	0.01	0.02	0.01	0.002	0.02
AK	0.01	0.01	0.03	0.01	0.02	0.01	0.03	0.02	−0.03 *	−0.002	−0.01	−0.01
B	0.03 *	0.04 *	0.01	0.04 *	0.03	0.05 *	0.02	0.05 **	−0.02	−0.03	−0.02	−0.03 *
Cu	0.02	0.02	0.02	0.01	0.001	0.002	0.002	0	0.02	0.02	0.02	0.02
Fe	0.02	0.02	0.01	0.02	0.02	0.03 *	0.01	0.03 *	−0.01	−0.02	−0.004	−0.01

Table 4. *Cont.*

Soil Properties	β-sor				β-sim				β-nes			
	0–20 cm	20–40 cm	40–60 cm	60–100 cm	0–20 cm	20–40 cm	40–60 cm	60–100 cm	0–20 cm	20–40 cm	40–60 cm	60–100 cm
Se	0.02	0.0004	0.01	0.01	0.01	0.001	0.003	0.01	0.002	−0.1	−0.001	−0.01
Cd	0.15 ***	0.02	0.01	0.01	0.13 ***	0.01	0.01	0.02	−0.04 *	−0.009	−0.001	−0.01
Cr	−0.0002	0.03	−0.003	0.002	−0.01	0.04	−0.0004	−0.003	0.02	−0.02	−0.001	−0.004
Ni	0.08 **	0.12 **	0.03 *	0.02	0.09 **	0.11 **	0.03	0.03	−0.05 **	−0.03	−0.02	−0.02
Hg	0.01	0.06 **	0.05 **	0.02	0.01	0.05 **	0.002	0.004	−0.0004	−0.01	0.09 **	0.01

Notes: p-value < 0.001, ***; p-value < 0.01, **; p-value < 0.05, *.

4. Discussion

4.1. The Predicted Contribution of Different Soil Depths to the β-Diversity of Nanling Forest Communities

Extensive studies on β-diversity of plant communities have helped to explain the mechanisms of community assemblage and diversity maintenance [6,36,37]. Through our study of Nanling ecosystems, it was found that plant species composition of three plant formations in the Nanling Mountains varies (mean β_{SOR} 0.92) along different community types, and the species composition dissimilarity of shrubs and herbs was greater than that of trees. As a mountain ecosystem with a maximum altitude of only 1902m, the plant β-diversity patterns of the Nanling mountain are similar to that of alpine ecosystems in China [37,38]. β-diversity was decomposed into species turnover and species nestedness components, and the species turnover component was dominant, which was consistent with the research in other subtropical forests [38–40]. It is suggested that the β-diversity pattern of plant communities in the Nanling Mountains may be mainly caused by the species turnover or community development in different spaces. In addition, the dissimilarity of plant community species composition increased with the geographical distance. Significant spatial differences exist in the species composition of Nanling forest communities, suggesting that plant β-diversity patterns may be affected by dispersal constraints [40,41].

In this study, the GDM model showed 11, 9, 8, and 8 variables were related to the β-diversity of trees in 0–20 cm, 20–40 cm, 40–60 cm, and 60–100 cm soil depths, respectively. The soil element contents of 0–20 cm accounted for 63.5% of the β-diversity of trees. There were 7, 13, 5, and 5 variables significantly explaining the shrub β-diversity in 0–20 cm, 20–40 cm, 40–60 cm, and 60–100 cm depth soil layers, respectively, and 0–20 cm soil element contents were also the best explainers, up to 30.1% of the β-diversity. Regarding the herb β-diversity, there were 7, 10, 8, and 9 explainers in 0–20 cm, 20–40 cm, 40–60 cm and 60–100 cm soil layers, respectively. The 0–20 cm also had the highest explanatory ability, accounting for 45.5% of the herbs' β-diversity. The distribution of trees (200 species) showed a more sensitive response to the soil element content heterogeneity than shrubs (46 species) and herbs (68 species). Previous studies suggested that GDM explained variation in species turnover for more widespread species, since, by definition, rare species are not shared by many sites [31,42]. We also speculate that the contribution of predictive variables fitted by GDM is closely related to the species pool in the study area, but further studies are needed to verify this [43].

4.2. Relative Effects of Environmental Filtration and Dispersal Limitation on the β-Diversity of Plant Communities

Many studies have confirmed that environmental filtration and dispersal limitation are important processes affecting species composition and community assemblage, but their relative roles in different ecosystems have not been generally concluded [6,44]. Based on GDM model, the species diversity of plant communities in Nanling showed a spatial increasing spatial pattern with the geographical distance. Environmental factors at different soil depths had different effects on the β-diversity of trees, shrubs, and herbs. This suggests that the environmental filtering process indicated by factors such as soil elements, and dispersal limitation characterized by geographical distance, both influenced the plant β-diversity and its components in the Nanling Mountains, which is consistent with previous studies in other mountains in southern China [38,40,45]. However, the relative contribution of environmental filtering and dispersal limitation on β-diversity in Nanling notably differs from the studies mentioned above, while according to our model, the dispersal limitation explained 8.78%, 16.64%, and 5.38% of the β-diversity of trees, shrubs, and herbs, while environmental factors such as altitude and soil explained significantly more. This seems to be indicative of ecosystem dynamics strongly driven by environmental heterogeneity [39,46], implying that in the Nanling Mountains ecosystems, environmental heterogeneity plays the dominant role in changes in species composition.

4.3. Effects of Environmental Factors on Different Components of β-Diversity

There are also some differences in the main factors affecting the species composition in different plant forms in Nanling. In this study, geographical distance showed a stronger effect on β-diversity components of shrubs than that of trees and herbs, which might be related to the heterogeneous dispersal ability of different plant forms. As an environmental factor reflecting the changes of water and heat conditions in mountain ecosystems, altitude usually has a significant correlation with species diversity [38,47]. In this study, altitude had a more important contribution to the species turnover components of tree and herb species, but a smaller influence on the β-diversity component of shrubs. Due to the nurse influence from trees, environment heterogeneity in the understory is low, so changes in precipitation, temperature and light radiation caused by altitude might have a lesser impact on the habitats of shrubs in the understory. Consequently, the shrubs considered in this study may not be markedly sensitive to altitude [38,48,49].

With the increase in soil depth, the effects of soil elements on the β-diversity of trees, shrubs, and herbs gradually decreased, which may be related to changes of the plant fine root amount with the increase of soil depth [50–52]. The active roots that plants can absorb nutrient elements and water through are mostly rootlets, and 50%–80% of the rootlets are located in the soil layer of 0–30 cm depth, and the root numbers decrease sharply after 30 cm depth [53,54]. The 0–60 cm depth soil was the main root distribution area of the tree species, strongly affected tree species composition. Available nitrogen in the 20–60 cm soil layer had a significant positive correlation with the turnover components of tree β-diversity, and available phosphorus in the 0–20 cm soil layer had a significant positive effect with the nested components of tree β-diversity. The shrub β-diversity and its turnover components were mainly related to the geographical distance between communities and were affected by soil factors of 0–40 cm depth soil, such as available nitrogen, phosphorus, potassium, and other nutrient elements. Soil nitrogen and phosphorus are essential nutrient elements for plant growth, which participate in various physiological processes of plant growth. However, the available phosphorus content in southern China's soils is extremely low, which limits plant growth [22]. In this study area, the available phosphorus content in the 0–20 cm soil depth was relatively high (1.98 mg/kg on average), but decreased sharply below the 20 cm soil depth. The average available phosphorus content in the 20–40 cm soil layer was 0.84 C mg/kg. Therefore, phosphorus was the limiting resource for subtropical forest species' growth.

The boron, iron, and copper content in the 0–20 cm soil layer had a significant positive effect on the β-diversity of trees, and copper in the 20–40 cm soil layer had a positive correlation with shrub β-diversity. Boron is involved in plant cell wall construction and affects plant metabolic pathways by binding apoplastic proteins of cell walls and membranes and by interfering with manganese-dependent enzymatic reactions [55]. Iron is essential for the structure of chloroplasts and mitochondria and for both plant productivity and nutritional quality [56,57]. Copper is a micronutrient which plays a role in processes such as photosynthesis, respiration, antioxidant activity, cell wall metabolism and hormone perception [58,59]. These three rare elements in the topsoil notably affect the tree species composition where the main rootlets of trees are distributed [60]. The chemical elements of 0–40 cm soil depth showed a stronger impact on the β-diversity of herbs, which may be related to the root of the herbaceous layer mainly concentrated in the topsoil [61]. Total nitrogen, total phosphorus, cadmium and nickel had significant positive correlations with β-diversity and turnover components of herbs, while the nested components were mainly negatively affected by available potassium, cadmium and nickel in the 0–20 cm topsoil. Heavy metals, such as cadmium and nickel, have been reported as phytotoxic at elevated concentrations, causing growth reduction, yield depression and disorder in plant metabolism and physiology [62,63].

5. Conclusions

This study provides three findings that might help to understand the species diversity maintenance mechanisms of different plant forms. First, the contribution of environmental factors and the dispersal process varied with plant forms, and soil factors had a stronger influence on the β-diversity of trees than that of shrubs and herbs. Secondly, with the increase of soil depth, the explanation rate of β-diversity of trees and herbs explained by soil chemical elements gradually decreased. Finally, the chemical elements of 0–60 cm soil mainly affected tree diversity, while the main soil factors affecting the β-diversity of shrubs and herbs were in 0–40 cm depth. We conclude that the deeper soil factors have broaden our understanding of influencing factors that shape plant communities. We suggest that models need to consider a broader spectrum of environmental factors that might affect plant community composition. We also suggest that the species composition and diversity patterns of the Nanling Mountains system need to be further studied.

Supplementary Materials: The following supporting information can be downloaded at: https://www.mdpi.com/article/10.3390/f13122184/s1, Figure S1: Partial response for 0–20 cm edaphic factors on trees' GDM; Figure S2: Partial response for 20–40 cm edaphic factors on trees' GDM; Figure S3: Partial response for 40–60 cm edaphic factors on trees' GDM; Figure S4: Partial response for 60–100 cm edaphic factors on trees' GDM; Figure S5: Partial response for 0–20 cm edaphic factors on shrubs' GDM; Figure S6: Partial response for 20–40 cm edaphic factors on shrubs' GDM; Figure S7: Partial response for 40–60 cm edaphic factors on shrubs' GDM; Figure S8: Partial response for 60–100 cm edaphic factors on shrubs' GDM; Figure S9: Partial response for 0–20 cm edaphic factors on herbs' GDM; Figure S10: Partial response for 20–40 cm edaphic factors on herbs' GDM; Figure S11: Partial response for 40–60 cm edaphic factors on herbs' GDM; Figure S12: Partial response for 60–100 cm edaphic factors on herbs' GDM; Table S1: Species composition and dominant species of different vegetation community at Nanling Mountains.

Author Contributions: Conceptualization, W.X. and P.Z.; methodology, Z.L.; validation, P.Y.; investigation, Z.T.; writing—review and editing, M.Á.G.-R.; project administration, P.Z. All authors have read and agreed to the published version of the manuscript.

Funding: This research was funded by the GDAS' Special Project of Science and Development [2021GDASYL-20210103005], the Sub-project of National Key Research and Development projects [2021FY100702], Biodiversity maintenance mechanism and regional biosecurity innovation in Lingnan region [2022GDASZH-2022010106] and Ecosystem and Biodiversity Monitoring in Nanling National Park [LC-2021124].

Conflicts of Interest: The authors declare no conflict of interest.

References

1. Rahbek, C. The role of spatial scale and the perception of large-scale species-richness patterns. *Ecol. Lett.* **2005**, *8*, 224–239. [CrossRef]
2. Pollock, L.J.; O'Connor, L.M.J.; Mokany, K.; Rosauer, D.F.; Talluto, M.V.; Thuiller, W. Protecting Biodiversity (in All Its Complexity): New Models and Methods. *Trends Ecol. Evol.* **2020**, *35*, 1119–1128. [CrossRef] [PubMed]
3. Magurran, A.E.; Dornelas, M.; Moyes, F.; Henderson, P.A. Temporal β diversity—A macroecological perspective. *Glob. Ecol. Biogeogr.* **2019**, *28*, 1949–1960. [CrossRef]
4. Legendre, P.; De Cáceres, M. Beta diversity as the variance of community data: Dissimilarity coefficients and partitioning. *Ecol. Lett.* **2013**, *16*, 951–963. [CrossRef]
5. Mori, A.S.; Isbell, F.; Seidl, R. Beta-Diversity, Community Assembly, and Ecosystem Functioning. *Trends Ecol. Evol.* **2018**, *33*, 549–564. [CrossRef]
6. Qian, H.; Jin, Y.; Leprieur, F.; Wang, X.; Deng, T. Geographic patterns and environmental correlates of taxonomic and phylogenetic beta diversity for large-scale angiosperm assemblages in China. *Ecography* **2020**, *43*, 1706–1716. [CrossRef]

7. Mori, A.S.; Shiono, T.; Koide, D.; Kitagawa, R.; Ota, A.T.; Mizumachi, E. Community assembly processes shape an altitudinal gradient of forest biodiversity. *Glob. Ecol. Biogeogr.* **2013**, *22*, 878–888. [CrossRef]
8. Svenning, J.C.; Fløjgaard, C.; Baselga, A. Climate, history and neutrality as drivers of mammal beta diversity in Europe: Insights from multiscale deconstruction. *J. Anim. Ecol.* **2011**, *80*, 393–402. [CrossRef]
9. König, C.; Weigelt, P.; Kreft, H. Dissecting global turnover in vascular plants. *Glob. Ecol. Biogeogr.* **2017**, *26*, 228–242. [CrossRef]
10. Baselga, A. Partitioning the turnover and nestedness components of beta diversity. *Glob. Ecol. Biogeogr.* **2010**, *19*, 134–143. [CrossRef]
11. Martins, P.M.; Poulin, R.; Goncalves-Souza, T. Drivers of parasite beta-diversity among anuran hosts depend on scale, realm and parasite group. *Philos. T. R. Soc. B* **2021**, *376*, 20200367. [CrossRef] [PubMed]
12. Chase, J.M. Stochastic community assembly causes higher biodiversity in more productive environments. *Science* **2010**, *328*, 1388–1391. [CrossRef] [PubMed]
13. Zhang, L.; Mi, X.; Harrison, R.D.; Yang, B.; Man, X.; Ren, H.; Ma, K. Resource Heterogeneity, Not Resource Quantity, Plays an Important Role in Determining Tree Species Diversity in Two Species-Rich Forests. *Fron. Ecol. Evol.* **2020**, *8*, 224. [CrossRef]
14. Chesson, P. Mechanisms of maintenance of species diversity. *Annu. Rev. Ecol. Syst.* **2000**, *31*, 343–366. [CrossRef]
15. Levine, J.M.; HilleRisLambers, J. The importance of niches for the maintenance of species diversity. *Nature* **2009**, *461*, 254–257. [CrossRef]
16. Xue, W.; Bezemer, T.M.; Berendse, F. Soil heterogeneity and plant species diversity in experimental grassland communities: Contrasting effects of soil nutrients and pH at different spatial scales. *Plant Soil.* **2019**, *442*, 497–509. [CrossRef]
17. Stein, A.; Gerstner, K.; Kreft, H. Environmental heterogeneity as a universal driver of species richness across taxa, biomes and spatial scales. *Ecol. Lett.* **2014**, *17*, 866–880. [CrossRef]
18. Kallimanis, A.S.; Mazaris, A.D.; Tzanopoulos, J.; Halley, J.M.; Pantis, J.D.; Sgardelis, S.P. How does habitat diversity affect the species–area relationship? *Glob. Ecol. Biogeogr.* **2008**, *17*, 532–538. [CrossRef]
19. Tukiainen, H.; Bailey, J.J.; Field, R.; Kangas, K.; Hjort, J. Combining geodiversity with climate and topography to account for threatened species richness. *Conserv. Biol.* **2017**, *31*, 364–375. [CrossRef]
20. Bailey, J.J.; Boyd, D.S.; Field, R. Models of upland species' distributions are improved by accounting for geodiversity. *Landscape Ecol.* **2018**, 33, 2071–2087. [CrossRef]
21. Hulshof, C.M.; Spasojevic, M.J. The edaphic control of plant diversity. *Glob. Ecol. Biogeogr.* **2020**, *29*, 1634–1650. [CrossRef]
22. Zhang, L.; Mi, X.; Shao, H.; Ma, K. Strong plant-soil associations in a heterogeneous subtropical broad-leaved forest. *Plant Soil* **2011**, *347*, 211–220. [CrossRef]
23. Hall, J.S.; McKenna, J.J.; Ashton, P.M.S.; Gregoire, T.G. Habitat characterizations underestimate the role of edaphic factors controlling the distribution of Entandrophragma. *Ecology* **2004**, *85*, 2171–2183. [CrossRef]
24. Berendse, F. Competition between plant populations with different rooting depths. *Oecologia* **1979**, *43*, 19–26. [CrossRef]
25. Bolte, A.; Villanueva, I. Interspecific competition impacts on the morphology and distribution of fine roots in European beech (*Fagus sylvatica* L.) and Norway spruce (*Picea abies* (L.) Karst.). *Eur. J. Forest Res.* **2006**, *125*, 15–26. [CrossRef]
26. Brassard, B.W.; Chen, H.Y.; Cavard, X.; Laganiere, J.o.; Reich, P.B.; Bergeron, Y.; Pare, D.; Yuan, Z. Tree species diversity increases fine root productivity through increased soil volume filling. *J. Ecol.* **2013**, *101*, 210–219. [CrossRef]
27. Jackson, R.B.; Canadell, J.; Ehleringer, J.R.; Mooney, H.; Sala, O.; Schulze, E.-D. A global analysis of root distributions for terrestrial biomes. *Oecologia* **1996**, *108*, 389–411. [CrossRef]
28. Nanjing Institute of Soil Science. *Analysis of Soil Physical and Chemical Properties*; Shanghai Science and Technology Publishing House: Shanghai, China, 1978.
29. Team, R.C. *R: A Language and Environment for Statistical Computing*; R Foundation for Statistical Computing: Vienna, Austria, 2021.
30. Ferrier, S.; Manion, G.; Elith, J.; Richardson, K. Using generalized dissimilarity modelling to analyse and predict patterns of beta diversity in regional biodiversity assessment. *Divers. Distrib.* **2007**, *13*, 252–264. [CrossRef]
31. Overton, J.M.; Barker, G.M.; Price, R. Estimating and conserving patterns of invertebrate diversity: A test case of New Zealand land snails. *Divers. Distrib.* **2009**, *15*, 731–741. [CrossRef]
32. Ashcroft, M.B.; Gollan, J.R.; Faith, D.P.; Carter, G.A.; Lassau, S.A.; Ginn, S.G.; Bulbert, M.W.; Cassis, G. Using generalised dissimilarity models and many small samples to improve the efficiency of regional and landscape scale invertebrate sampling. *Ecol. Inform.* **2010**, *5*, 124–132. [CrossRef]
33. Leaper, R.; Hill, N.A.; Edgar, G.J.; Ellis, N.; Lawrence, E.; Pitcher, C.R.; Barrett, N.S.; Thomson, R. Predictions of beta diversity for reef macroalgae across southeastern Australia. *Ecosphere* **2011**, *2*, 1–18. [CrossRef]
34. Oksanen, J.; Simpson, G.L.; Blanchet, F.G.; Kindt, R.; Legendre, P.; Minchin, P.R.; O'Hara, R.B.; Solymos, P.; Stevens, M.H.H.; Eduard, S.; et al. *Package 'vegan'. Community Ecology Package*; R Package Version 2.6-4; R Foundation for Statistical Computing: Vienna, Austria, 2022.
35. Baselga, A.; Orme, D.; Villeger, S.; Bortoli, J.D.; Leprieur, F.; Logez, M. *betapart: Partitioning Beta Diversity into Turnover and Nestedness Components, R package Version 1.5.6*; R Foundation for Statistical Computing: Vienna, Austria, 2022.
36. Zellweger, F.; Roth, T.; Bugmann, H.; Bollmann, K. Beta diversity of plants, birds and butterflies is closely associated with climate and habitat structure. *Glob. Ecol. Biogeogr.* **2017**, *26*, 898–906. [CrossRef]

37. Nanda, S.A.; Haq, M.-u.; Singh, S.; Reshi, Z.A.; Rawal, R.S.; Kumar, D.; Bisht, K.; Upadhyay, S.; Upreti, D.; Pandey, A. Species richness and β-diversity patterns of macrolichens along elevation gradients across the Himalayan Arc. *Sci. Rep.* **2021**, *11*, 20155. [CrossRef] [PubMed]
38. Tang, Z.; Fang, J.; Chi, X.; Feng, J.; Liu, Y.; Shen, Z.; Wang, X.; Wang, Z.; Wu, X.; Zheng, C. Patterns of plant beta-diversity along elevational and latitudinal gradients in mountain forests of China. *Ecography* **2012**, *35*, 1083–1091. [CrossRef]
39. Kouba, Y.; Martínez-García, F.; de Frutos, Á.; Alados, C.L. Plant β-diversity in human-altered forest ecosystems: The importance of the structural, spatial, and topographical characteristics of stands in patterning plant species assemblages. *Eur. J. Forest Res.* **2014**, *133*, 1057–1072. [CrossRef]
40. Shi, W.; Wang, Y.Q.; Xiang, W.S.; Li, X.K.; Cao, K.F. Environmental filtering and dispersal limitation jointly shaped the taxonomic and phylogenetic beta diversity of natural forests in southern China. *Ecol. Evol.* **2021**, *11*, 8783–8794. [CrossRef] [PubMed]
41. Jiang, L.; Lv, G.; Gong, Y.; Li, Y.; Wang, H.; Wu, D. Characteristics and driving mechanisms of species beta diversity in desert plant communities. *PLoS ONE* **2021**, *16*, e0245249. [CrossRef] [PubMed]
42. Latombe, G.; Hui, C.; McGeoch, M.A. Multi-site generalised dissimilarity modelling: Using zeta diversity to differentiate drivers of turnover in rare and widespread species. *Methods Ecol. Evol.* **2017**, *8*, 431–442. [CrossRef]
43. Krasnov, B.R.; Shenbrot, G.I.; Vinarski, M.M.; Korallo-Vinarskaya, N.P.; Khokhlova, I.S.; McGeoch, M. Multi-site generalized dissimilarity modelling reveals drivers of species turnover in ectoparasite assemblages of small mammals across the northern and central Palaearctic. *Glob. Ecol. Biogeogr.* **2020**, *29*, 1579–1594. [CrossRef]
44. Myers, J.A.; Chase, J.M.; Jiménez, I.; Jørgensen, P.M.; Araujo-Murakami, A.; Paniagua-Zambrana, N.; Seidel, R. Beta-diversity in temperate and tropical forests reflects dissimilar mechanisms of community assembly. *Ecol. Lett.* **2013**, *16*, 151–157. [CrossRef]
45. Legendre, P.; Mi, X.; Ren, H.; Ma, K.; Yu, M.; Sun, I.-F.; He, F. Partitioning beta diversity in a subtropical broad-leaved forest of China. *Ecology* **2009**, *90*, 663–674. [CrossRef] [PubMed]
46. Jankowski, J.E.; Ciecka, A.L.; Meyer, N.Y.; Rabenold, K.N. Beta diversity along environmental gradients: Implications of habitat specialization in tropical montane landscapes. *J. Anim. Ecol.* **2009**, *78*, 315–327. [CrossRef] [PubMed]
47. Ficetola, G.F.; Mazel, F.; Thuiller, W. Global determinants of zoogeographical boundaries. *Nat. Ecol. Evol.* **2017**, *1*, 1–7. [CrossRef] [PubMed]
48. Barnes, P.W.; Archer, S. Influence of an overstorey tree (Prosopis glandulosa) on associated shrubs in a savanna parkland: Implications for patch dynamics. *Oecologia* **1996**, *105*, 493–500. [CrossRef] [PubMed]
49. Wang, J.; Wang, Y.; Li, M.; He, N.; Li, J. Divergent roles of environmental and spatial factors in shaping plant β-diversity of different growth forms in drylands. *Glob. Ecol. Conserv.* **2021**, *26*, e01487. [CrossRef]
50. Wang, Y.; Dong, X.; Wang, H.; Wang, Z.; Gu, J. Root tip morphology, anatomy, chemistry and potential hydraulic conductivity vary with soil depth in three temperate hardwood species. *Tree Physiol.* **2016**, *36*, 99–108. [CrossRef]
51. Maeght, J.-L.; Gonkhamdee, S.; Clement, C.; Isarangkool Na Ayutthaya, S.; Stokes, A.; Pierret, A. Seasonal patterns of fine root production and turnover in a mature rubber tree (Hevea brasiliensis Müll. Arg.) stand-differentiation with soil depth and implications for soil carbon stocks. *Front. Plant Sci.* **2015**, *6*, 1022. [CrossRef]
52. Schwinning, S.; Litvak, M.; Pockman, W.; Pangle, R.; Fox, A.; Huang, C.-W.; McIntire, C. A 3-dimensional model of Pinus edulis and Juniperus monosperma root distributions in New Mexico: Implications for soil water dynamics. *Plant Soil* **2020**, *450*, 337–355. [CrossRef]
53. Krišāns, O.; Samariks, V.; Donis, J.; Jansons, Ā. Structural Root-plate characteristics of wind-thrown Norway spruce in hemiboreal forests of Latvia. *Forests* **2020**, *11*, 1143. [CrossRef]
54. Awé, D.V.; Noiha, N.V.; Vroh, B.T.A.; Zapfack, L. Root Distribution of Four Tree Species Planted in Living Hedges according to Two Types of Soil and Three Agroforestry Technologies in the Sudano-Sahelian Zone of Cameroon. *Open J. Agric. Res.* **2021**, *1*, 74–83.
55. Blevins, D.G.; Lukaszewski, K.M. Boron in plant structure and function. *Annu. Rev. Plant Biol.* **1998**, *49*, 481–500. [CrossRef] [PubMed]
56. Briat, J.-F.; Duc, C.; Ravet, K.; Gaymard, F. Ferritins and iron storage in plants. *BBA–Gen. Subj.* **2010**, *1800*, 806–814. [CrossRef] [PubMed]
57. López-Millán, A.F.; Duy, D.; Philippar, K. Chloroplast iron transport proteins–function and impact on plant physiology. *Front. Plant sci.* **2016**, *7*, 178. [CrossRef] [PubMed]
58. Pilon, M.; Abdel-Ghany, S.E.; Cohu, C.M.; Gogolin, K.A.; Ye, H. Copper cofactor delivery in plant cells. *Curr. Opin. Plant biol.* **2006**, *9*, 256–263. [CrossRef] [PubMed]
59. Yruela, I. Copper in plants: Acquisition, transport and interactions. *Funct. Plant Biol.* **2009**, *36*, 409–430. [CrossRef] [PubMed]
60. Schenk, H.J.; Jackson, R.B. The global biogeography of roots. *Ecol. Monogr.* **2002**, *72*, 311–328. [CrossRef]
61. Li, J.; He, B.; Chen, Y.; Huang, R.; Tao, J.; Tian, T. Root distribution features of typical herb plants for slope protection and their effects on soil shear strength. *TCSAE* **2013**, *29*, 144–152.
62. Sheoran, I.; Aggarwal, N.; Singh, R. Effects of cadmium and nickel on in vivo carbon dioxide exchange rate of pigeon pea (*Cajanus cajan* L.). *Plant Soil* **1990**, *129*, 243–249. [CrossRef]
63. Papazoglou, E.G.; Serelis, K.G.; Bouranis, D.L. Impact of high cadmium and nickel soil concentration on selected physiological parameters of *Arundo donax* L. *Eur. J. Soil Biol.* **2007**, *43*, 207–215. [CrossRef]

forests

MDPI

Review

Precipitation and Temperature Influence the Relationship between Stand Structural Characteristics and Aboveground Biomass of Forests—A Meta-Analysis

Yingdong Ma [1], Anwar Eziz [2], Ümüt Halik [1,*], Abdulla Abliz [3] and Alishir Kurban [2,*]

[1] College of Ecology and Environment, Ministry of Education Key Laboratory of Oasis Ecology, Xinjiang University, Ürümqi 830046, China
[2] Xinjiang Institute of Ecology and Geography, Chinese Academy of Sciences, Ürümqi 830011, China
[3] College of Tourism, Key Laboratory of Sustainable Development of Xinjiang's Historical and Cultural Tourism, Xinjiang University, Ürümqi 830046, China
* Correspondence: halik@xju.edu.cn (Ü.H.); alishir@ms.xjb.ac.cn (A.K.)

Abstract: Forest aboveground biomass (AGB) is not simply affected by a single factor or a few factors, but also by the interaction between them in complex ways across multiple spatial scales. Understanding the joint effect of stand structural characteristics and climate factors on AGB on large scales is critical for accurate forest carbon storage prediction and sustainable management. Despite numerous attempts to clarify the relationships between stand structural characteristics (tree density/TD, diameter at breast height/DBH, basal area/BA), climate factors (mean annual temperature/MAT, mean annual precipitation/MAP), and AGB, they remain contentious on a large scale. Therefore, we explored the relationships between stand structural characteristics, climate factors, and AGB at a biome level by meta-analyzing datasets contained in 40 articles from 25 countries, and then answered the questions of how stand structural characteristics influence AGB at the biome level and whether the relationships are regulated by climate on a large scale. Through using regression analysis and the establishment of a structural equation model, the results showed that the influence of basal area on AGB at the biome level was more substantial than that of tree density and DBH, and the significant relationship between basal area and AGB was relatively stable regardless of biome variation, but the effects of tree density and DBH was non-negligible within the biome. Climatic factors (e.g., temperature and precipitation), should be considered. Our meta-analysis illustrated the complicated interactions between climate factors, stand structural characteristics, and the AGB of forests, highlighting the importance of climate effects on regulating stand structural characteristics and AGB relationships. We suggest that basal area be preferred and considered in forest sustainable management practice to optimize stand structure for increasing carbon storage potential, with close attention to local climate conditions. Overall, our meta-analysis will crucially aid forest management and conservation in the context of global environmental changes, and provide novel insights and a scientific reference to lead to future carbon storage research on large scales.

Keywords: stand structural characteristic; aboveground biomass; temperature; precipitation; influence

Citation: Ma, Y.; Eziz, A.; Halik, Ü.; Abliz, A.; Kurban, A. Precipitation and Temperature Influence the Relationship between Stand Structural Characteristics and Aboveground Biomass of Forests—A Meta-Analysis. *Forests* **2023**, *14*, 896. https://doi.org/10.3390/f14050896

Academic Editor: José Aranha

Received: 30 March 2023
Revised: 13 April 2023
Accepted: 24 April 2023
Published: 27 April 2023

1. Introduction

Changes in carbon storage in terrestrial ecosystems have far-reaching implications for the carbon cycle of the global ecosystem and climate change [1,2]. Forest ecosystems play a vital role in the global carbon cycle. Furthermore, mitigation of the detrimental impact of global warming has great significance in ecosystem stability and function [3,4]. Thus, accurate forest biomass estimation is necessary and has been the main area of focus in climate change research, forest management, and sustainable development [5–13]. Forest degradation and deforestation caused by human activities and climate change pose a significant challenge to ecosystem function and sustainable development. An estimated

420 million ha of forest has been lost worldwide through deforestation since 1990, and the total carbon stock in forests decreased from 668 gigatonnes in 1990 to 662 gigatonnes in 2020 [14]. Thus, increasingly widespread concern about global climate change has led to an interest in reducing carbon emissions through quantifying carbon sequestration by forests [15]. Biomass is a valuable indicator in forest stand structure assessment and forest carbon stock estimation [16]. Aboveground biomass (AGB) and carbon storage are commonly derived from tree inventory data (diameter and height) [17,18]. In contrast, tree density, species diversity, and other forest structural attributes are interlinked with carbon storage in the forest ecosystem [19–21]. Along with variety, stand structure is equally critical in deciphering management implications [22].

It is widely accepted that current increases in forest biomass result partially from the positive effects of climate change and changes in forest management [23–26]. Additionally, numerous studies have suggested that mean annual temperature (MAT) and mean annual precipitation (MAP) are the main factors influencing forest biomass across broad geographic scales and climatic gradients, e.g., [27–29]. However, whether and to what extent climate influences forest biomass changes remain debatable [30]. For example, as carbon sinks, tropical forests are prone to carbon losses caused by variations in stand structure and species composition [31–34]. The positive effects of climate change on forest biomass may be offset by increasing climatic variability and extreme climate events, such as intense drought and extremely low temperatures [35]. Thus, temperature and precipitation can constrain AGB through seasonality and extended dry periods, such as by regulating the relationships between functional traits, including stand structure, and AGB at a community level through the length of the growing and the dry seasons [36–38]. Specifically, AGB is directly associated with stand structure as it is influenced by climate.

Previous studies have indicated that lower TD and BA affected forest biomass in tropical forests [31,39,40]. Furthermore, in subtropical or pantropical forests, community composition and stand structure (mainly as it relates to tree height) significantly impact AGB [41,42]. In temperate forests, TD, DBH, and BA tend to determine biomass allocation and AGB [43,44]. Future forest carbon sinks could be affected by large-scale changes in mortality and growth rate due to climate, stand structure, and their interactions [45,46]. In addition, these drivers, which are used to predict climate change's effects on forest biomass, can also interact in complex ways across various spatial scales [47]. These studies on small scales contribute to our understanding of how climate change impacts forest dynamics and ecosystem services, which are crucial for managing forests and conserving biodiversity in the context of global environmental changes. However, the present relationships between the factors mentioned above (e.g., stand structural characteristics, climate factors, and aboveground biomass) remain contentious at a large scale. Therefore, an enhanced understanding of how stand structural characteristics and climate factors influence AGB in forests at large scales is essential for estimating future potentials in forest carbon storage. In particular, for sustainable forest management, the deeper comprehension is needed because the optimized stand structure will affect seedling regeneration in forests, and the regeneration will ultimately determine tree species composition of forest stands and, hence, the AGB. Here, the objective of our meta-analysis was to examine the relationships between stand structural characteristics and AGB, and further make it clear how climate (e.g., temperature, precipitation) affects this relationship. More specifically, we sought to answer the following scientific questions:

(1) What effect do stand structural characteristics have on AGB, and how do such trait effects compare?

(2) How do climate factors affect the relationships between stand structural characteristics and AGB?

We hypothesized that temperature and precipitation would affect the stand structural characteristics–AGB relationships. To test the hypothesis, we conducted a meta-analysis on the effect of MAT and MAP on the relationships between TD, DBH, BA, and AGB through a systematic literature review. We firstly evaluated the relationship variation between stand structural characteristics and AGB in each biome (e.g., boreal forest, temperate seasonal forest, tropical rain forest, and tropical seasonal forest). We then explored the joint influence of stand structural characteristics and climate factors on AGB across all biomes. Finally, we investigated the multiple relationships between stand structural characteristics, climate factors, biome, and AGB.

2. Materials and Methods

2.1. Literature Search and Data Compilation

A systematic literature search was conducted using the subject headings "forest structure" AND "stand structure" AND "aboveground biomass" AND "forest biomass" to search published articles in the Web of Science, Google Scholar, Open Access Library, and CNKI, while Google Scholar and the Open Access Library were used to download articles that are not accessible in the Web of Science. The articles that were shortlisted for analysis had to be published in English and exclusively focused on stand structure–AGB relationships. Hence, the relevant literature up to December 2022 were identified and filtered on the basis of the following strict criteria: (1) each experiment must cover one or more sample plots; (2) studies must present stand structural characteristics, including DBH, TD, and BA in each site; (3) studies must be based on stand-level field research, not greenhouse or growth chamber pot experiments; and (4) the forest stands must cover either managed or non-managed natural and secondary forest. Following the criteria for the literature selection, 40 articles from 25 countries (Supplementary Figure S1) were ultimately included in our meta-analysis. With regard to data compilation, most data were directly obtained from the original articles, whereas a small portion of data were extracted from figures and tables using GetData Graph Digitizer 2.2 (Free Software Foundation, Boston, MA, USA). Finally, 227 entries of DBH, BA, TD, and AGB data (Supplementary Table S1) were included in our datasets for analysis. Here, these values of stand structural characteristics and AGB refer to a stand level. To analyze the interactive effects of climate factors (MAT, MAP) on AGB, MAT and MAP data of the sampling points in our datasets were extracted directly from the 40 articles. Meanwhile, the sampling points from the literature were classified into four biomes (e.g., boreal forest, temperate seasonal forest, tropical rainforest, and tropical seasonal forest) based on the MAP and MAT of each point using Whittaker's biome diagram [48].

2.2. Data Analysis

In the meta-analysis, we aimed to evaluate the relationship between stand structural characteristics and AGB within biomes and, further, the impact of climate on these relationships. To analyze the relationship of stand structural characteristics and AGB, the log transformation was first undertaken for each variable to ensure that they conformed to a normal distribution [49]. Then, the regression relationship between stand structural characteristics and AGB were described by the following allometric growth equation (Equation (1)):

$$\lg y = \alpha (\lg x) + \beta \quad (1)$$

where y and x represent the AGB values and different structural characteristics in the stand, α is the allometric exponent (the slope of this equation), and β is the allometric constant (the intercept of this equation); the slope is compared against $\alpha = 1$, representing the isometric relationship between two variables when the slope is exactly $\alpha = 1$, and the allometric relationship when it does not ($\alpha < 1$ or $\alpha > 1$) [50–52]. Then, the approach of standard major axis regression analysis (SMA) was used to assess the parameters α and β, and the allometric relationship between structural characteristics and AGB was tested through a comparison to 1.0; if the slope α is significantly different from 1.0, this indicates an allometric relationship between stand structural characteristics and AGB; otherwise, they are isometric [49,53]. To explore the joint effects of MAP, MAT, and stand structural characteristics on AGB across all biomes, multiple linear regression analysis (MLR) was carried out, and the variance inflation factor (VIF > 10) was used to estimate the collinearity between variables in the analysis [54,55]; meanwhile, a Pearson correlation analysis was conducted to analyze the correlation between stand structural characteristics and climate factors.

Structural equation modeling (SEM) was used to assess the multiple effects of MAT, MAP, TD, DBH, BA, and the biome on AGB, and, furthermore, the relationships in the multivariate data and the complex relationships and causality between variables were identified. For the evaluation of the most effective SEM, several statistical parameters including the chi-square (χ^2) test, goodness-of-fit index (GFI), comparative fit index (CFI), and standardized root mean square residual (SRMR) were used in the model. Here, we used the χ^2 test to critically evaluate the model fit and selection of SEM. These indicators for a strong model fit to the data contained an insignificant ($p > 0.05$) χ^2 test statistic, SRMR < 0.05, and GFI and CFI > 0.90, and the lowest AIC was selected as our final model among all acceptable models [19,56]. The SEM explained that a combination of factors, namely MAT, MAP, TD, DBH, BA, and the biome, would result in direct and indirect effects on AGB. All data processing, statistical analysis, and graphing were conducted using MS Excel, R 4.2.3 (https://www.r-project.org/, accessed on 20 March 2023), Amos 26.0 (https://www.ibm.com/spss, accessed on 24 December 2022), Origin 9.4 (OriginLab, Northampton, MA, USA), and ArcGIS 10.7.

3. Results and Analysis

3.1. Allometric Relationship of Stand Structural Characteristics and AGB

An SMA was performed to determine the allometric relationship between variables (Figure 1). The results showed that the allometric relationships were different across all biomes. Firstly, significant allometric associations of TD ($\alpha = -0.417$; CI = -0.640, -0.272; $p < 0.001$) and DBH ($\alpha = 0.575$, CI = 0.389, 0.851; $p < 0.05$), respectively, with AGB were observed, indicating that the growth rate in AGB exceeded TD and DBH in boreal forest (Figure 1a). In terms of temperate seasonal forest, the relationship of TD and AGB was significantly allometric ($\alpha = -0.76$, CI = -0.908, -0.637; $p < 0.05$), whereas the relationships were not found for DBH nor BA ($p > 0.05$), indicating an isometric growth of DBH and BA, respectively, and AGB in this biome (Figure 1b). Moreover, significant allometric relationships of BA ($\alpha = 1.309$, CI = 1.135, 1.510; $p < 0.001$) and DBH ($\alpha = 1.524$, CI = 1.122, 2.069; $p < 0.05$), respectively, with AGB were observed in tropical rainforest (Figure 1c), whereas those of TD and AGB non-significant ($p > 0.05$). Meanwhile, significant allometric relationships of BA ($\alpha = 1.145$, CI= 1.026, 1.279; $p < 0.05$) and TD ($\alpha = 0.608$, CI = 0.471, 0.785; $p < 0.001$), respectively, with AGB were further found in tropical seasonal forest, whereas there was an isometric relationship between DBH and AGB here ($p > 0.05$) (Figure 1d).

3.2. Interrelation of Stand Structural Characteristics and Climate Factors

As shown in Table 1, correlation analysis indicated the significant correlation ($p < 0.05$) between stand structural characteristics and climate factors in boreal forest, which stated that the considerable influence of BA on AGB was related to climate factors. In addition, BA was negatively correlated with TD ($r = -0.450$, $p < 0.05$), whereas it was positively correlated with DBH ($r = 0.743$, $p < 0.01$), indicating that the increase in DBH rather than TD was suitable for BA in boreal forest. Significant correlations of TD and climate factors, such as MAT ($r = 0.262$, $p < 0.01$) and MAP ($r = 0.231$, $p < 0.05$), were observed in temperate seasonal forest; meanwhile, a negative correlation of TD with DBH ($r = -0.205$, $p < 0.05$) and positive correlation with BA ($r = 0.246$, $p < 0.05$) were also found, indicating that the relationship of TD and AGB was not simply affected by climate factors, but associated with DBH. This further revealed that increasing TD will promote BA in temperate seasonal forest. Moreover, MAP was positively correlated with BA ($r = 0.499$, $p < 0.01$) in tropical rainforest, indicating the reciprocal effect of MAP and BA on AGB. However, BA was not significantly correlated with any stand or climate factors in tropical seasonal forest ($p > 0.05$), whereas TD ($p < 0.05$) and DBH ($p < 0.01$), respectively, were associated with climate factors.

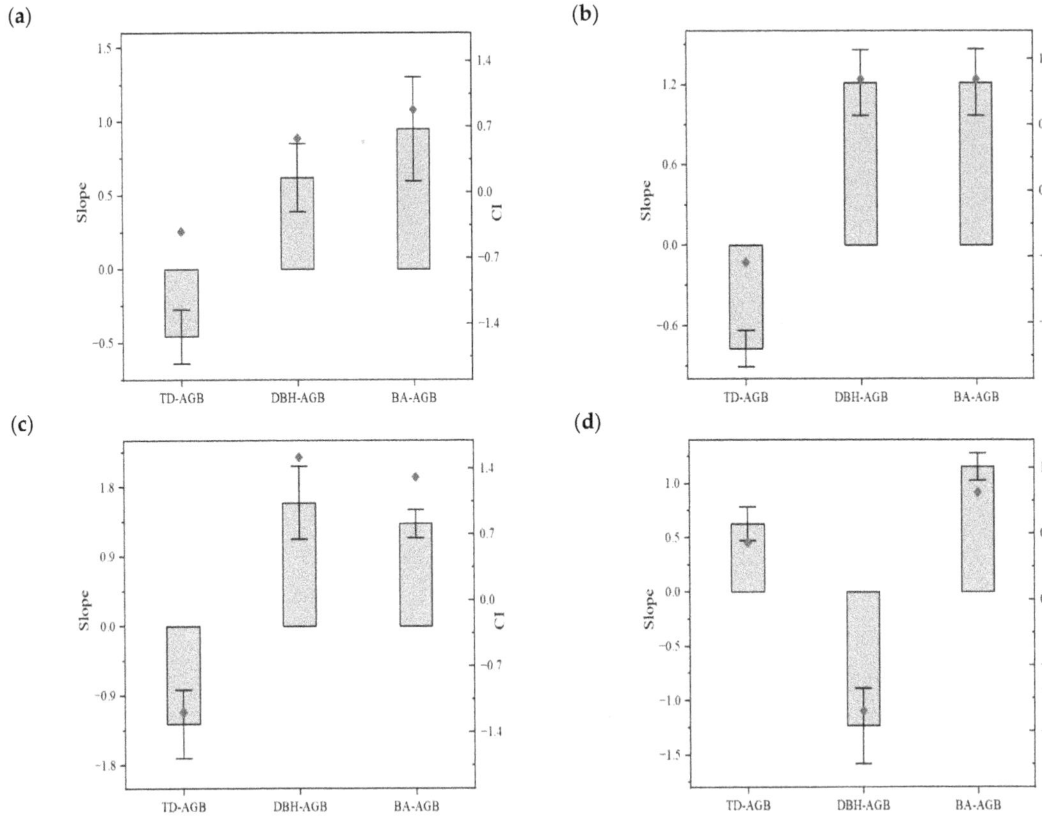

Figure 1. Allometric relationships of stand structural characteristics and AGB in boreal forest (**a**), temperate seasonal forest (**b**), tropical rainforest (**c**), and tropical seasonal forest (**d**). The diamond's position relative to the box indicates the slope, and the two short horizontal lines denote the confidence interval (CI). DBH: diameter at breast height; TD: tree density; BA: basal area.

Table 1. Relationships of stand structural characteristics and climate factors.

Biome	Variables	MAT	MAP	TD	DBH
Boreal forest	MAP	0.951 **			
	TD	−0.473 *	−0.599 **		
	DBH	0.781 **	0.841 **	−0.594 **	
	BA	0.816 **	0.803 **	−0.450 *	0.743 **
Temperate seasonal forest	MAP	0.263 **			
	TD	0.262 **	0.231 *		
	DBH	0.12	−0.12	−0.205 *	
	BA	0.06	0.15	0.246 *	0.13
Tropical rainforest	MAP	0.09			
	TD	0.786 **	−0.08		
	DBH	−0.32	0.24	−0.31	
	BA	−0.12	0.449 **	−0.07	0.10
Tropical seasonal forest	MAP	−0.302 *			
	TD	0.03	−0.311 *		
	DBH	−0.691 **	0.559 **	−0.25	
	BA	−0.24	−0.10	0.07	−0.07

* indicates a significant correlation at the 0.05 level, and ** represents significance at the 0.01 level. MAP: mean annual precipitation; MAT: mean annual temperature; DBH: diameter at breast height; TD: tree density; BA: basal area.

3.3. *Joint Influences of Stand Structural Characteristics and Climate Factors on AGB*

Multiple regression analysis indicated that AGB was jointly influenced by stand structural characteristics and climate factors (Figure 2). In terms of boreal forest, stand factors jointly influenced AGB, and a significant effect of BA on AGB was observed ($r = 0.803$, $p < 0.05$) (Figure 2a). However, BA ($r = 0.647$, $p < 0.001$) and TD ($r = -0.387$, $p < 0.001$) significantly affected AGB ($r = 1.041$, $p < 0.05$) in temperate seasonal forest, and the influence of BA was stronger than TD; meanwhile, a significant effect of MAT on AGB was found ($r = -0.226$, $p < 0.01$), showing a joint influence of a stand structural characteristic and climate factor on AGB (Figure 2b). Similarly, BA significantly affected AGB in tropical rainforest ($r = 0.665$, $p < 0.001$); meanwhile, a significant effect of DBH ($r = 0.216$, $p < 0.05$) and MAP ($r = 0.226$, $p < 0.05$) on AGB was observed here (Figure 2c). In addition, a significant effect of BA on AGB was found in tropical seasonal forest ($r = 0.804$, $p < 0.001$) (Figure 2d). According to the relative contribution of stand structural characteristics and climate factors to AGB across all biomes, the results revealed that TD, DBH, and BA explained 37.8% of the variation in AGB in boreal forest. Concerning the temperate seasonal forest, the contribution of stand structural characteristics and climate factors to AGB totaled 51%. A higher contribution of stand and climate factors was found in tropical rainforest (78.6%) and seasonal forest (64.9%), indicating a more substantial influence of stand structural characteristics and climate factors on AGB in tropical forest than in temperate and boreal forest (Figure 3).

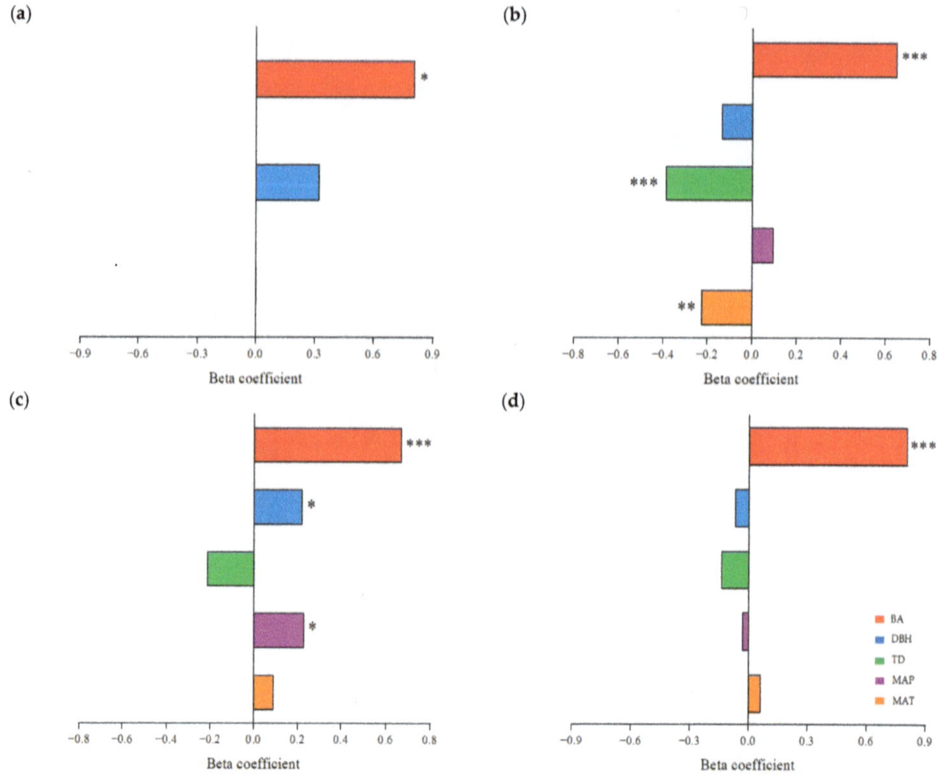

Figure 2. Multiple regression relationships of stand structural characteristics, climate factors, and AGB in boreal forest (**a**), temperate seasonal forest (**b**), tropical rainforest (**c**), and tropical seasonal forest (**d**). The significant results are indicated with an asterisk (* $p < 0.05$; ** $p < 0.01$; *** $p < 0.001$). MAP: mean annual precipitation; MAT: mean annual temperature; DBH: diameter at breast height; TD: tree density; BA: basal area.

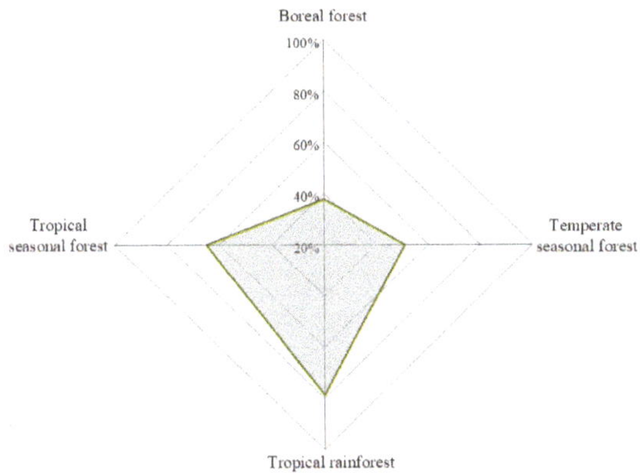

Figure 3. Total contribution of stand structural characteristics and climate factors to AGB.

3.4. The Establishment of SEM among Observed Variables

Our results (Figure 4) revealed that the final selected and best-fit SEM with the different paths for the interrelationships between climate factors, stand structural characteristics, and biome explained 43% of the variation in AGB. Meanwhile, the climate factors explained 3%, 7%, and 2% of the variation, respectively, in TD, DBH, and BA. An important role of climate factors was found when stand structural characteristics influenced AGB. However, at this point, the effects of climate factors were indirect. The SEM showed that BA had a significant impact on AGB ($p < 0.001$), which was promoted directly by MAT and MAP. TD had a non-significant impact on AGB, but MAP significantly and directly affected TD ($p < 0.05$). Meanwhile, DBH significantly affected AGB ($p < 0.05$), which was related to the direct effects of MAT ($p < 0.001$) and MAP ($p < 0.01$). Despite climate factors being associated with biome ($p < 0.001$), it was shown that the biome did not have a direct or indirect effect on stand structure. However, the biome significantly and directly affected AGB ($p < 0.05$). Furthermore, a significant influence of climate factors on AGB was observed in the SEM ($p < 0.001$), which was related to the biome ($p < 0.001$).

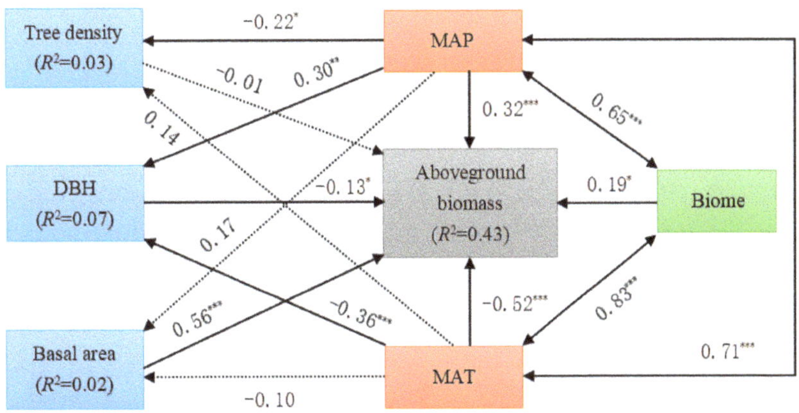

Model goodness fit: χ^2/df = 3.49, SRMR = 0.049, GFI = 0.974, CFI = 0.974, AIC = 64.95.

Figure 4. The final best-fit structural equation model (SEM) for evaluating multiple influences of climate factors (MAT, MAP), stand structural characteristics (tree density, DBH, basal area), and biome on AGB. "−" are negative relationships while "+" are positive ones, and the dotted line indicates insignificant relationships (* $p < 0.05$; ** $p < 0.01$; *** $p < 0.001$). R^2 indicates the total variation in a dependent variable that is explained by the combined independent variables. MAP: mean annual precipitation; MAT: mean annual temperature.

4. Discussion

4.1. Differences between Stand Structural Characteristics in Terms of Their Influence on AGB within Biomes

As the main characteristics of stand structure, TD, DBH, and BA strongly influenced AGB. However, owing to different abiotic and biotic conditions, the influence of stand structural characteristics on AGB differed across all biomes [57,58]. Alexander's finding [43] showed that DBH and BA significantly influenced AGB in boreal forest, and the significance of BA on AGB was also found on a large scale in our study. However, Wang et al. [59] revealed that TD significantly affected AGB in boreal forest (e.g., Korean pine forests). Our findings further showed that the relationship between TD and AGB was allometric in boreal forest, which may be associated with regeneration after a fire and human disturbance [60,61]. Previous results revealed that AGB in *Larix gmelinii* stands remained constant at decreasing TD due to fire and nutrient limitations triggered by competition [62], indicating the important effects of self-thinning on AGB at the biome level [63]. The relationships between TD and AGB in competitive environments were mediated by tree crowns, since TD significantly influenced crown structure, including branches and leaves, by regulating crown growth in order to obtain plenty of sunshine [64–66]. However, tree death from repeated ground fires and slow recovery due to nutrient limitations changed self-thinning trajectories [67]. Particularly in even-aged stands that regenerated after fire, since the relationship between competition and habitat conditions are more complicated, the association of TD and AGB might be affected by insect outbreak or windthrow [68]. TD significantly affected AGB in temperate forest, consistent with the findings of Yang et al. [69]; meanwhile, an allometric relationship of TD and AGB was found within this biome. Most likely, community type and species diversity should be considered, such as in broadleaf and conifer forests or mixed forests [70], since trees' biological traits (e.g., leaf area, thickness) may regulate different nutrient uptake and partitioning. Moreover, due to the resource complementarity effect, inter-species interactions drive trees to make efficient use of light, water, and soil nutrients [71,72].

Our study revealed that DBH significantly influenced AGB in tropical rainforest, likely because dominant tree species regulated the aboveground resource allocation via luxuriant branches, which have a positive feedback on carbon storage [73]. The relationship may also be explained by increased light capture and light use efficiencies in association with complex tree-sized structures [74,75], showing different responses of different-sized trees to light efficiency. Several authors have confirmed that large trees increased AGB in the natural forests [76–78], since large trees could obtain and utilize more nutrients through a well-developed root system and crown structure, and govern carbon sequestration [79]. However, the effects of small trees in a stand should not be ignored, since in forests with dense small DBH trees in larger numbers, and many sparse large trees, AGB was stable [80]. Our finding also showed a strong correlation between TD and AGB in boreal and temperate forests. Most likely, there is a competitive exclusion mechanism in forests; for example, large trees lead to species competition in different ways, and, hence, directly lead to a reduction in TD and indirectly to AGB [81]. A significant allometric relationship of TD and DBH, respectively, with AGB was found in tropical forests, where it has been confirmed that the relationships were related to site condition, tree age, tree species diversities, and origin of the forest [15,82–85].

BA significantly influenced AGB across all biomes in our results. Conversely, Alexander et al. [43] and Taylor et al. [86] concluded that a non-significant relationship occurred in the boreal forest of interior Alaska and central Canada, indicating the significance of the geographical environment. However, in southern Finland, the significant association between BA and AGB resulted from forest management [87], such as felling [88], which indicated that harvesting may alter the influence of BA on AGB regardless of stand age and species diversity. In this study, we showed a correlation between TD, DBH, and BA in boreal forest, which means that if felling reduced TD, fewer trees may affect DBH and BA, and, hence, AGB. It is perhaps for this reason that DBH and AGB were allometric.

Similar management may also influence the relationship between BA and AGB in tropical forests [89,90]. Nevertheless, in unmanaged forests, BA in the upper trees significantly affected AGB; therefore, the vital role of dominant trees should be noted [91]. Furthermore, numerous authors have suggested that the relationship of BA and AGB in the tropics was involved in tree species, elevation, tree age, and soil properties [17,92–94], which may explain the potential reason for the allometry between them. In temperate forest, it was found that the significant influence of BA on AGB was consistent with previous studies [43,95]; however, the influence was not absolute due to dominant species. Even if the BA of different sizes of trees was a strong driver of AGB, overstory trees dominated the relationship and played a more critical role than the understory trees [96]. Moreover, owing to the dominant species, the effect of BA on AGB was stable with the change in altitude [97]. In our results, we did not find a significant correlation between BA and DBH in spite of the substantial impact of BA on AGB in temperate forest. However, other findings showed that BA in medium–small DBH trees had the most important effect on AGB, whereas BA in large trees slightly affected AGB, which may be related to tree species diversity [98].

4.2. Determinants of Co-Driving AGB among Climate, Stand Structural Characteristics, and Biome

Previous studies have proposed factors that govern forest AGB, highlighting several climatic variables, including MAT and MAP [28,99]. A consensus has yet to be reached on which climate variables are more important in terms of affecting the relationships on different spatial scales [86]. Our results revealed that both climate factors and stand structural characteristics jointly influenced AGB across all biomes, whereas there was a significant influence on AGB mainly in temperate and tropical rainforest. A possible explanation was that climate factors regulated stand structure in forests, and, hence, AGB. Specifically, temperature drives the trees' utilization of light energy within the photosynthetic uptake processes to increase AGB [100]. In previous studies, the linear relationship between absorbed photosynthetic radiation and tree biomass has been confirmed for different tree species [101]. The temperature, which regulated the relationship of stand structure and AGB, could also have an adverse impact on AGB. For instance, the increasing temperature may lead to an increase in fire risk in the future [102], especially in boreal and temperate forests, since the risk of extreme fire events was expected to trigger great AGB loss [103]. Our results revealed that precipitation and BA together influenced AGB in tropical rainforest, and BA had a significant correlation with the precipitation, indicating that precipitation may influence BA and, thus, AGB. Similarly, the interrelations between stand structure characteristics and precipitation were also observed in tropical seasonal forest. Nevertheless, the climate factors and BA did not jointly affect AGB in this biome, which has been revealed in other studies due to influences of the topography [104–106], soil properties, management protocols, and species diversity [107–110]. Additionally, we found that the contribution of climate and stand factors to AGB in tropical and temperate forests was greater than that in boreal forest, but the potential ecological reasons remained unclear on a large scale.

Either a direct or indirect effect on AGB for climate, stand structure, and biome was further observed through the SEM. It was indicated that both temperature and precipitation had a direct effect on stand structure and AGB, which was not entirely consistent with the previous results showing the more important influence of temperature on AGB than other climatic variables [111]. This was likely associated with diverse biomes in our study on a large scale. Although our finding did not reveal any direct or indirect effects of biome on stand structure on a large scale, previous findings have confirmed that climate factors influenced stand structure due to biome on a local scale [112,113]. Most likely, the mutual effects of biome and climate factors reflected trees' different growth strategies in adaptation to favorable climatic conditions [67,114,115]. This indicated that the effect of biome on the relationship between stand structure and AGB should be considered regardless of scales. Previous findings revealed that climate factors alone cannot explain the patterns of AGB on the regional scale [116], showing the interaction between temperature and precipitation in the influence on AGB, which was further confirmed in our findings. Moreover, we found that temperature and precipitation together had a direct effect on stand structural characteristics, especially on DBH and, thus, AGB; however, climate factors alone significantly and directly affected TD and BA and, hence, AGB, which means there were no fixed patterns in the effects of climate variables on the relationship between stand structure and AGB.

4.3. Study Limitations and Weaknesses

Our review highlights an important issue related to carbon storage across all biomes in recent years. Unfortunately, there were still potential research shortcomings and biases due to there being fewer datasets from limited studies with fewer biome types and fewer countries included. Although temperature and precipitation directly affected stand structural characteristics and, hence, AGB on a large scale, the common effects of climate factors and other stand structural characteristics such as tree height and crown breadth rather than only TD, DBH, and BA may affect AGB at the biome level, particularly in tropical and temperate forests. In the process of carrying out the meta-analysis, we found that fewer original studies directly resulted in smaller datasets, and, hence, affected the research results. For instance, our meta-analysis indicated that basal area had a more substantial influence on AGB than that of tree density and DBH in the boreal, temperate seasonal, tropical rain, and tropical seasonal forests; however, the relationships may be changed with variation in tree age, regional scale, and biodiversity, which were not fully revealed due to inadequate data on tree age in the original articles. Owing to the fewer original studies on the relationship between stand structural characteristics and AGB, including an incomplete data set or fewer parameters on stand structure (e.g., only one or two indicators used in characterizing the stand structure) in the literature, a comprehensive and detailed explanation of the complicated relation between climate, stand structure, and AGB could not be presented in the meta-analysis. Furthermore, this may give rise to several large and mostly unexplained relationships between stand structural characteristics and AGB in different aged stands, especially in forests of the underrepresented regions, or in undiscovered and complex biomes. In addition, an ecological mechanism of stand structural characteristics–AGB relationships could not be understood in depth due to the limited datasets from fewer original studies. In fact, our initial idea was to explore the relationships between stand structural characteristics of different sized trees and AGB in different temperature and precipitation conditions through a comparison of plantation forests and natural or secondary forests, which, however, was replaced due to the low number of original studies on the relationship between stand structural characteristics and AGB in plantation forests. However, it is also possible that several original articles were not found due to the limited databases. Nevertheless, we agree that temperature and precipitation will likely be a major determinant in the environmental components of regulating stand structural characteristics and AGB relationships, and large datasets are essential for detailed meta-analysis research on a large scale.

5. Conclusions

Understanding the relationships of stand structural characteristics and AGB on a large scale is vital to both predicting the rate and potential of forest carbon storage and guiding future multifunctional forest management in the face of global environmental changes. The effects of climate factors and stand structural characteristics on AGB were discussed based on a comparative analysis of the datasets used in our meta-analysis. The results showed that the important effects of stand structural characteristics and climate factors in increasing AGB should be considered for sustainable forest management. Different stand structural characteristics had different impacts on AGB within biomes; through comparison between stand structural characteristics, the significant influences of basal area on AGB were found to be more substantial than those of tree density and DBH in the boreal, temperate seasonal, tropical rain, and seasonal forest. Therefore, the main stand structural characteristic affecting AGB was undoubtedly basal area; however, the influence of tree density and DBH should not be ignored at the biome level. This study sheds light on the complicated interactions between climate factors, stand structural characteristics, and AGB in forests, and highlights that both temperature and precipitation affect the relationship between stand structural characteristics and AGB on a large scale, hence supporting our hypothesis. Therefore, to increase AGB and forest carbon storage, we suggest that basal area be preferred and considered in forest management practice to optimize stand structure for promoting AGB growth, with close attention to local climate conditions. Moreover, in terms of a meta-analysis on the potential for forest carbon storage on a large scale, future research should concentrate on clarifying the relationship between other stand structural characteristics and AGB in natural forests under hydrothermal conditions, and could also attempt an extended study of plantation forests using an adequate dataset.

Supplementary Materials: The following supporting information can be downloaded at: https://www.mdpi.com/article/10.3390/f14050896/s1, Figure S1: Sampling points from 25 countries; Table S1: Datasets from 25 countries.

Author Contributions: Y.M. designed and performed the research framework, collected and analyzed the data, and wrote and prepared the original draft. Ü.H. and A.K. designed and supervised the study, reviewed the manuscript, and approved the final draft. A.E. participated in data analysis and reviewed the manuscript draft. A.A. reviewed the manuscript with critical comments and language proofreading. All authors have read and agreed to the published version of the manuscript.

Funding: This research work was supported by the National Natural Science Foundation of China (Grant Nos.: 32260285, 32071655), the Third Xinjiang Scientific Expedition and Research Program (Grant No: 2022xjkk0301), and Enterprise research project funded by Tarim River Basin Authority (Grant No: TGJGLJJJG2021ZXFW0007).

Data Availability Statement: The data presented in this study are available upon reasonable request from the author team.

Conflicts of Interest: The authors declare no conflict of interest.

References

1. Zhou, L.; Liu, H.; Zhou, G.; Zhou, X.; Hong, Y.; Li, C.; Lu, C.; He, Y.; Shao, J.; Sun, X.; et al. Responses of biomass allocation to multi-factor global change: A global synthesis. *Agric. Ecosyst. Environ.* **2020**, *304*, 107115. [CrossRef]
2. Tian, L.; Tao, Y.; Fu, W.X.; Li, T.; Ren, F.; Li, M.Y. Dynamic simulation of land use/cover change and assessment of forest ecosystem carbon storage under climate change Scenarios in Guangdong Province, China. *Remote Sens.* **2022**, *14*, 2330. [CrossRef]
3. Payne, N.J.; Cameron, D.A.; Leblanc, J.D.; Morrison, I.K. Carbon storage and net primary productivity in Canadian boreal mixedwood stands. *J. For. Res.* **2019**, *30*, 1667–1678. [CrossRef]
4. Li, X.; Du, H.; Mao, F.; Zhou, G.; Han, N.; Xu, X.; Liu, Y.; Zhu, D.; Zheng, J.; Dong, L.; et al. Assimilating spatiotemporal MODIS LAI data with a particle filter algorithm for improving carbon cycle simulations for bamboo forest ecosystems. *Sci. Total Environ.* **2019**, *694*, 133803. [CrossRef] [PubMed]
5. De Castilho, C.V.; Magnusson, W.E.; De Araújo, R.N.O.; Luizão, F.J. Short-term temporal changes in tree live biomass in a central Amazonian forest Brazil. *Biotropica* **2010**, *42*, 95–103. [CrossRef]
6. Furley, P. Tropical savannas: Biomass, plant ecology, and the role of fire and soil on vegetation. *Prog. Phys. Geog.* **2010**, *34*, 563–585. [CrossRef]
7. Keith, H.; van Gorsel, E.; Jacobsen, K.L.; Cleugh, H.A. Dynamics of carbon exchange in a Eucalyptus forest in response to interacting disturbance factors. *Agric. For. Meteorol.* **2012**, *153*, 67–81. [CrossRef]
8. Lu, D.; Chen, Q.; Wang, G.; Moran, E.; Batistella, M.; Zhang, M.; Laurin, G.V.; Saah, D. Aboveground forest biomass estimation with Landsat and LiDAR data and uncertainty analysis of the estimates. *Int. J. For. Res.* **2012**, *1*, 436537. [CrossRef]
9. Brahma, B.; Nath, A.J.; Deb, C.; Sileshi, G.W.; Sahoo, U.K.; Das, A.H. A critical review of forest biomass estimation equations in India. *Tree For. People* **2021**, *5*, 100098. [CrossRef]
10. Fang, J.Y.; Guo, Z.; Hu, H.; Kato, T.; Muraoka, H.; Son, Y. Forest biomass carbon sinks in East Asia, with special reference to the relative contributions of forest expansion and forest growth. *Glob. Chang. Biol.* **2014**, *20*, 2019–2030. [CrossRef]
11. McEwan, R.W.; Lin, Y.; Sun, I.; Hsieh, C.; Su, S.; Chang, L.; Song, G.M.; Wang, H.; Hwong, J.; Lin, K.; et al. Topographic and biotic regulation of aboveground carbon storage in subtropical broad-leaved forests of Taiwan. *For. Ecol. Manag.* **2011**, *262*, 1817–1825. [CrossRef]

12. Pan, Y.; Birdsey, R.A.; Fang, J.; Houghton, R.; Kauppi, P.E.; Kurz, W.A.; Phillips, O.L.; Shvidenko, A.; Lewis, S.L.; Canadell, J.G.; et al. A large and persistent car bon sink in the world's forests. *Science* **2011**, *333*, 988–993. [CrossRef] [PubMed]
13. Ou, G.; Li, C.; Lv, Y.; Wei, A.; Xiong, H.; Xu, H.; Wang, G. Improving aboveground biomass estimation of *Pinus densata* forests in Yunnan using Landsat 8 Imagery by incorporating age dummy variable and method comparison. *Remote Sens.* **2019**, *11*, 738. [CrossRef]
14. FAO. Global Forest Resources Assessment 2020: Main Report. Rome: FAO. 2020. Available online: https://www.fao.org/forest-resources-assessment/en/ (accessed on 20 September 2022).
15. Azad, M.S.; Kamruzzaman, M.; Osawa, K. Quantification and understanding of above and belowground biomass in Medium Saline Zone of the Sundarbans, Bangladesh: The relationships with forest attributes. *J. Sustain. For.* **2019**, *39*, 331–345. [CrossRef]
16. Soares, M.L.G.; Schaeffer-Novelli, Y. Aboveground biomass of mangrove species. I. Analysis of models. *Estuar. Coast. Shelf Sci.* **2005**, *65*, 1–18. [CrossRef]
17. Gross, J.; Flores, E.E.; Schwendenmann, L. Stand structure and aboveground biomass of a *Pelliciera rhizophorae* mangrove forest, Gulf of Monitjo Ramsar Site, Pacific Coast, Panama. *Wetlands* **2014**, *34*, 55–65. [CrossRef]
18. Clough, B.J.; Russell, M.B.; Domke, G.M.; Woodall, C.W. Quantifying allometric model uncertainty for plot-level live tree biomass stocks with a data-driven, hierarchical framework. *For. Ecol. Manag.* **2016**, *372*, 175–188. [CrossRef]
19. Ali, A.; Yan, E.R.; Chen, H.Y.H.; Chang, S.X.; Zhao, Y.T.; Yang, X.D.; Xu, M.S. Stand structural diversity rather than species diversity enhances aboveground carbon storage in secondary subtropical forests in Eastern China. *Biogeosciences* **2016**, *13*, 4627–4635. [CrossRef]
20. Wang, W.; Lei, X.; Ma, Z.; Kneeshaw, D.D.; Peng, C. Positive relationship between aboveground carbon stocks and structural diversity in spruce-dominated forest stands in New Brunswick, Canada. *For. Sci.* **2011**, *57*, 506–515. [CrossRef]
21. Dănescu, A.; Albrecht, A.T.; Bauhus, J. Structural diversity promotes productivity of mixed, uneven-aged forests in southwestern Germany. *Oecologia* **2016**, *182*, 319–333. [CrossRef]
22. Kaushal, S.; Baishya, R. Stand structure and species diversity regulate biomass carbon stock under major Central Himalayan forest types of India. *Ecol. Process.* **2021**, *10*, 14. [CrossRef]
23. Huang, C.; Liang, Y.; He, H.S.; Wu, M.M.; Liu., B.; Ma, T.X. Sensitivity of aboveground biomass and species composition to climate change in boreal forests of Northeastern China. *Ecol. Model.* **2021**, *445*, 109472. [CrossRef]
24. Hember, R.A.; Kurz, W.A.; Metsaranta, J.M.; Black, T.A.; Guy, R.D.; Coops, N.C. Accelerating regrowth of temperate-maritime forests due to environmental change. *Glob. Chang. Biol.* **2012**, *18*, 2026–2040. [CrossRef]
25. Peng, J.; Dan, L.; Huang, M. Sensitivity of global and regional terrestrial carbon storage to the direct CO2 effect and climate change based on the CMIP5 model intercomparison. *PLoS ONE* **2014**, *9*, e95282. [CrossRef]
26. Luyssaert, S.; Ciais, P.; Piao, S.L.; Schulze, E.D.; Jung, M.; Zaehle, S.; Schelhaas, M.J.; Reichstein, M.; Churkina, G.; Papale, D.; et al. The European carbon balance, Part 3: Forests. *Glob. Chang. Biol.* **2010**, *16*, 1429–1450. [CrossRef]
27. Chen, X.; Luo, M.Y.; Larjavaara, M. Effects of climate and plant functional types on forest above-ground biomass accumulation. *Carbon Balance Manag.* **2023**, *18*, 305. [CrossRef]
28. Sankaran, M.; Hanan, N.P.; Scholes, R.J.; Ratnam, J.; Augustine, D.J.; Cade, B.S.; Gignoux, J.; Higgins, S.I.; Le Roux, X.; Ludwig, F.; et al. Determinants of woody cover in African savannas. *Nature* **2005**, *438*, 846–849. [CrossRef] [PubMed]
29. Zhang, T.; Ding, G.J.; Zhang, J.P.; Qi, Y.J. Stand, plot characteristics, and tree species diversity jointly dominate the recruitment biomass of subtropical forests. *For. Ecol. Manag.* **2023**, *531*, 120814. [CrossRef]
30. McMahon, S.M.; Parker, G.G.; Miller, D.R. Evidence for a recent increase in forest growth. *Proc. Natl. Acad. Sci. USA* **2010**, *107*, 3611–3615. [CrossRef] [PubMed]
31. Osuri, A.M.; Kumar, V.S.; Sankaran, M. Altered stand structure and tree allometry reduce carbon storage in evergreen forest fragments in India's Western Ghats. *For. Ecol. Manag.* **2014**, *329*, 375–383. [CrossRef]
32. Phillips, O.L.; Aragão, L.E.; Lewis, S.L. Drought sensitivity of the Amazon rainforest. *Science* **2009**, *323*, 1344–1347. [CrossRef]
33. Calvo-Rodriguez, S.; Sánchez-Azofeifa, G.A.; Durán, S.M.; Espírito-Santo, M.M.D.; Nunes, Y.R.F. Dynamics of Carbon Accumulation in Tropical Dry Forests under Climate Change Extremes. *Forests* **2021**, *12*, 106. [CrossRef]
34. Hiltner, U.; Huth, A.; Fischer, R. Importance of the forest state in estimating biomass losses from tropical forests: Combining dynamic forest models and remote sensing. *Biogeosciences* **2022**, *19*, 1891–1911. [CrossRef]
35. Zhao, M.; Running, S.W. Drought-induced reduction in global terrestrial net primary production from 2000 through 2009. *Science* **2010**, *329*, 940–943. [CrossRef] [PubMed]
36. Ali, A.; Lin, S.L.; He, J.K.; Kong, F.M.; Yu, J.H.; Jiang, H.S. Elucidating space, climate, edaphic and biodiversity effects on aboveground biomass in tropical forests. *Land Degrad. Dev.* **2019**, *30*, 918–927. [CrossRef]
37. Michaletz, S.T.; Kerkhoff, A.J.; Enquist, B.J. Drivers of terrestrial plant production across broad geographical gradients. *Glob. Ecol. Biogeogr.* **2018**, *27*, 166–174. [CrossRef]
38. Malhi, Y.; Wood, D.; Baker, T.R.; Wright, J.; Phillips, O.L.; Cochrane, T. The regional variation of aboveground live biomass in old-growth Amazonian forests. *Glob. Chang. Biol.* **2006**, *12*, 1107–1138. [CrossRef]
39. Lee, L.; Lee, J.; Kim, S.J.; Roh, Y.J.; Salim, K.A.; Lee, W.K.; Son, Y.H. Forest structure and carbon dynamics of an intact lowland mixed Dipterocarp forest in Brunei Darussalam. *J. For. Res.* **2018**, *29*, 199–203. [CrossRef]
40. Gao, L.S.; Zhang, X.L. Above-ground biomass estimation of plantation with complex forest stand structure using multiple features from airborne laser scanning point cloud data. *Forests* **2021**, *12*, 1713. [CrossRef]

41. Feldpausch, T.R.; Lloyd, J.; Lewis, S.L.; Brienen, R.J.W.; Gloor, M.; Mendoza, A.M. Tree height integrated into pantropical forest biomass estimates. *Biogeosciences* **2012**, *9*, 3381–3403. [CrossRef]
42. Wang, X.; Huang, X.; Wang, Y.; Yu, P.; Guo, J. Impacts of site conditions and stand structure on the biomass allocation of single trees in larch plantations of Liupan Mountains of Northwest China. *Forests* **2022**, *13*, 177. [CrossRef]
43. Alexander, H.D.; Mack, M.C.; Goetz, S.; Beck, P.S.A.; Belshe, E.F. Implications of increased deciduous cover on stand structure and aboveground carbon pools of Alaskan boreal forests. *Ecosphere* **2012**, *3*, 45. [CrossRef]
44. Mantgem, P.J.; Stephenson, N.L.; Byrne, J.C.; Daniels, L.D.; Franklin, J.F.; Fule, P.Z.; Harmon, M.E.; Larson, A.J.; Smith, J.M.; Taylor, A.H.; et al. Widespread increase of tree mortality rates in the western United States. *Science* **2009**, *323*, 521–524. [CrossRef]
45. Ruiz-Benito, P.; Lines, E.R.; Go´mez-Aparicio, L.; Zavala, M.A.; Coomes, D.A. Patterns and drivers of tree mortality in Iberian forests: Climatic effects are modified by competition. *PLoS ONE* **2013**, *8*, e56843. [CrossRef]
46. Ruiz-Benito, P.; Madrigal-Gonza´lez, J.; Ratcliffe, S.; Coomes, D.A.; Kandler, G.; Lehtonen, A.; Wirth, C.; Zavala, M.A. Stand structure and recent climate change constrain stand basal area change in European forests: A comparison across boreal, temperate, and mediterranean biomes. *Ecosystems* **2014**, *17*, 1439–1454. [CrossRef]
47. Holdaway, R.J.; Easdale, T.A.; Carswell, F.E.; Richardson, S.J.; Peltzer, D.A.; Mason, N.W.H.; Brandon, A.M.; Coome, D.A. Nationally representative plot network reveals contrasting drivers of net biomass change in secondary and old-growth forests. *Ecosystems* **2017**, *20*, 944–959. [CrossRef]
48. Westman, W.E.; Peet, R.K.; Robert, H. Whittaker (1920–1980): The man and his work. *Vegetatio* **1982**, *48*, 97–122. [CrossRef]
49. Warton, D.I.; Wright, I.J.; Falster, D.S.; Westoby, M. Bivariate line-fitting methods for allometry. *Biol. Rev.* **2006**, *81*, 259–291. [CrossRef]
50. Mankou, G.S.; Ligot, G.; Panzou, G.J.L.; Boyemba, F. Tropical tree allometry and crown allocation, and their relationship with species traits in central Africa. *For. Ecol. Manag.* **2021**, *493*, 119262. [CrossRef]
51. Eziz, A.; Yan, Z.B.; Tian, D.; Han, W.X.; Tang, A.Y.; Fang, J.Y. Drought effect on plant biomass allocation: A meta-analysis. *Ecol. Evol.* **2017**, *7*, 11002–11010. [CrossRef]
52. Li, J.L.; Wang, M.T.; Li, H.S.; Chen, X.P.; Sun, J.; Zhong, Q.L.; Cheng, D.L. Effects of Canopy Height on the Relationship Between Individual Leaf Mass and Leafing Intensity of 69 Broad Leaved Trees in Jiangxi Province. *Scientia Silvae Sinicae* **2021**, *57*, 62–71. [CrossRef]
53. Niklas, K.J.; Cobb, E.D.; Niinemets, U.; Reich, P.B.; Sellin, A.; Shipley, B.; Wright, I.J. "Diminishing returns" in the scaling of functional leaf traits across and within species groups. *Proc. Natl. Acad. Sci. USA* **2007**, *104*, 8891–8896. [CrossRef] [PubMed]
54. Gomez, R.S.; Perez, J.G.; Martin, M.D.L.; Garcia, C.G. Collinearity diagnostic applied in ridge estimation through the variance inflation factor. *J. Appl. Stat.* **2016**, *43*, 1831–1849. [CrossRef]
55. Kim, J.H. Multicollinearity and misleading statistical results. *Korean J. Anesthesiol.* **2019**, *72*, 558–569. [CrossRef]
56. Malaeb, Z.; Summers, J.; Pugesek, B. Using structural equation modeling to investigate relationships among ecological variables. *Environ. Ecol. Stat.* **2000**, *7*, 93–111. [CrossRef]
57. Gandhi, D.S.; Sundarapandian, S. Large-scale carbon stock assessment of woody vegetation in tropical dry deciduous forest of Sathanur reserve forest, Eastern Ghats, India. *Environ. Monit. Assess.* **2017**, *189*, 187. [CrossRef]
58. Becknell, J.M.; Kucek, L.K.; Powers, J.S. Aboveground biomass in mature and secondary seasonally dry tropical forests: A literature review and global synthesis. *For. Ecol. Manag.* **2012**, *276*, 88–95. [CrossRef]
59. Wang, X.M.; Guo, Z.W.; Guo, X.; Wang, X.P. The relative importance of succession, stand age and stand factors on carbon allocation of Korean Pine Forests in the Northern Mt. Xiaoxing'anling, China. *Forests* **2020**, *11*, 512. [CrossRef]
60. Kamara, M.; Said, S.M. Estimation of aboveground biomass, stand density, and biomass growth per year in the past using stand reconstruction technique in black spruce and Scotch pine in boreal forest. *Polar Sci.* **2022**, 33, 100787. [CrossRef]
61. Vygodskaya, N.N.; Schulze, E.D.; Tchebakova, N.M.; Karpachevskii, L.O.; Kozlov, D.; Sidorov, K.N.; Panfyorov, M.I.; Abrazko, M.A.; Shaposhnikov, E.S.; Solnzeva, O.N.; et al. Climatic control of stand thinning in unmanaged spruce forests of the southern taiga in European Russia. *Tellus B* **2002**, *54*, 443–461. [CrossRef]
62. Schulze, E.D.; Schulze, W.; Kelliher, F.M.; Vygodskaya, N.N.; Ziegler, W.; Koba, K.I.; Koch, H.; Arneth, A.; Kusnetsova, W.A.; Sogachev, A.; et al. Aboveground biomass and nitrogen nutrition in a chronosequence of pristine Dahurian Larix stands in Eastern Siberia. *Can. J. For. Res.* **1995**, *25*, 943–960. [CrossRef]
63. Mund, M.; Kummetz, E.; Hein, M.; Bauer, G.A.; Schulze, E.D. Growth and carbon stocks of a spruce forest chronosequence in central Europe. *For. Ecol. Manag.* **2002**, *171*, 275–296. [CrossRef]
64. Ruiz-Benito, P.; Madrigal-González, J.; Young, S.; Mercatoris, P.; Cavin, L.; Huang, T.J.; Chen, J.C.; Jump, A.S. Climatic stress during stand development alters the sign and magnitude of age-related growth responses in a subtropical mountain pine. *PLoS ONE* **2015**, *10*, e0126581. [CrossRef] [PubMed]
65. Sitcha, S.; Smithb, B.; Prenticec, I.C.; Arnethc, I.; Bondeaua, I.A.; Cramera, W.; Kapland, J.O.; Levisc, S.; Luchta, W.; Sykesb, M.T.; et al. Evaluation of ecosystem dynamics, plant geography and terrestrial carbon cycling in the LPJ dynamic global vegetation model. *Glob. Chang. Biol.* **2003**, *9*, 161–185. [CrossRef]
66. Li, Q.C.; Liu, Z.L.; Jin, G.Z. Impacts of stand density on tree crown structure and biomass: A global meta-analysis. *Agric. For. Meteorol.* **2022**, *326*, 109181. [CrossRef]
67. Wirth, C.; Schulze, E.D.; Luhker, B.; Grigoriev, S.; Siry, M.; Hardes, G.; Ziegler, W.; Backor, M.; Bauer, G.; Vygodskaya, N.N. Fire and site type effects on the long-term carbon balance in pristine Siberian Scots pine forest. *Plant Soil.* **2002**, *242*, 41–63. [CrossRef]

68. Franc, A. Bimodality for plant sizes and spatial pattern in cohorts: The role of competition and site conditions. *Theor. Popul. Biol.* **2001**, *60*, 117–132. [CrossRef]
69. Yang, B.Y.; Ali, A.; Xu, M.S.; Guan, M.S.; Li, Y.; Zhang, X.N.; He, X.M.; Yang, Y.D. Large plants enhance aboveground biomass in arid natural forest and plantation along differential abiotic and biotic conditions. *Front. Plant Sci.* **2022**, *13*, 999793. [CrossRef]
70. Fernandez-Nunez, E.; Rigueiro-Rodriguez, A.; Mosquera-Losada, M.R. Silvopastoral systems established with *Pinus radiata D. Don* and *Betula pubescens Ehrh.*: Tree growth, understorey biomass and vascular plant biodiversity. *Forestry* **2014**, *87*, 512–524. [CrossRef]
71. Morin, X.; Fahse, L.; Scherer-Lorenzen, M.; Bugmann, H. Tree species richness promotes productivity in temperate forests through strong complementarity between species. *Ecol. Lett.* **2011**, *14*, 1211–1219. [CrossRef]
72. Forrester, D.I.; Ammer, C.; Annighöfer, P.J.; Barbeito, I.; Bielak, K.; BravoOviedo, A.; Coll, L.; del Río, M.; Drössler, L.; Heym, M.; et al. Effects of crown architecture and stand structure on light absorption in mixed and monospecific fagus sylvatica and pinus sylvestris forests along a productivity and climate gradient through Europe. *J. Ecol.* **2018**, *106*, 746–760. [CrossRef]
73. Lutz, J.A.; Larson, A.J.; Swanson, M.E.; Freund, J.A.; Ben, B.L. Ecological importance of Large-diameter trees in a temperate mixed-conifer forest. *PLoS ONE* **2012**, *7*, e36131. [CrossRef]
74. Yachi, S.; Loreau, M. Does complementary resource use enhance ecosystem functioning? A model of light competition in plant communities. *Ecol. Lett.* **2007**, *10*, 54–62. [CrossRef] [PubMed]
75. Zhang, Y.; Chen, H.Y.H. Individual size inequality links forest diversity and above-ground biomass. *J. Ecol.* **2015**, *103*, 1245–1252. [CrossRef]
76. Slik, J.F.; Paoli, G.; McGuire, K.; Amaral, I.; Barroso, J.; Bastian, M.; Blanc, L.; Bongers, F.; Boundja, P.; Clark, C.; et al. Large Trees drive forest aboveground biomass variation in moist lowland forests across the tropics. *Glob. Ecol. Biogeogr.* **2013**, *22*, 1261–1271. [CrossRef]
77. Bastin, J.F.; Barbier, N.; Réjou-Méchain, M.; Fayolle, A.; Gourlet-Fleury, S.; Maniatis, D.; de Haulleville, T.; Baya, F.; Beeckman, H.; Beina, D.; et al. Seeing central African forests through their largest trees. *Sci. Rep.* **2015**, *5*, 13156. [CrossRef]
78. Lutz, J.A.; Furniss, T.J.; Johnson, D.J.; Davies, S.J.; Allen, D.; Alonso, A.; Anderson-Teixeira, K.J.; Andrade, A.; Baltzer, J.; Becker, K.M.L.; et al. Global importance of large-diameter trees. *Glob. Ecol. Biogeogr.* **2018**, *27*, 849–864. [CrossRef]
79. Mensah, S.; du Toit, B.; Seifert, T. Diversity–biomass relationship across forest layers: Implications for niche complementarity and selection effects. *Oecologia* **2018**, *187*, 783–795. [CrossRef]
80. Boucher, Y.; Auger, I.; Arseneault, D.; Elzein, T.; Sirois, L. Long-term (1925–2015) forest structure reorganization in an actively managed temperate-boreal forest region of eastern North America. *For. Ecol. Manag.* **2021**, *481*, 118744. [CrossRef]
81. Ouyang, S.; Xiang, W.H.; Wang, X.P.; Xiao, W.F.; Chen, L.; Li, S.D.; Sun, H.; Deng, X.W.; Forrester, D.I.; Zeng, L.X.; et al. Effects of stand age, richness and density on productivity in subtropical forests in China. *J. Ecol.* **2019**, *107*, 2266–2277. [CrossRef]
82. Fotis, A.T.; Murphy, S.J.; Ricart, R.D.; Krishnadas, M.; Whitacre, J.; Wenze, J.W.; Queenborough, S.A.; Comita, L.A. Above-ground biomass is driven by mass-ratio effects and stand structural attributes in a temperate deciduous forest. *J. Ecol.* **2018**, *106*, 561–570. [CrossRef]
83. Alves, L.F.; Vieira, S.A.; Scaranello, M.A.; Camargo, P.B.; Santos, F.A.M.; Joly, C.A.; Martinelli, L.A. Forest structure and live aboveground biomass variation along an elevational gradient of tropical Atlantic moist forest (Brazil). *For. Ecol. Manag.* **2010**, *260*, 679–691. [CrossRef]
84. Zhang, J.W.; Fiddler, G.O.; Young, D.H.; Shestak, C.; Carlson, R. Allometry of tree biomass and carbon partitioning in ponderosa pine plantations grown under diverse conditions. *For. Ecol. Manag.* **2021**, *497*, 119526. [CrossRef]
85. Cysneiros, V.C.; Souza, F.C.D.; Gaui, T.D.; Pelissari, A.L.; Orso, G.A.; Machado, S.D.A.; Carvalho, D.S.D.; Silveira-Filho, T.B. Integrating climate, soil and stand structure into allometric models: An approach of site-effects on tree allometry in Atlantic Forest. *Ecol. Indic.* **2021**, *127*, 107794. [CrossRef]
86. Taylor, A.R.; Gao, B.L.; Chen, H.Y.H. The effect of species diversity on tree growth varies during forest succession in the boreal forest of central Canada. *For. Ecol. Manag.* **2020**, *455*, 117641. [CrossRef]
87. Wittkea, S.; Yu, X.W.; Karjalainena, M.; Hyyppä, J.; Puttonen, E. Comparison of two-dimensional multitemporal Sentinel-2 data with three-dimensional remote sensing data sources for forest inventory parameter estimation over a boreal forest. *Int. J. Appl. Earth Obs.* **2019**, *76*, 167–178. [CrossRef]
88. Xu, W.R.; He, H.S.; Luo, X.; Tang, Z.Q.; Liu, K.; Cong, Y.; Gu, X.L.; Zong, S.W.; Du, H.B. Long-term effects of commercial harvest exclusion on forest structure and aboveground biomass in the Great Xing'an Mountains. China. *Acta Ecol. Sin.* **2018**, *38*, 1203–1215. [CrossRef]
89. Chave, J.; Condit, R.; Lao, S.; Caspersen, J.P.; Foster, R.B.; Hubbell, S.P. Spatial and temporal variation of biomass in a tropical forest: Results from a large census plot in Panama. *J. Ecol.* **2003**, *91*, 240–252. [CrossRef]
90. Urquiza-Haas, T.; Dolman, P.M.; Peres, C.A. Regional scale variation in forest structure and biomass in the Yucatan Peninsula, Mexico: Effects of forest disturbance. *For. Ecol. Manag.* **2007**, *247*, 80–90. [CrossRef]
91. Kenin, L.; Elferts, D.; Baders, E.; Jansons, A. Carbon pools in a hemiboreal over-mature Norway Spruce stands. *Forests* **2018**, *9*, 435. [CrossRef]
92. Azmana, M.S.; Sharma, S.; Shaharuddin, M.A.M.; Hamzah, M.L.; Adibah, S.N.; Zakaria, R.M.; MacKenzie, R.A. Stand structure, biomass and dynamics of naturally regenerated and restored mangroves in Malaysia. *For. Ecol. Manag.* **2021**, *482*, 118852. [CrossRef]

93. Schedlbauer, J.S.; Finegan, B.; Kavanagh, K.L. Rain Forest Structure at Forest-Pasture Edges in Northeastern Costa Rica. *Biotropica* **2007**, *39*, 578–584. [CrossRef]
94. Clark, D.B.; Clark, D.A. Landscape-scale variation in forest structure and biomass in a tropical rain forest. *For. Ecol. Manag.* **2000**, *137*, 185–198. [CrossRef]
95. Burrow, W.H.; Henry, B.K.; Back, P.V.; Hoffmann, M.B.; Tait, L.J. Growth and carbon stock change in eucalypt woodlands in northeast Australia: Ecological and greenhouse sink implications. *Clim. Chang. Biol.* **2002**, *8*, 769–784. [CrossRef]
96. Larsary, M.K.; Pourbabaei, H.; Sanaei, A.; Salehi, A.; Yousefpour, R.; Ali, A. Tree-size dimension inequality shapes aboveground carbon stock across temperate forest strata along environmental gradients. *For. Ecol. Manag.* **2021**, *496*, 119482. [CrossRef]
97. Aiba, S.; Hillb, D.A.; Agetsuma, N. Comparison between old-growth stands and secondary stands regenerating after clear-felling in warm-temperate forests of Yakushima, southern Japan. *For. Ecol. Manag.* **2001**, *140*, 163–175. [CrossRef]
98. Mensah, S.; Noulekoun, F.; Ago, E.E. Aboveground tree carbon stocks in West African semi-arid ecosystems: Dominance patterns, size class allocation and structural drivers. *Glob. Ecol. Conserv.* **2020**, *24*, e01331. [CrossRef]
99. Raich, J.W.; Russell, A.E.; Kitayama, K.; Parton, W.J.; Vitousek, P.M. Temperature influences carbon accumulation in moist tropical forests. *Ecology* **2006**, *87*, 76–87. [CrossRef] [PubMed]
100. Cerny, J.; Pokorny, R.; Vejpustkova, M.; Sramek, V.; Bednar, P. Air temperature is the main driving factor of radiation use efficiency and carbon storage of mature Norway spruce stands under global climate change. *Int. J. Biometeorol.* **2020**, *64*, 1599–1611. [CrossRef]
101. Larcher, W. *Physiological Plant Ecology: Ecophysiology and Stress Physiology of Functional Groups*; Springer Science & Business Media: Berlin/Heidelberg, Germany, 2003.
102. Silva, P.S.; Bastos, A.; Libonati, R.; Rodrigues, J.A.; DaCamara, C.C. Impacts of the 1.5 °C global warming target on future burned area in the brazilian cerrado. *For. Ecol. Manag.* **2019**, *446*, 193–203. [CrossRef]
103. Silva, P.; Bastos, A.; DaCamara, C.C.; Libonati, R. Future projections of fire occurrence in brazil using EC-earth climate model. *Rev. Bras. Meteorol.* **2016**, *31*, 288–297. [CrossRef]
104. Blundo, C.; Malizia, A.; Malizia, L.R.; Lichstein, J.W. Forest biomass stocks and dynamics across the subtropical Andes. *Biotropica* **2021**, *53*, 170–178. [CrossRef]
105. Chi, C.H.; McEwan, R.W.; Chang, C.T.; Zheng, C.Y.; Yang, Z.J.; Chiang, J.M.; Lin, T.C. Typhoon disturbance mediates elevational patterns of forest structure, but not species diversity, in humid monsoon Asia. *Ecosystems* **2015**, *18*, 1410–1423. [CrossRef]
106. Xiong, X.Y.; Zhu, J.L.; Li, S.; Fan, F.; Cai, Q.; Ma, S.H.; Su, H.J.; Ji, C.J.; Tang, Z.Y.; Fang, J.Y. Aboveground biomass and its biotic and abiotic modulators of a main food bamboo of the giant panda in a subalpine spruce-fir forest in southwestern China. *J. Plant Ecol.* **2022**, *15*, 1–12. [CrossRef]
107. Yang, X.Q.; Blagodatsky, S.; Liu, F.; Beckschäferf, P.; Xu, J.C.; Cadisch, G. Rubber tree allometry, biomass partitioning and carbon stocks in mountainous landscapes of sub-tropical China. *For. Ecol. Manag.* **2017**, *404*, 84–99. [CrossRef]
108. Paoli, G.D.; Curran, L.M.; Slik, J.W.F. Soil nutrients affect spatial patterns of aboveground biomass and emergent tree density in southwestern Borneo. *Oecologia* **2008**, *155*, 287–299. [CrossRef]
109. Gao, Y.; Skutsch, M.; Rodríguez, D.L.J.; Solórzano, J.V. Identifying variables to discriminate between conserved and degraded forest and to quantify the differences in Biomass. *Forests* **2020**, *11*, 1020. [CrossRef]
110. Bennett, A.C.; Penman, T.D.; Arndt, S.K.; Roxburgh, S.H.; Bennett, L.T. Climate more important than soils for predicting forest biomass at the continental scale. *Ecography* **2020**, *43*, 1692–1705. [CrossRef]
111. Huang, X.; Huang, C.; Teng, M.; Zhou, Z.; Wang, P. Net primary productivity of Pinus massoniana dependence on climate, soil and forest characteristics. *Forests* **2020**, *11*, 404. [CrossRef]
112. Coomes, D.A.; Flores, O.; Holdaway, R.; Jucker, T.; Lines, E.R.; Vanderwel, M.C. Wood production response to climate change will depend critically on forest composition and structure. *Glob. Chang. Biol.* **2014**, *20*, 3632–3645. [CrossRef]
113. Stegen, J.C.; Swenson, N.G.; Enquist, B.J.; White, E.P.; Phillips, O.L.; Jorgensen, P.M.; Weiser, M.D.; Mendoza, A.M.; Vargas, P.N. Variation in above-ground forest biomass across broad climatic gradients. *Glob. Ecol. Biogeogr.* **2011**, *20*, 744–754. [CrossRef]
114. Chu, C.J.; Bartlett, M.; Wang, Y.S.; He, F.L.; Weiner, J.; Chave, J. Does climate directly influence NPP globally? *Glob. Chang. Biol.* **2016**, *22*, 12–24. [CrossRef]
115. Lohbeck, M.; Bongers, F.; Martinez-Ramos, M.; Poorter, L. The importance of biodiversity and dominance for multiple ecosystem functions in a human-modified tropical landscape. *Ecology* **2016**, *97*, 2772–2779. [CrossRef]
116. Sichone, P.; De Cauwer, V.; Chissungui, A.V.; Goncalves, F.M.P.; Finckh, M.; Revermann, R. Patterns of above-ground biomass and its environmental drivers: An analysis based on plot-based surveys in the dry tropical forests and woodlands of southern Africa. *Biodivers. Ecol.* **2018**, *6*, 309–316. [CrossRef]

forests

MDPI

Article

The Influence of Stand Structure on Understory Herbaceous Plants Species Diversity of *Platycladus orientalis* Plantations in Beijing, China

Ranran Cui [1], Shi Qi [1,*], Bingchen Wu [2], Dai Zhang [1], Lin Zhang [1], Piao Zhou [1], Ning Ma [1] and Xian Huang [1]

1 School of Soil and Water Conservation, Beijing Forestry University, Beijing 100083, China
2 Jiangxi Academy of Water Science and Engineering, Nanchang 330029, China
* Correspondence: qishi@bjfu.edu.cn; Tel.: +86-135-2204-6290

Abstract: Species diversity is a crucial index used to evaluate the stability and complexity of forest ecosystems. Studying the relationship between stand structure and understory herbaceous plants species diversity is useful for managers to formulate the best forest structure optimization method with the goal of improving herbaceous species diversity. In this research, *Platycladus orientalis* plantations in Beijing were taken as the research object. Pearson's correlation analysis was used to explore the single-factor correlation between stand structure and understory herbaceous plants species diversity; furthermore, a typical correlation analysis and multiple linear regression were used to explore the multi-factor correlation and analyze the dominant stand structure parameters affecting understory herbaceous plants species diversity. In the range of stand structures studied, the results showed that canopy density was negatively correlated with the Shannon–Wiener index and Simpson index ($p < 0.01$), and tree density was negatively correlated with the Shannon–Wiener index ($p < 0.05$). In terms of stand spatial structure, the mingling degree was positively correlated with the Shannon–Wiener index, Simpson index, Margalef richness index and Pielou evenness index ($p < 0.05$), while the uniform angle was negatively correlated with the Pielou evenness index ($p < 0.05$). The correlation coefficient of the first group of typical variables in the typical correlation analysis was 0.90 ($p < 0.05$); from this group of typical variables, it can be concluded that canopy density is the most influential indicator affecting the comprehensive index of understory herbaceous plants species diversity, with a load of −0.690, and the Shannon–Wiener index and Simpson index are the most responsive indicators of changes in the comprehensive index of stand structure, with loads of 0.871 and 0.801, respectively. In the process of the management of *Platycladus orientalis* plantations under a low altitude, south slope, thin soil layer and hard soil parent material, in order to improve the herbaceous species diversity, the canopy density of the overstory and tree density should be appropriately reduced. Additionally, it is necessary to regulate the horizontal spatial structure of stands. When the trees are randomly distributed and the mingling degree is high, the species diversity of herbs can be increased.

Keywords: stand structure; understory vegetation; species diversity; canonical correlation analysis

Citation: Cui, R.; Qi, S.; Wu, B.; Zhang, D.; Zhang, L.; Zhou, P.; Ma, N.; Huang, X. The Influence of Stand Structure on Understory Herbaceous Plants Species Diversity of *Platycladus orientalis* Plantations in Beijing, China. *Forests* **2022**, *13*, 1921. https://doi.org/10.3390/f13111921

Academic Editor: Bruno Foggi

Received: 19 October 2022
Accepted: 13 November 2022
Published: 15 November 2022

1. Introduction

Species diversity has long been a prevalent issue in the area of forest ecology [1–3], and species diversity, representing one of the key indicators used to evaluate the complexity and stability of forest ecosystems, and it can help to maintain the function of the forest ecosystem. Many studies have shown that the increase in species diversity can improve the stability and herbaceous plants productivity of plantations [4,5]. As an important component of forest communities, understory vegetation plays an indispensable role in improving the physical and chemical properties of soil, preventing soil erosion, and maintaining ecosystem functions [6,7]. The understory herbaceous plants species diversity is influenced by factors such as stand structure, site conditions and climate, among which the variability of water, heat and light, and soil nutrients in understory vegetation caused by different stand structures is a key factor affecting the understory herbaceous plants species diversity on a small scale [8,9]. Therefore, it is important to investigate factors of stand structure that affect understory herbaceous plants species diversity.

Stand structure is divided into the following two parts: the stand non-spatial structure and stand spatial structure. Currently, most of the research on the factors affecting species diversity still focus on environmental factors [10–13], and relatively few studies have been conducted on the effects of stand spatial structure and non-spatial structure on species diversity. When environmental factors do not seriously disturb the community, stand structure influences the number and composition of species in the understory. The stand structure indices such as canopy density and tree density have an effect on species richness and diversity [14,15]. It is important to understand how stand structure affects understory herbaceous plants species diversity, as it is of great significance for the rational management and nurturing of plantations. Stand spatial structure is a description of the spatial pattern between trees within a forest ecosystem, reflecting spatial characteristic information of stands in the arbor layer in the horizontal and vertical directions [16], and it is more focused on reflecting the spatial distribution characteristics of stands. The stand spatial structure cannot only determine the competition intensity between neighboring trees, but also the spatial niche among trees [17]. Changes in the spatial structure characteristics of the arbor layer contribute to the spatial heterogeneity of the stand; thus, the understory vegetation habitat changes, which ultimately has an impact on the understory herbaceous plants species diversity [18,19]. Therefore, exploring the reasonable stand structure is of great significance for improving the species diversity of the herbaceous layer, enhancing the stability of plantations and allowing the functions of forests to proceed.

As an ecological construction project, the Beijing–Tianjin Sand-storm Source Control II Project is of great strategic significance in China [20]. Its main aim is to protect and improve vegetation coverage, and the project adjusts stand structure as a way to improve the understory herbaceous plants species diversity. Following the implementation of the Beijing–Tianjin Sand-storm Source Control II Project from 2013 to 2021, the total area of the project in Beijing now amounts to about 732.5 km^2, of which the area of *Platycladus orientalis* plantations is about 380.08 km^2, accounting for 52% of the total area of the project. The *Platycladus orientalis* plantations of the project are mainly located in low altitude, south slope, thin soil layer and hard soil parent material areas, because *Platycladus orientalis* tends to grow towards the sun in a low altitude, and it can adapt to the barren environment. In this study, in order to explore methods of forest management which can be used to effectively improve the understory herbaceous plants species diversity, 16 sample plots located in areas of a low altitude, south slope, thin soil layer and hard soil parent material were selected based on the Beijing–Tianjin Sand-storm Source Control II Project to investigate the relationship between the stand structure and understory herbaceous plants species diversity. The aim was to provide a basis for the adjustment of the stand structure of *Platycladus orientalis* plantations so as to improve the stability of the plantations.

2. Materials and Methods

2.1. Overview of the Study Area

The study area was located in the mountainous area of Beijing (Figure 1), and its geographical coordinates are 39°28′~41°05′ N, 115°25′~117°30′ E. The area has a warm, temperate semi-humid and semi-arid monsoon climate, with four distinct seasons. Its average annual temperature is 13–14 °C, with a frost-free period of 180–200 days, and annual average precipitation of 470–655 mm. Along the elevation gradient, from high to low, the soil types are Eutric Cambisols, Chromic Cambisols and Gleyic Cambisols. The vegetation types are mainly coniferous forests and broad-leaved forests, among which *Pinus tabuliformis* and *Platycladus orientalis* are the dominant species in coniferous forests. The forest in this study belongs to *Platycladus orientalis* plantations, the soil is Chromic

Cambisols, and the tree species mainly include *Platycladus orientalis*, *Armeniaca sibirica* and *Cotinus coggygria*.

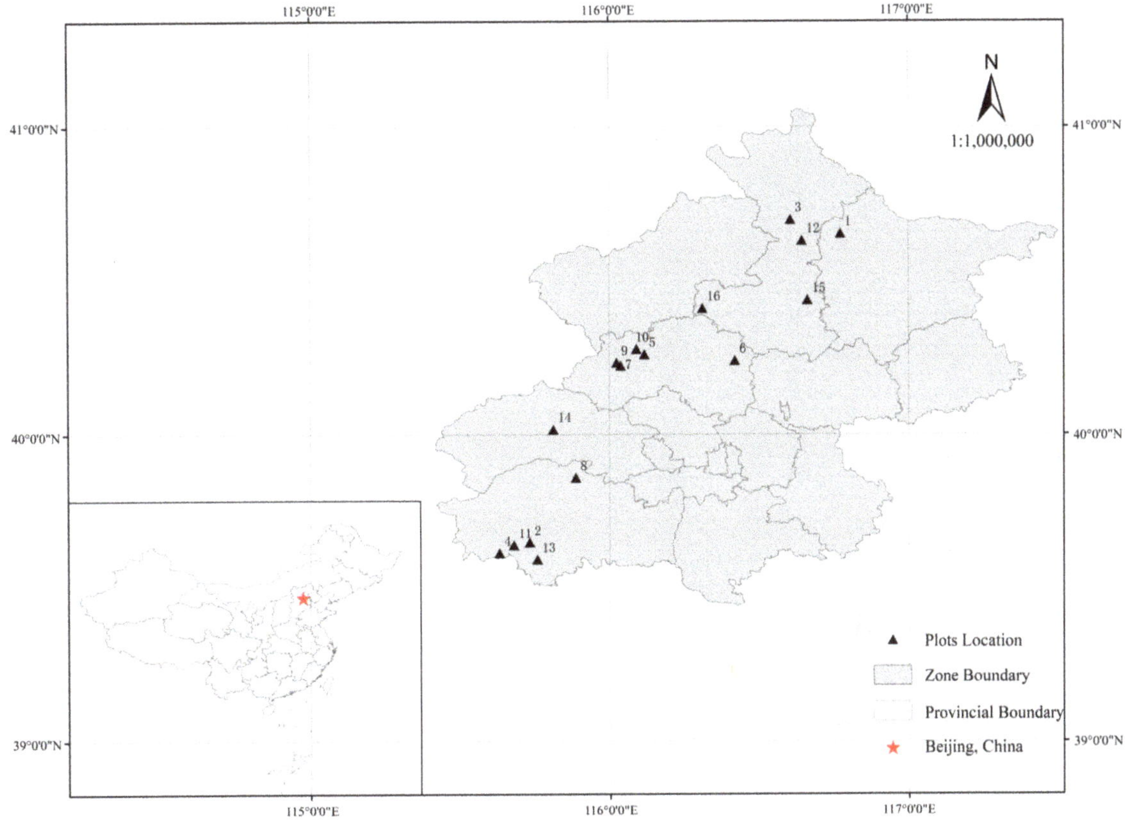

Figure 1. Overview of the study area.

2.2. Sample Plot Setting

The Beijing–Tianjin Sand-storm Source Control II Project mainly includes low-quality and inefficient forests within its transformation project, alongside a closed forests project and difficult site afforestation project. The forests selected in this study were closed from 2018 to 2021, and were middle-aged and near-mature forests planted in the last century through tree planting. The closure measures mainly include setting up closure fences, closure signs and arranging forest patrollers, while forestry measures adopted artificial means to promote natural regeneration. Of the 80 sub-compartments in the same condition, 20% were selected to set up sample plots. The 16 sample plots were established in 2021 in the sub-compartments of the closed forest of the project. Based on the results of available forestry surveys, sample plots with a similar altitude, slope, aspect, soil thickness and soil texture were selected. Sample plots were located in areas with a low altitude (100–300 m), south slope (SS), thin soil layer (0–30 cm) and hard soil parent material. Sixteen standard plots with *Platycladus orientalis* as the dominant tree species were selected, and the accompanying tree species were *Armeniaca sibirica* and *Cotinus coggygria*, with an area of 20 m × 20 m. Each tree in the sample plot was measured, and the investigation factors included tree species composition, DBH, tree height, the coordinates of each tree and other factors. Among them, canopy density was measured using the sample line method, the sample line was set along the diagonal of the sample plot to calculate the total length of the canopy falling on the sample line, and the canopy density was equal to the ratio of the total length of the canopy to the total length of the sample lines. The tree density can be obtained from the data of the forest survey. The study procedures are as follows: set five 1 m × 1 m herbaceous samples in the corners and the middle of each plot, and record the species names, the number of plants and the topographic information of each plot, including altitude, slope and aspect. The basic information of each sample plot is provided in Table 1.

Table 1. Basic information of each sample plot.

Plot	Altitude (m)	Slope (°)	Aspect	Tree Height (m)	DBH (cm)	Canopy Density	Tree Density (Stem·ha^{-1})	Age (a)	Proportion of Dominant Tree Species (%)
1	272	23	SS	6.1	7.1	0.8	1125	49	56
2	223	16	SS	5.2	7.6	0.5	1000	45	50
3	203	24	SS	4.0	6.7	0.6	975	45	97
4	297	19	SS	4.8	9.6	0.6	1175	60	72
5	120	16	SS	7.4	11.0	0.8	1500	80	57
6	181	22	SS	3.6	6.0	0.6	1575	40	65
7	225	21	SS	7.2	10.9	0.7	1425	78	79
8	285	24	SS	4.0	7.2	0.7	1100	53	61
9	166	25	SS	7.1	8.0	0.8	2125	69	93
10	125	17	SS	7.4	12.1	0.7	1300	80	85
11	155	24	SS	4.9	9.6	0.5	1550	65	89
12	100	18	SS	6.2	8.4	0.8	1175	50	68
13	160	22	SS	4.8	9.0	0.5	1300	52	81
14	229	22	SS	4.6	6.5	0.6	1250	34	56
15	199	24	SS	5.5	8.9	0.8	1275	76	96
16	239	23	SS	3.9	6.1	0.6	1200	36	96

2.3. *Research Methods*

2.3.1. Stand Structure Parameters

Stand structure is mainly divided into stand spatial structure and stand non-spatial structure parameters. According to the coordinates and DBH of trees in the sample plot, the method of Winklemass [21] and Excel were used to calculate the stand spatial structure parameters in various plots.

In this study, in terms of stand non-spatial structure parameters, the following four indices were adopted: tree height, DBH, canopy density and tree density, and in terms of stand spatial structure parameters, the following four indices were adopted: uniform angle, neighborhood comparison, mingling degree and opening degree. The stand spatial structure is based on the stand spatial structure unit. Each tree in the forest and its adjacent trees (n) can form a stand spatial structure unit. To avoid the adjacent trees of the object tree on the boundary falling outside the sample plots and thus affecting the results, a 5 m buffer zone was set, the tree in the buffer zone could not be set as the object tree, and the number of adjacent trees n was 4. The schematic diagram of the sample plot is shown in the Figure 2.

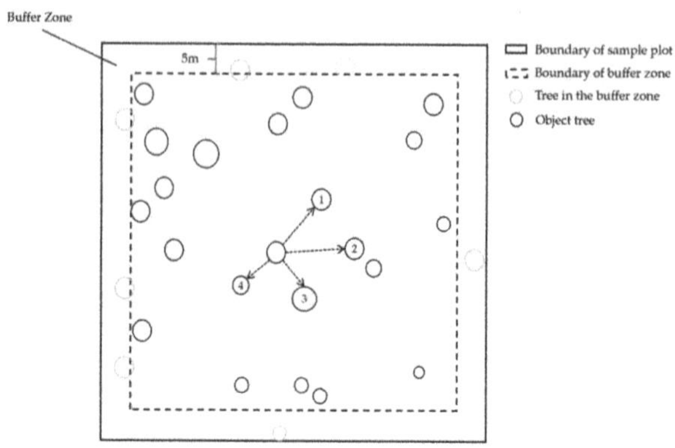

Figure 2. Schematic diagram of the sample plot. Note: Adjacent trees n = 4, Buffer zone = 5 m.

The formula and the meaning of each parameter are as follows:

(1) Uniform angle is an index reflecting the horizontal spatial distribution pattern of trees [22], and it was calculated according to Equation (1):

$$W_i = \frac{1}{n}\sum_{j=1}^{n} Z_{ij} \quad (1)$$

where W_i is the uniform angle of object tree i, n is the number of adjacent trees. When the angle of adjacent trees is less than the standard angle, Z_{ij} = 1; otherwise, Z_{ij} = 0, and the standard angle is 360°/(n + 1).

(2) Neighborhood comparison is an index reflecting the degree of differentiation of the object tree (DBH, tree height and crown width, etc.) [23], in order to express the spatial dominance of the stand, and it was calculated using Equation (2):

$$U_i = \frac{1}{n}\sum_{j=1}^{n} K_{ij} \quad (2)$$

where U_i is the neighborhood comparison of object tree i, and n is the number of adjacent trees. When the DBH of the object tree i is smaller than the adjacent tree j, K_{ij} = 0; otherwise, K_{ij} = 1.

(3) The mingling degree is an index reflecting the degree of isolation between trees [24], and it was calculated using Equation (3):

$$M_i = \frac{1}{n}\sum_{j=1}^{n} V_{ij} \quad (3)$$

where M_i is the mingling degree of the object tree i, and n is the number of adjacent trees. When the object tree i is not the same species as the adjacent tree j, V_{ij} = 1; otherwise, V_{ij} = 0.

(4) The opening degree is an index reflecting the light transmission conditions in the forest [25], and it was calculated using Equation (4):

$$K_i = \frac{1}{n}\sum_{j=1}^{n} \frac{D_{ij}}{H_{ij}} \quad (4)$$

where K_i is the opening degree of object tree i, n is the number of adjacent trees. D_{ij} is the horizontal distance between the object tree i and the adjacent tree j, and H_{ij} is the tree height of the adjacent tree j.

2.3.2. Indices of Understory Herbaceous Plants Species Diversity

In this study, the Simpson index, Shannon–Wiener index, Margalef richness index and Pielou evenness index [26] were selected to reflect the understory herbaceous plants species diversity.

(1) Margalef richness index

The Margalef richness index is an index reflecting the number of species in a community, and it was calculated using Equation (5):

$$R' = \frac{S-1}{\ln N} \quad (5)$$

(2) Simpson index

The Simpson index, indicates the probability that two individuals are randomly selected from the community and not put back, and that two herbaceous individuals belong to the same species. The greater the probability that two herbaceous individuals belong to the same species, the higher the concentration of herbaceous individuals and the lower the species diversity. It was calculated by using Equation (6):

$$D = 1 - \sum_{i=1}^{s} P_i^2 \tag{6}$$

(3) Shannon–Wiener index

The Shannon–Wiener index applies an information-theoretic approach to predict which species the next collected individual belongs to; and if the community species diversity is higher, the greater the uncertainty in predicting the next individual is. It was calculated by using Equation (7):

$$H' = -\sum_{i=1}^{s} P_i \ln P_i \tag{7}$$

(4) Pielou evenness index

The Pielou evenness index reflects the distribution of individuals of all species in a community, and the higher the species evenness index, the more evenly the number of individuals of each species is distributed. It was calculated by using Equation (8):

$$J_{sw} = -\sum_{i=1}^{s} \frac{P_i \ln P_i}{\ln S} \tag{8}$$

where S is the number of species, N is the total number of individuals of all species, and P_i is the ratio of the number of the ith species to the overall number of species.

2.3.3. Data Processing

In this paper, we adopted the Pearson correlation analysis, multiple linear regression and typical correlation analysis to explore the relationships between stand structure and understory herbaceous plants species diversity with SPSS.

(1) Pearson's correlation analysis

The Pearson correlation coefficient [27] was used to calculate the univariate correlation between stand structure and understory herbaceous plants species diversity with the following Equation (9):

$$r = \frac{\Sigma(x - \overline{x})(y - \overline{y})}{\Sigma(x - \overline{x})^2 \Sigma(y - \overline{y})^2} \tag{9}$$

where r is correlation coefficient, x and y are samples, and $\overline{x}$ and $\overline{y}$ are the average of samples.

(2) Canonical Correlation Analysis

Canonical Correlation Analysis (CCA) [28] is a mathematical method used to measure the correlation between two sets of multiple variables, where one set of variables is the stand structure indices and the other is understory herbaceous plants species diversity indices, assuming that the two sets of variables are as follows:

$$\begin{cases} X = (x_1, x_2, \ldots, x_n) \\ Y = (y_1, y_2, \ldots, y_n) \end{cases} \tag{10}$$

One or more representative comprehensive variables, Z_i and V_i, were selected among the two sets of variables. This comprehensive variable can be used to replace the original variable. The equation is as follows:

$$\begin{cases} Z_i = a_1x_1 + a_2x_2 + \ldots + a_nx_n \\ V_i = b_1x_1 + b_2x_2 + \ldots + b_nx_n \end{cases} \tag{11}$$

The correlation between X and Y was explored using the relationship between linear combinations of Z_i and V_i to find the best vectors *a* and *b* to maximize the correlation coefficient between Z_i and V_i.

(3) Multiple linear regression

Stepwise regression was conducted to establish a multiple linear regression model with species diversity indices as the dependent variable and stand structure indices as the independent variables, and the equation was as follows.

$$y = \beta_0 + \beta_1 x_1 + \beta_2 x_2 + \ldots + \beta_p x_p + \mu \quad (12)$$

where y is the dependent variable, x_p is the independent variable, β_p is a constant, and μ is residual.

3. Results and Analysis

3.1. Basic Characteristics of Stand Structure and Species Diversity

There were 56 herbaceous plants species in the selected sample plots, involving 26 families and 50 genera, among which *Asteraceae* was the most abundant, accounting for 27% of the total number of herbaceous plants species, followed by *Poaceae* and *Rosaceae.* In terms of herbaceous plants abundance, the top five species were *Artemisia carvifolia* (relative abundance is 8.8), *Carex hancockiana* (relative abundance is 8.6), *Setaria viridis* (relative abundance is 5.5), *Cleistogenes caespitosa* (relative abundance is 5.4) and *Chenopodium album* (relative abundance is 5.2), which belonged to *Asteraceae, Cyperaceae, Poaceae, Poaceae,* and *Chenopodiaceae,* respectively.

The basic information of stand structure parameters (stand non-spatial structure and stand spatial structure) and the understory herbaceous plants species diversity index sample plots is shown in Table 2, from which it can be seen that the coefficient of variation of most of the indices was large. Additionally, the coefficient of variation of the mingling degree was the largest at 67.9%. The variation in canopy density, uniform angle and neighborhood comparison was small among the sample plots, and the coefficients of variation which were 16.7%, 6.9% and 4.9%, respectively.

Table 2. Basic information of stand structure and species diversity index of the sampling plots.

Category	Index	Mean ± Standard Deviation	Minimum	Maximum	Coefficient of Variation (%)
Stand non-spatial structure	Tree height (m)	5.4 ± 1.3	3.6	7.4	24.6%
	DBH (cm)	8.4 ± 1.9	6.0	12.1	21.9%
	Canopy density	0.7 ± 0.1	0.5	0.8	16.7%
	Tree density (stem·ha^{-1})	1315 ± 280	975	2125	21.3%
Stand spatial structure	Uniform angle	0.5053 ± 0.0347	0.4186	0.5391	6.9%
	Neighborhood comparison	0.5044 ± 0.0247	0.4571	0.5379	4.9%
	Mingling degree	0.3075 ± 0.2087	0.0278	0.6667	67.9%
	Opening degree	0.5945 ± 0.2747	0.2683	1.4281	46.2%
Species diversity	Shannon–Wiener index	0.8220 ± 0.2748	0.2752	1.2636	33.4%
	Simpson index	0.4853 ± 0.1390	0.1980	0.6342	28.6%
	Margalef richness index	0.7537 ± 0.2678	0.3913	1.3815	35.5%
	Pielou evenness index	0.7825 ± 0.1602	0.3970	0.9518	20.5%

3.2. Univariate Correlation Analysis between Stand Structure and Understory Herbaceous Plants Species Diversity

The univariate correlation analysis between stand structure and understory herbaceous plants species diversity was performed using the Pearson correlation coefficient. The result, as shown in Figure 3, indicates that tree height and DBH were not significantly correlated with understory herbaceous plants species diversity ($p > 0.05$). In Figures 3 and 4, in the range of the stand structure studied, the canopy density was significantly negatively correlated with the Shannon–Wiener index and Simpson index ($p < 0.01$), and tree density was significantly negatively correlated with the Shannon–Wiener index ($p < 0.05$). In terms of stand spatial structure, the neighborhood comparison and opening degree were not significantly correlated with understory herbaceous plants species diversity ($p > 0.05$). The mingling degree was significantly positively correlated with all four indices, the Shannon–Wiener index, Simpson index, Margalef richness index and Pielou evenness index ($p < 0.05$), and the uniform angle was significantly negatively correlated with the Pielou index ($p < 0.05$).

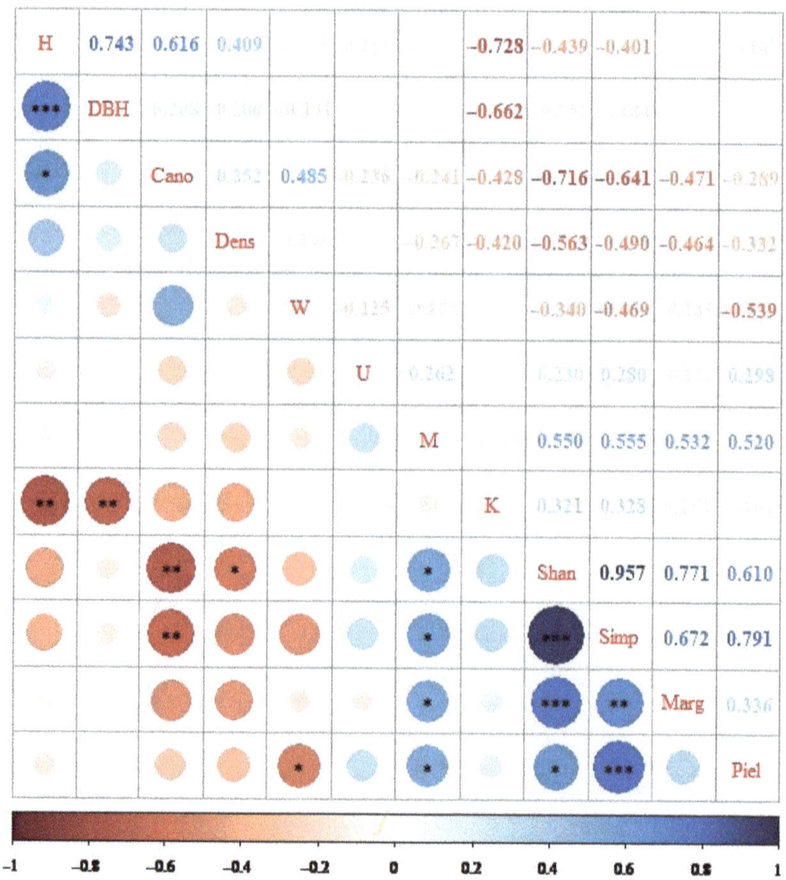

Figure 3. Correlation coefficients between stand structure and understory herbaceous plants species diversity. Note: ***: $p < 0.001$; **: $p < 0.01$; *: $p < 0.05$. H—Tree Height; Cano—Canopy density; Dens—Tree density; W—Uniform angle; U—Neighborhood comparison; M—Mingling degree; K—Opening degree; Shan—Shannon–Wiener index; Simp—Simpson index; Marg—Margalef richness index; Piel—Pielou evenness index.

3.3. *Typical Correlation Analysis between Stand Structure and Understory Herbaceous Plants Species Diversity*

The four stand structure indices that significantly influenced the understory herbaceous plants species diversity index were derived from a Pearson correlation analysis, namely the canopy density, tree density, uniform angle and mingling degree. The four stand structure indices were considered as one set of variables (U_1), and the four species diversity indices were considered as another set of variables (V_1). The linear relationship between the two sets of variables was obtained by performing a typical correlation analysis.

It is shown in Table 3 that the typical correlation coefficient between the first group of typical variables was 0.907 and the results passed the typical correlation coefficient test at the $\alpha = 0.05$ level, indicating that there was a strong correlation between the two groups of variables; therefore, the group of variables consisting of the four indices of stand structure could be used to explain the group of understory herbaceous plants species' diversity variables. The correlation coefficients between the typical variables of the second, third and fourth group were 0.724, 0.453 and 0.137. The four indices (canopy density, tree density, uniform angle and mingling degree) were considered in all three groups; however, all three groups of typical variables did not pass the significance test. Therefore, only the first group of typical variables was analyzed.

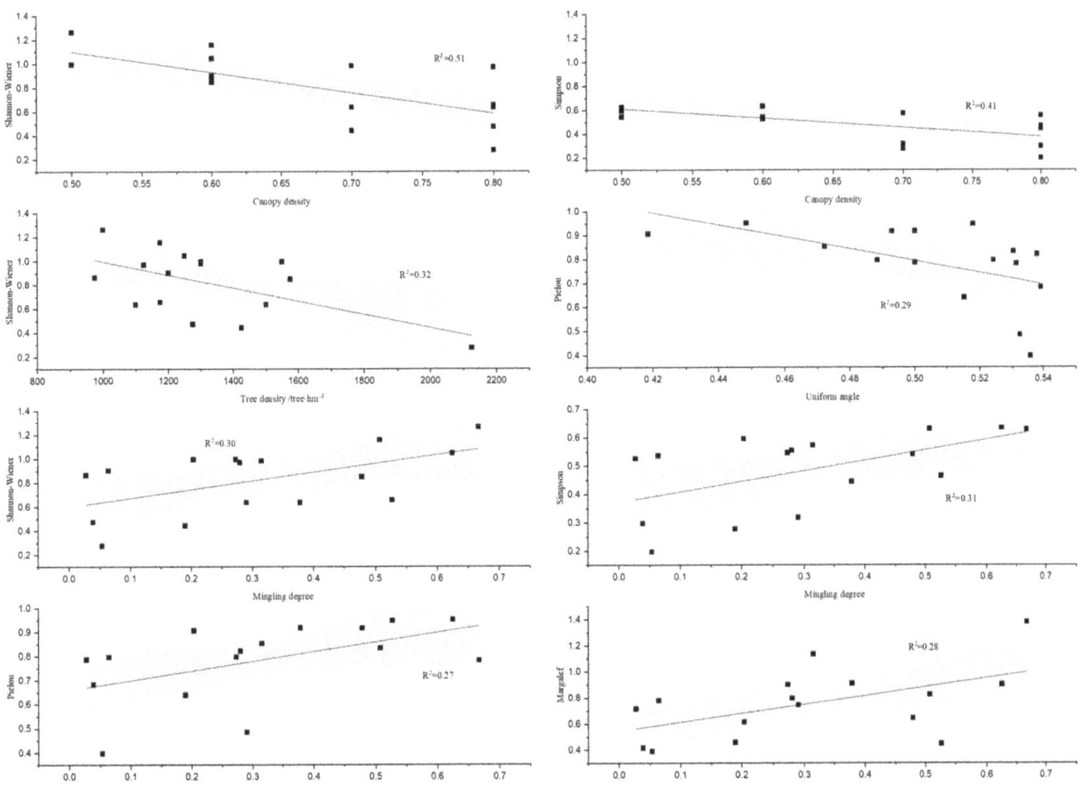

Figure 4. Correlation plots for cases that had a significant correlation between stand structure and understory herbaceous plants species diversity.

Table 3. Canonical correlation coefficient and its significance test.

Typical Variable Group	Canonical Correlation Coefficient	*p*
1	0.907	0.033
2	0.724	0.327
3	0.453	0.627
4	0.137	0.656

As can be seen from the standardized coefficients in Table 4, the equation for the first group of typical variables was as follows:

$$\begin{cases} Z_1 = -0.536X_1 - 0.473X_2 - 0.022X_3 + 0.388X_4 \\ V_1 = 2.827Y_1 - 2.331Y_2 - 0.107Y_3 + 0.733Y_4 \end{cases} \quad (13)$$

Table 4. Standardization coefficient and cross-load coefficient of typical variables.

Variable		X_1	X_2	X_3	X_4	Y_1	Y_2	Y_3	Y_4
Standardization coefficient	Z_1	−0.536	−0.473	−0.022	0.388	-	-	-	-
	V_1	-	-	-	-	2.827	−2.331	−0.107	0.733
Cross-load coefficient	Z_1	−0.690	−0.644	−0.246	0.587	-	-	-	-
	V_1	-	-	-	-	0.871	0.801	0.682	0.525

Note: X_1—canopy density; X_2—Density; X_3—Uniform angle; X_4—Mingling degree; Y_1—Shannon–Wiener index; Y_2—Simpson index; Y_3—Margalef richness index; Y_4—Pielou evenness index.

In the linear combination Z_1, the loads of X_1 (canopy density) and X_2 (tree density) were larger, at −0.536 and −0.473, respectively, indicating that two stand non-spatial structure indices, canopy density and tree density, were the dominant factors in the comprehensive index of stand structure. In the linear combination V_1, Y_1 (Shannon–Wiener index) and Y_2 (Simpson index) had larger loads of 2.827 and −2.331, respectively, indicating that the Shannon–Wiener index and Simpson index played a decisive role in the comprehensive index of understory herbaceous plants species diversity. Additionally, Y_3 (Margalef richness index) had the smallest load of −0.107, indicating that the species richness index had a small weight in the understory herbaceous plants species diversity comprehensive index.

From the cross-loading coefficients in Table 4, it can be seen that the loads of X_1 (canopy density), X_2 (tree density) and X_4 (mingling degree) in Z_1 were larger, at −0.690, −0.644 and 0.587, respectively. The loads of Y_1 (Shannon–Wiener index) and Y_2 (Simpson index) in V_1 were the largest, at 0.871 and 0.801, respectively. It indicated that the canopy density and tree density of the stand non-spatial structure played a key role in affecting understory herbaceous plants species diversity, the mingling degree of stand spatial structure had a greater influence on understory herbaceous plants species diversity, and the Shannon–Wiener index and Simpson index are the most sensitive to the changes to the stand structure.

3.4. Multiple Stepwise Regression Results

Regression equations for each species diversity index were obtained using stepwise regression, reflecting the stand structure indices which were significantly associated with understory herbaceous plants species diversity, and four main explanatory variables affecting understory herbaceous plants species diversity were retained. Each regression equation passed the significance test ($p < 0.05$).

As can be seen from Table 5, among the stand non-spatial structure indices, the regression equation of the Shannon–Wiener index ($R^2 = 0.765$) retained two factors, canopy density and tree density, indicating that canopy density and tree density played a key role in affecting the Shannon–Wiener index. The regression equation of the Simpson index ($R^2 = 0.581$) retained the canopy density with its standardized coefficient of −0.539, which had a considerable effect on the Simpson index. Canopy density and tree density were excluded from the stepwise regression equations of the Margalef richness index and Pielou evenness index. Tree height and DBH were excluded from the stepwise regression equations of all species' diversity indices, indicating that the effect of DBH and tree height on the understory herbaceous plants species diversity was small.

Among the stand spatial structure indices, the mingling degree could be used in the regression equations of the Shannon–Wiener index, Simpson index, Margalef richness index and Pielou evenness index, indicating that the mingling degree had a significant effect on the diversity, richness and evenness of understory vegetation species. The uniform angle was chosen in the regression equation of the Pielou evenness index (R^2 = 0.486), indicating that the uniform angle index influenced the Pielou evenness index, and the standardized coefficient of the uniform angle in the regression equation was −0.471. The neighborhood comparison and the opening degree were excluded from the regression equations of all species diversity indices, indicating that the species diversity indices were not correlated with the neighborhood comparison and the opening degree.

Table 5. Stepwise regression equation of species diversity index.

Species Diversity Index	Regression Equation	R^2	p	AIC
Shannon–Wiener	y = −0.552 × Canopy − 0.336 × Density + 0.327 × M	0.765	0.000	−16.176
Simpson	y = −0.539 × Canopy + 0.425 × M	0.581	0.004	−8.951
Margalef	y = 0.532 × M	0.283	0.034	−2.353
Pielou	y = −0.471 × W + 0.448 × M	0.486	0.013	−5.687

Note: Canopy—Canopy density; Density—Tree density; W—Uniform angle; M—Mingling degree. AIC is the akaike information criterion.

4. Discussion

As an important part of forest ecosystem communities, understory vegetation plays an indispensable role in plantations due to its high rate of nutrient accumulation and biological turnover; furthermore, intense competition exists between trees and understory vegetation [29,30]. The study concluded from the univariate correlation analysis, typical correlation analysis and multiple linear regression that the canopy density and tree density of the stand non-spatial structure were significantly and negatively correlated with understory herbaceous plants species diversity. The factors of canopy density and tree density had large correlation coefficients with the linear combination V_1 (species diversity variable group), of −0.690 and −0.644, respectively, indicating that the combined effect of canopy density and tree density on understory herbaceous plants species diversity was the largest. The coefficients of the canopy density in equations of the Shannon–Wiener index and Simpson index were −0.552 and −0.539, respectively. The main reason was that understory vegetation can receive less light when the canopy density is high; therefore, the species diversity is limited by light conditions. Some studies [31,32] showed that there is a decrease in understory herbaceous plants species diversity with increased canopy density, which is consistent with the results of this paper. The coefficient of tree density in the regression equation of the Shannon–Wiener index was −0.336, indicating that tree density was negatively correlated with species diversity in the herbaceous layer. Some studies also found that the number of understory vegetation species increased with rising tree density until a certain threshold [15,33]. The main reason is that when the tree density exceeds the ideal density, a competitive situation between stands forms for subsistence resources, which inhibits the growth and development of understory vegetation, resulting in a decrease in diversity. The results indicate that the range of tree density selected exceeded the threshold.

The heterogeneity of the stand spatial structure led to the formation of different community composition, light and soil conditions in the understory. The opening degree is the vertical spatial structure parameter of the stand in the study, and the opening degree was not significantly correlated with the understory herbaceous plants species diversity index in the Pearson correlation analysis ($p > 0.05$), which is consistent with the results of Zhu et al. [21]. The uniform angle, neighborhood comparison and mingling degree were all spatial structure parameters on the horizontal plane, among which the neighborhood comparison had no significant effect on understory herbaceous plants species diversity ($p > 0.05$). The uniform angle was only significantly and negatively correlated ($p < 0.05$) with the Pielou evenness index with a correlation coefficient of −0.539, and a correlation coefficient of −0.471 in the Pielou index regression equation. The uniform angle reflected the horizontal distribution pattern of the stand. The stand distribution was random when the uniform angle was small, and it was aggregated when the uniform angle was large [34]. When the distribution of trees was aggregated, the size of the forest gap varied, leading to a decrease in species evenness. The uniform angle affects the size and location of the forest gap in the tree layer, which directly affects the value of ultraviolet radiation [35]. The appropriate forest gap size helps maintain the forest microclimate and understory microenvironment and facilitates the improvement of understory species evenness. In contrast, when the forest gap is too large, it also leads to a reduction in environmental heterogeneity within the forest gap, which results in a decrease in species evenness. The study showed that the mingling degree was significantly and positively correlated with all four understory herbaceous plants species diversity indices ($p < 0.05$), and the mingling degree was incorporated into the regression equation of each herbaceous species diversity index as an independent variable. Additionally, in the results of the typical correlation analysis, the loading coefficient of the mingling degree was 0.587 in the variable group of the stand structure index, which was the index with the greatest influence on the comprehensive index of understory herbaceous plants species diversity in terms of the stand spatial structure. The effect of the mingling degree on understory herbaceous plants species diversity was reflected in the competition among trees, leading to mutual suppression between trees and the release of nutrient space for understory vegetation. Both understory vegetation and *Platycladus orientalis* are shallow-rooted plants, while *Armeniaca sibirica* and *Cotinus coggygria* both have well-developed root systems [36,37]. Root systems of different lengths crisscross the soil, which improves the permeability and looseness of the soil, activates soil microbial activities, and thus accelerates the circulation of soil nutrients [38]. The distribution characteristics of soil nutrients explained the uneven distribution of understory vegetation [39,40]. At the same time, the increase in the mingling degree improved the heterogeneity of the forest environment, which was more conducive to the improvement of the uniformity of understory vegetation species.

The interrelationship between stand structure and understory herbaceous plants species diversity can be better reflected by the description of one-to-one, one-to-many, and many-to-many relationships in the study. In the later stage of forest management work, an improvement of the understory herbaceous plants species diversity can be achieved by adjusting the canopy density and tree density in the stand non-spatial structure, and the horizontal spatial structure (uniform angle and mingling degree) in the stand spatial structure. The three models used in this study are linear models, which cannot accurately predict the relationship between stand structure and understory herbaceous plants species diversity. To further study the response of understory herbaceous plants species diversity to stand structure, we can start from the perspective of nonlinear models and increase the number of samples in the future.

5. Conclusions

Stand structure had a significant effect on understory herbaceous plants species diversity, among which the canopy density and tree density had a considerable effect on the understory herbaceous plants species diversity in terms of the stand non-spatial structure, and the uniform angle and mingling degree had a great effect on understory herbaceous plants species diversity in terms of stand spatial structure. In order to increase understory herbaceous plants species diversity to enhance the plantation ecosystem function, in the subsequent management of *Platycladus orientalis* plantations, the canopy density and tree density of the forest can be improved to provide more sunlight, nutrients and space for herbaceous species. The canopy density of the overstory and tree density should be appropriately reduced. Within the range of 0.5–0.8, canopy density can be set at the value of 0.5; in the range of tree densities studied, the value of around 1000 tree·hm^{-2} was found to be the most advantageous. In terms of the stand spatial structure, when trees are randomly distributed and the mingling degree is large, understory herbaceous plants species diversity is higher. Therefore, the horizontal spatial structure of the stand should be adjusted by increasing the mingling degree and disrupting the distribution pattern of trees. The results have certain limitations, however, due to the narrow range of stand structures studied and the use of a linear model. Nonlinear models can be used to investigate the correlation between stand structure and understory herbaceous plants species diversity with increased precision in the future.

Author Contributions: Conceptualization, S.Q.; methodology, R.C.; validation, R.C.; formal analysis, R.C.; investigation, R.C., B.W., D.Z., L.Z., P.Z., N.M., X.H.; writing—original draft, R.C.; writing—review and editing, R.C.; visualization, R.C.; supervision S.Q.; funding acquisition S.Q. All authors have read and agreed to the published version of the manuscript.

Funding: This research was funded by the Forestry Monitoring and Evaluation of the 2020 Beijing–Tianjin Sand-Storm Source Control II Project, grant number 2020-SYZ-01-17JC05.

Data Availability Statement: Not applicable.

Acknowledgments: The authors are thankful to the Beijing Gardening and Greening Bureau for providing forestry engineering design data.

Conflicts of Interest: The authors declare no conflict of interest.

References

1. Zheng, L.T.; Chen, H.Y.H.; Hautier, Y.; Bao, D.F.; Xu, M.S.; Yang, B.Y.; Zhao, Z.; Zhang, L.; Yan, E.R. Functionally diverse tree stands reduce herbaceous diversity and productivity via canopy packing. *Funct. Ecol.* **2022**, *36*, 950–961. [CrossRef]
2. Han, X.; Xu, Y.; Huang, J.H. Species Diversity Regulates Ecological Strategy Spectra of Forest Vegetation Across Different Climatic Zones. *Front. Plant Sci.* **2022**, *13*, 807369. [CrossRef] [PubMed]
3. Liang, W.J.; Wei, X. Relationships between ecosystems above and below ground including forest structure, herb diversity and soil properties in the mountainous area of Northern China. *Glob. Ecol. Conserv.* **2020**, *24*, 1–10. [CrossRef]
4. Hou, L.Y.; Zhang, Y.Q.; Li, Z.C.; Shao, G.D.; Song, L.G.; Sun, Q.W. Comparison of Soil Properties, Understory Vegetation Species Diversities and Soil Microbial Diversities between Chinese Fir Plantation and Close-to-Natural Forest. *Forests* **2021**, *12*, 632. [CrossRef]
5. Rota, E.; Tancredi, C.; Bargagli, R. Community structure, diversity and spatial organization of enchytraeids in Medi-terranean urban holm oak stands. *Eur. J. Soil Biol.* **2014**, *62*, 83–91. [CrossRef]
6. Landuyt, D.; De, L.E.; Perring, M.P. The functional role of temperate forest understorey vegetation in a changing world. Global change biology. *Glob. Chang. Biol.* **2019**, *25*, 3625–3641. [CrossRef]
7. Lü, X.T.; Yin, J.X.; Tang, J.W. Diversity and composition of understory vegetation in the tropical seasonal rain forest of Xishuangbanna. *Rev. Biol. Trop.* **2011**, *59*, 455–463. [CrossRef]
8. Ali, A.; Yan, E.R.; Chen, H.Y.H.; Chang, S.X.; Zhao, Y.T.; Yang, X.D.; Xu, M.S. Stand structural diversity rather than species diversity enhances aboveground carbon storage in secondary subtropical forests in Eastern China. *Biogeosciences* **2016**, *13*, 4627–4635. [CrossRef]
9. Deal, R.L. Management strategies to increase stand structural diversity and enhance biodiversity in coastal rainforests of Alaska. *Biol. Conserv.* **2007**, *137*, 520–532. [CrossRef]
10. Do, T.V.; Sato, T.; Saito, S.; Kozan, O.; Yamagawa, H.; Nagamatsu, D.; Manabe, T. Effects of micro-topographies on stand structure and tree species diversity in an old-growth evergreen broad-leaved forest, southwestern Japan. *Glob. Ecol. Conserv.* **2015**, *4*, 185–196. [CrossRef]
11. Chapungu, L.; Nhamo, L.; Gatti, R.C.; Chitakira, M. Quantifying Changes in Plant Species Diversity in a Savanna Ecosystem Through Observed and Remotely Sensed Data. *Sustainability* **2020**, *12*, 2345. [CrossRef]
12. Sreelekshmi, S.; Bijoy Nandan, S.; Sreejith, V.K.; Radhakrishnan, C.K.; Suresh, V.R. Mangrove species diversity, stand structure and zonation pattern in relation to environmental factors—A case study at Sundarban delta, east coast of India. *Reg. Stud. Mar. Sci.* **2020**, *35*, 101111. [CrossRef]
13. Song, X.; Cao, M.; Li, J.; Kitching, R.L.; Nakamura, A.; Laidlaw, M.J.; Yang, J. Different environmental factors drive tree species diversity along elevation gradients in three climatic zones in Yunnan, southern China. *Plant Divers.* **2021**, *43*, 433–443. [CrossRef]
14. Chevaux, L.; Marell, A.; Baltzinger, C.; Boulanger, V.; Cadet, S.; Chevalier, R.; Paillet, Y. Effects of stand structure and ungulates on understory vegetation in managed and unmanaged forests. *Ecol. Appl.* **2022**, *32*, e2531. [CrossRef]
15. Ahmad, B.; Wang, Y.H.; Hao, J.; Liu, Y.H.; Bohnett, E.; Zhang, K.B. Optimizing stand structure for trade-offs between overstory timber production and understory plant diversity: A case-study of a larch plantation in northwest China. *Land Degrad. Dev.* **2018**, *29*, 2998–3008. [CrossRef]
16. Liu, D.; Zhou, C.F.; He, X.; Zhang, X.H.; Feng, L.Y.; Zhang, H.R. The Effect of Stand Density, Biodiversity, and Spatial Structure on Stand Basal Area Increment in Natural Spruce-Fir-Broadleaf Mixed Forests. *Forests* **2022**, *13*, 162. [CrossRef]
17. Gao, W.Q.; Lei, X.D.; Liang, M.W.; Larjavaara, M.; Li, Y.T.; Gao, D.L.; Zhang, H.R. Biodiversity increased both productivity and its spatial stability in temperate forests in northeastern China. *Sci. Total Environ.* **2021**, *780*, 146674. [CrossRef]
18. Li, Y.; Zhao, Z.; Hui, G.; Hu, Y.; Ye, S. Spatial structural characteristics of three hardwood species in Korean pine broad-leaved forest—Validating the bivariate distribution of structural parameters from the point of tree population. *For. Ecol. Manag.* **2014**, *314*, 17–25. [CrossRef]

19. Stoll, P.; Newbery, D.M. Evidence of species-specific neighborhood effects in the dipterocarpaceae of a Bornean rain forest. *Ecology* **2005**, *86*, 3048–3062. [CrossRef]
20. China National Forestry and Grassland Administration. Phase II of Beijing Tianjin Sandstorm Source Control Project. Available online: http://hdl.handle.net/20.500.11822/29776 (accessed on 8 November 2022).
21. Zhu, G.Y.; Xu, Q.G.; Lü, Y. Effects of stand spatial structure on species diversity of shrubs in Quercus spp. natural secondary forests in Hunan Province. *Acta Ecol. Sin.* **2018**, *38*, 5404–5412.
22. Hui, G.Y.; Gadow, K. The neighbourhood pattern-a new structure parameter for describing distribution of foerst tree position. *Sci. Silvae Sin.* **1999**, *35*, 37–42.
23. Hui, G.Y.; Gadow, K.; Albert, M. A new parameter for stand spatial structure—Neighbourhood comparison. *For. Res.* **1999**, *1*, 4–9.
24. Hui, G.Y.; Hu, Y.B. Measuring species spatial isolation in mixed forests. *For. Res.* **2001**, *14*, 23–27.
25. Wang, P.; Jia, L.M.; Wei, S.P.; Wang, Q.F. Analysis of stand spatial structure of *Platycladus orientalis* recreational forest based on Voronoi diagram method. *J. Beijing For. Univ.* **2013**, *35*, 39–44.
26. Peng, Y.; Fan, M.; Song, J.Y.; Cui, T.T.; Li, R. Assessment of plant species diversity based on hyperspectral indices at a fine scale. *Sci. Rep.* **2018**, *8*, 4776. [CrossRef]
27. Zhang, L.J.; Du, H.; Yang, Z.Q.; Song, T.Q.; Zeng, F.P.; Peng, W.X.; Huang, G.Q. Topography and Soil Properties Determine Biomass and Productivity Indirectly via Community Structural and Species Diversity in Karst Forest, Southwest China. *Sustainability* **2022**, *14*, 7644. [CrossRef]
28. Hamayun, S.; Shujaul, M.K.; David, M.H.; Zahid, U.; Rizwana, A.Q. Species Diversity, Community Structure, and Distribution Patterns in Western Himalayan Alpine Pastures of Kashmir, Pakistan. *Mt. Res. Dev.* **2011**, *31*, 153–159.
29. Pan, P.; Zhao, F.; Ning, J.; Zhang, L.; Ouyang, X.; Zang, H. Impact of understory vegetation on soil carbon and nitrogen dynamic in aerially seeded Pinus massoniana plantations. *PLoS ONE* **2018**, *13*, e0191952. [CrossRef]
30. Zhang, J.W.; Young, D.H.; Oliver, W.W.; Fiddler, G.O. Effect of overstorey trees on understorey vegetation in California (USA) ponderosa pine plantations. *Forestry* **2016**, *89*, 91–99. [CrossRef]
31. Barbier, S.; Gosselin, F.; Balandier, P. Influence of tree species on understory vegetation diversity and mechanisms involved—A critical review for temperate and boreal forests. *For. Ecol. Manag.* **2008**, *254*, 1–15. [CrossRef]
32. Berger, A.L.; Puettmann, K.J. Overstory composition and stand structure influence herbaceous plant diversity in the mixed aspen forest of northern Minnesota. *Am. Midl. Nat.* **2000**, *143*, 111–125. [CrossRef]
33. VanderSchaaf, C.L. Estimating understory vegetation response to multi-nutrient fertilization in Douglas-fir and ponderosa pine stands. *J. For. Res.* **2008**, *13*, 43–51. [CrossRef]
34. Dong, L.B.; Bettinger, P.; Liu, Z.G. Optimizing neighborhood-based stand spatial structure: Four cases of boreal forests. *For. Ecol. Manag.* **2022**, *506*, 119965. [CrossRef]
35. Chen, L.; Han, W.Y.; Liu, D.; Liu, G.H. How forest gaps shaped plant diversity along an elevational gradient in Wolong National Nature Reserve? *J. Geogr. Sci.* **2019**, *29*, 1081–1097. [CrossRef]
36. Cai, L.; Zhu, W.R.; Wang, H.T.; Zhang, G.C.; Liu, X.; Yang, J.H.; Wang, Y.P. Root Morphology of Six Tree Species in Mountain Area of Middle South Shandong. *Sci. Soil Water Conserv.* **2015**, *13*, 83–91.
37. Chen, Y.; Wang, X.W. Vertical Distribution of Root System of Several Plants in Nenjiang Sand Land. *Prot. For. Sci. Technol.* **2007**, 30–32.
38. Wan, X.H.; Huang, Z.Q.; He, Z.M.; Yu, Z.P.; Wang, M.H.; Davis, M.R.; Yang, Y.S. Soil C:N ratio is the major determinant of soil microbial community structure in subtropical coniferous and broadleaf forest plantations. *Plant Soil* **2015**, *387*, 103–116. [CrossRef]
39. Ma, T.S.; Deng, X.W.; Chen, L.; Xiang, W.H. The soil properties and their effects on plant diversity in different degrees of rocky desertification. *Sci. Total Environ.* **2020**, *736*, 139667. [CrossRef]
40. Mureva, A.; Ward, D. Soil microbial biomass and functional diversity in shrub-encroached grasslands along a precipitation gradient. *Pedobiologia* **2017**, *63*, 37–45. [CrossRef]

forests

MDPI

Article

Effects of Precipitation and Soil Moisture on the Characteristics of the Seedling Bank under *Quercus acutissima* Forest Plantation in Mount Tai, China

Longmei Guo [1], Ruiqiang Ni [1], Xiaoli Kan [1], Qingzhi Lin [1], Peili Mao [1,*], Banghua Cao [1], Peng Gao [1], Jinwei Dong [1], Wendong Mi [2] and Boping Zhao [2]

[1] Taishan Mountain Forest Ecosystem Research Station, Key Laboratory of State Forestry Administration for Silviculture of the Lower Yellow River, Shandong Agricultural University, Tai'an 271018, China; guolm3180@126.com (L.G.); wind0309@163.com (R.N.); lwsdau2021@163.com (X.K.); lin15634195020@163.com (Q.L.); caobanghua@126.com (B.C.); gaopengy@163.com (P.G.); jwdong@sdau.edu.cn (J.D.)

[2] Mount Taishan Management Committee, Tai'an 271000, China; miwendong888@163.com (W.M.); tszbp@163.com (B.Z.)

* Correspondence: maopl1979@163.com

Abstract: Natural regeneration is crucial for the development of sustainable forestry practices in light of the current global climate changes. In this paper, we compared the size distributions of *Quercus acutissima* seedlings in the understory of *Q. acutissima* forest plantations in Mount Tai in 2010 and 2017, studied the physiological and morphological responses of seedlings to the microenvironment, and explored the maintenance mechanisms of the seedling bank. The results showed that the density of understory seedlings in 2017 was only 61.63% of that in 2010, especially in the 20–40 cm height class. Between 2011 and 2016, the precipitation and soil water content were the highest in 2011, followed by 2013. The 2–4-year seedlings (height < 40 cm) were not significantly different in seedling biomass, biomass allocation, and root morphology (root total surface area, root volume, and root average diameter), and were significantly different in total root length, specific root length, specific root surface area, and nonstructural carbohydrate content of root, stem, and leaves. However, 5–6-year seedlings (height > 40 cm) showed the largest biomass. Principal component analysis indicated that altering root morphology, nonstructural carbohydrate, and biomass allocation played significant roles in the drought adaptation of seedlings in the understory. In conclusion, drought stress together with seedling adaptation influenced the dynamics of seedling bank in the understory of *Q. acutissima* plantations.

Keywords: seedling bank; seedling age; microenvironment; biomass allocation; nonstructural carbohydrates

Citation: Guo, L.; Ni, R.; Kan, X.; Lin, Q.; Mao, P.; Cao, B.; Gao, P.; Dong, J.; Mi, W.; Zhao, B. Effects of Precipitation and Soil Moisture on the Characteristics of the Seedling Bank under *Quercus acutissima* Forest Plantation in Mount Tai, China. *Forests* **2022**, *13*, 545. https://doi.org/10.3390/f13040545

Academic Editor: Ilona Mészáros

Received: 17 February 2022
Accepted: 29 March 2022
Published: 31 March 2022

Publisher's Note: MDPI stays neutral with regard to jurisdictional claims in published maps and institutional affiliations.

1. Introduction

Forest plantation has increased rapidly from 1990 to 2015 at the global scale, with a percentage increase from 4.06 to 6.95% of the total forest area between 1990 and 2015 [1]. The forest plantation plays an increasingly crucial role in timber production, environmental improvement, landscape rehabilitation, and climate change mitigation [2]. Natural regeneration is vital for sustainable forest management. Naturally regenerated forests have the advantages of better plant establishment, self-regenerated material, and high seedling densities [3,4]. However, improper management has made the planted forests unable to complete natural regeneration [5,6]. Stand structure [7], litter density and grass cover [8], seedling adaptation [9–11], stand management [12,13], and year-to-year variation in stand conditions [14] are considered to be vital factors affecting forest natural regeneration. The natural regeneration of plantations is a long-term and complex process affected by many factors. At present, the regeneration mechanisms of plantations are still poorly understood.

Sufficient seedling bank is the foundation of the success of natural forest regeneration [15–18]. The structure and composition of natural regeneration are closely related to environmental conditions under forest stands [8,19], including precipitation [20], air temperature [21], soil moisture [22], and soil temperature [14], which significantly affect the establishment and survival of tree seedlings. Long-term monitoring of tree seedlings revealed that the fraction of seedlings eventually reaching the sapling class is quite small due to high seedling mortality [14,23,24]. Therefore, Rey and Alcántara [24] suggested that seedling survival is a critical process for tree species recruitment. Abiotic factors, especially drought, are considered to be the main factors causing seedling death [24]. With the increase of seedling size, the proportion of non-photosynthetic tissues to total sapling biomass increases, as well as respiration costs [25–27]. The minimum light demands of tree species increased with increasing seedling size, which led to the decrease of shade tolerance [28,29]. Thus, the transformation from seedlings to saplings requires higher light intensity [29,30], which is one of the most important limiting factors under forest stands. In addition, Soto et al. [31] suggested that light and nitrogen interact to influence regeneration in old-growth Nothofagus-dominated forests in South-Central Chile. Therefore, the current understanding of the dynamics of seedling banks under forest remains very limited.

Nonstructural carbohydrates (NSC) storage is a fundamental process that allows organisms to meet variable demands for resources during their development and buffer environmental fluctuations in resource supply. NSC is mainly stored in the roots and stems of tree seedlings [32], and starch is its main component [25,33]. Under short-term drought stress, the NSC content of *Pinus massoniana* Lamb. seedlings increased rapidly, and the variation in NSC partitioning among organs was more highly significant than the variation in biomass partitioning [34]. However, the change of NSC in the roots of *Manihot esculenta* Crantz. was not significant [35]. In shaded forest understories, NSC in tree seedlings acts as a buffer during the periods of negative net carbon balance against herbivores and diseases [32], defoliation [36], and suddenly shade increases [37], etc. Under severe shading conditions, NSC in seedlings is greatly consumed [34,38], or little remains [38]. Similarly, reduced NSC reserve in the root system of *Populus tremuloides* Michx. and *P. balsamifera* L. seedlings during severe drought contributed to the root death of seedlings during the dormant season by compromising the frost tolerance of the root system [39]. Ivanov et al. [40] also found that critical exhaustion of the NSC reserves in the roots under water stress led to the greatest mortality of *Pinus sylvestris* L. seedlings. After the relief of drought stress, Tomasella et al. [41] suggested that preserving higher NSC content at the end of a drought can be important for the hydraulic resilience of trees. The survival of tree seedlings under the forest is a long-term process. With the continuous growth of tree seedlings, their NSC content gradually increases [28,42], which is considered conducive to their survival in the forest [36]. However, some studies have suggested that carbohydrate storage is not related to low-light survival for tree seedlings [42,43]. Myers and Kitajima [32] proposed that tree seedlings rely on NSC reserves to survive short-term periods of negative carbon balance under forest, thereby enabling net carbon gain over the long term. Therefore, the relationship between NSC and the survival of tree seedlings under forest has not yet achieved a unified understanding.

China has the largest forest plantation area in the world. For a long time, forest plantation management has paid much attention to short-term productivity and economic benefits but had ignored natural regeneration [2]. *Quercus acutissima* Carruth., a deciduous tree species belonging to *Quercus* in the Fagaceae family, is the main component tree of forest vegetation in warm temperate and subtropical regions of China. *Q. acutissima* is a pioneer tree in barren mountains and barren land, and an excellent tree species for soil and water conservation, with high ecological, economic, and landscape values. At present, there are few studies on the natural regeneration of *Q. acutissima* plantation, and the limiting mechanism is unclear. *Q. acutissima* has certain shade tolerance, and its seedlings rarely die under more than 12% full light [44]. Xue et al. [45] found there were a large number of seedlings in the secondary forest of *Q. acutissima*, but few saplings. To further explain the restriction mechanism of the natural regeneration of *Q. acutissima* plantations, we studied the pure forest plantation of *Q. acutissima*, the main forest type of Mount Tai, and with the following aims: (1) to describe changes of the regeneration structure of *Q. acutissima* seedlings under the forest across a long-term period; (2) to explore the effects of the environmental factors on regeneration structure and seedling establishment; (3) to clarify the morphological and physiological adaptation with the increase of seedling age.

2. Materials and Methods

2.1. Study Area

The study area of Mount Tai lies in the middle of Shandong province, China, between Jinan City and Tai'an City. The climate is classified as the warm temperate continental monsoon zone, the average annual temperature is 12.6 °C, the frost-free period is 196 days, and the accumulated temperature greater than or equal to 10 °C is 3821 °C. The average annual precipitation is 758 mm and is concentrated from June to September. The soil type is mainly brown loam with a pH value of about 6.0 and a soil depth of 15~60 cm. Most of the vegetation in Mount Tai was planted in the 1950s and 1960s, with a total forest area of about 9490 hm^2 and a forest coverage rate of 81.5%. The main tree species include *Platycladus orientalis* (L.) Franco, *Q. acutissima*, *Robinia pseudoacacia* L., *Pinus thunbergii* Parl., *Pinus densiflora* Sieb. et Zucc., *Pinus tabuliformis* Carr., etc. [46,47].

2.2. Investigation of *Q. acutissima* Plantation

The investigation of *Q. acutissima* plantation was conducted in the Mountain Tai Forest Ecosystem Research Station of State Forestry and Grassland Administration (117°06′48.2″ N, 36°20′05.3″ E). The forest was planted in 1960 using 1-year-old seedlings. The plantation is a pure forest with an age of 60 years and an average altitude of 730 m. The permanent monitoring plot was established in 2010 with an area of 0.48 hm^2 (60 m × 80 m). Twelve 20 m × 20 m quadrats were set up in the plot. In each quadrat, we measured the diameter at breast height (DBH) for all trees with heights ≥1.5 m, which were defined as adult trees. Saplings and seedlings below 1.5 m in height were divided into eight height classes: <20 cm, 20–40 cm, 40–60 cm, 60–80 cm, 80–100 cm, 100–120 cm, 120–140 cm, and 140–150 cm. Adult trees were classified into DBH sizes, with one diameter class for every 5 cm. The permanent monitoring plot was reviewed in May 2017. The sampling and investigation methods in this paper referred to the forestry industry standard of the People's Republic of China issued by the State Forestry Administration (State Forestry Administration, 2011).

2.3. Measurement of Seedling Traits of *Q. acutissima*

On May 10, 2016, 19 seedlings with heights between 9 and 46 cm were selected in the *Q. acutissima* plantation. The height, diameter, and number of leaves of the seedlings were measured and collected. Then, each seedling was carefully excavated from the soil, and all the seedlings were sealed and stored in a fresh-keeping bag and brought back to the laboratory.

The age of seedlings was determined by counting the number of annual rings. These seedlings were divided into four groups according to their age: 2-year (I), 3-year (II), 4-year (III), and 5–6-year (IV). There were 3–9 seedlings in each group.

In the laboratory, the seedlings were divided into leaves, stems, and roots. The area of each leaf of each seedling was measured with a CI-202 portable laser leaf area meter (CID Inc., Washington, USA). After rinsing with clean water, these roots were scanned by HP Scanjet 8200 scanner, and the scanned images were analyzed by root parameter analysis software (Delta-T Area Meter Type AMB2) to obtain the length and surface area of roots. The roots, stems, and leaves of seedlings were dried at 80 °C until maintaining a constant weight. Then, the dry weight was weighed and recorded. After the dried samples of roots, stems, and leaves were pulverized, their concentrations of soluble sugar and starch were determined by anthranone–H_2SO_4 colorimetry, respectively [48]. Starch and soluble sugars in each plant organ were added together to determine nonstructural carbohydrate concentration (NSC). All the seedling indices are shown in Table 1.

Table 1. Seedling characteristic indices, abbreviations, and units.

Seedling Characteristic Indices	Abbreviations	Units
Relative soil water content	RSWC	%
Seedling biomass (leaf + stem + root mass)	SB	g
Leaf mass ratio (leaf dry mass/total seedling dry mass)	LMR	$g\ g^{-1}$
Stem mass ratio (stem + petiole mass)/total plant mass	SMR	$g\ g^{-1}$
Root mass ratio (root dry mass/total seedling dry mass)	RMR	$g\ g^{-1}$
Specific leaf area (total leaf area/total leaf dry mass)	SLA	$cm^2\ g^{-1}$
Leaf area ratio (total leaf area/total seedling dry mass)	LAR	$cm^2\ g^{-1}$
Photosynthetic tissues/non-photosynthetic tissues (leaf mass/(stem + root mass))	P/NP	$g\ g^{-1}$
Total Root length	TRL	cm
Root volume	RV	cm^3
Root average diameter	RAD	mm
Root total surface area	RSA	cm^2
Specific root length (root length/dry root mass)	SRL	$cm\ g^{-1}$
Specific root area (root surface area/ dry root mass)	SRA	$cm^2\ g^{-1}$
Root soluble sugar	RSS	$mg\ g^{-1}$
Stem soluble sugar	SSS	$mg\ g^{-1}$
Leaf soluble sugar	LSS	$mg\ g^{-1}$
Root starch	RS	$mg\ g^{-1}$
Stem starch	SS	$mg\ g^{-1}$
Leaf starch	LS	$mg\ g^{-1}$

2.4. Monitoring Environmental Parameters under *Q. acutissima* Plantation

The environmental factors in the forest plot were measured from July 2011 to December 2016. The precipitation at 1.5 m in the forest was measured. The relative soil water content (RSWC) was measured at 10, 20, 30, and 40 cm depth of soil, respectively. The indices above were determined using a CR3000 automatic weather station (Campbell Scientific, Logan, UT, USA), which automatically recorded daily data every 10 min. Data were collected every month.

2.5. Statistical Analysis

The precipitation effect was tested by a two-way ANOVA, with year and month as the sources of variations. The relative soil water content was tested by a three-way ANOVA, with year, month, and depth as the sources of variations. Seedling indices were analyzed by a one-way ANOVA. The responses of functional traits to height were compared, and multiple comparisons were made. The linear regression analysis of height and age was carried out using the excavated 19 seedlings. At the same time, the correlation analysis of each index was carried out. All the seedling indices were analyzed by principal component analysis (PCA). The test level was $p = 0.05$. All statistical analyses were conducted using R 4.1.2 for Windows.

3. Results

3.1. The Diameter Structure of *Q. acutissima* Plantation

The density of *Q. acutissima* adult trees in 2010 stands was 634 N ha^{-1}, with trees in the DBH class 2–5 cm accounting for 19.70%, 5–10 cm for only 3.44%, and 15–35 cm for 70.94%. In 2017, the density of adult trees was 675 N ha^{-1}, with trees in the DBH class 2–5 cm accounting for 14.81%, 5–10 cm for 0, and 15–35 cm for 77.78%. In 2017, compared with 2010, the density of trees in the DBH 2–5 cm decreased by 4.89%, 5–10 cm trees disappeared, and the density of 15–35 cm trees increased by 6.84% (Figure 1A).

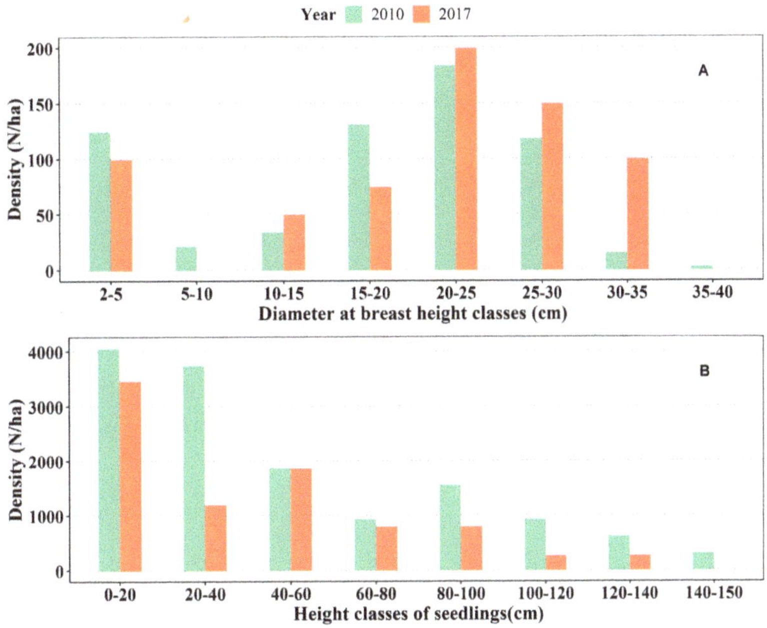

Figure 1. Distribution of diameter at breast height classes for adult trees (**A**) and distribution of height classes for seedlings (**B**) of *Quercus acutissima* in 2010 and 2017, respectively.

The height distribution of seedlings in 2010 and 2017 was pyramid-shaped (Figure 1B). The tree number reached a maximum in the height class 0–20 cm and 20–40 cm, and then gradually decreased as the height increased. Compared with 2010, the density of seedlings in 2017 was only 61.63% of that in 2010, showing a significant decrease. Moreover, seedlings of all height levels were reduced, especially at the height of 20–40 cm, which was only 32% of that in 2010. All seedlings at 140–150 cm disappeared in 2017. Regression analysis showed that there was a very significant linear relationship between height and age for seedlings with heights less than 46 cm (F = 35.85, $p < 0.01$, Figure 2).

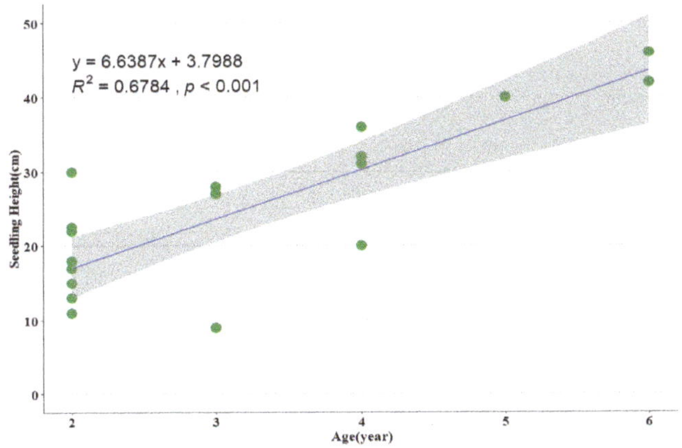

Figure 2. Linear regression between seedling height and age of *Quercus acutissima* seedlings.

3.2. *Characteristics of Environmental Factors under Q. acutissima Plantation*

3.2.1. Precipitation

There were significant differences in precipitation among different months ($F = 11.34$, $p < 0.01$) and years ($F = 2.84$, $p < 0.05$). With the increase of months, the precipitation first increased and then decreased, and was mainly concentrated between June and August each year (Figure 3). Among different years, precipitation in 2011 was the highest ($p < 0.01$), and precipitation in 2013 was significantly higher than that in 2016 ($p < 0.05$), and there was no significant difference in precipitation among other years ($p > 0.05$). During 2012 and 2016, precipitation in the same months fluctuated significantly among years (Figure 3).

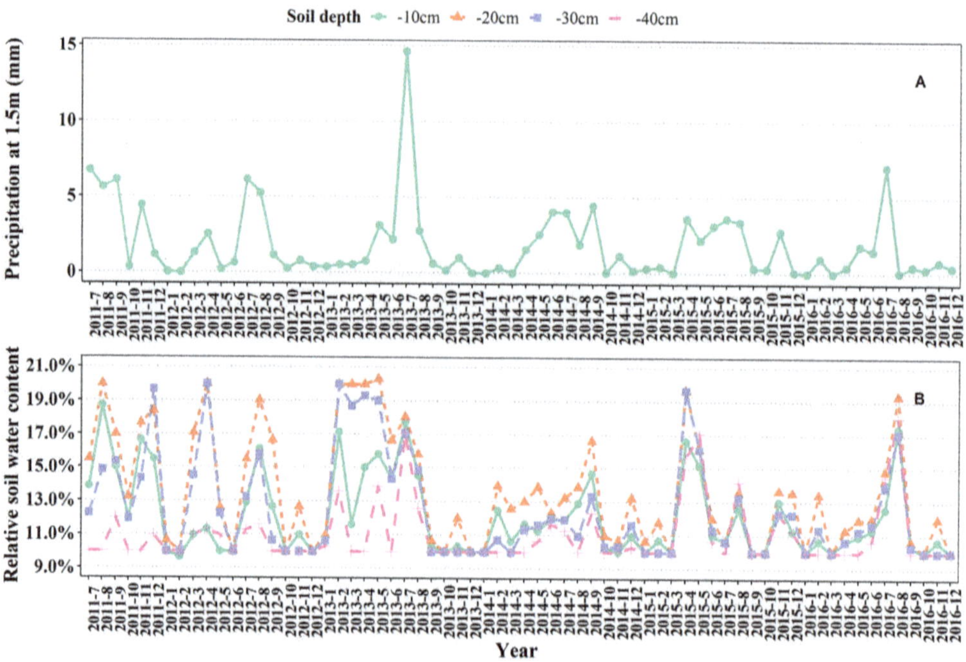

Figure 3. Variation of precipitation and relative soil water content in the *Quercus acutissima* plantation from 2011 to 2016. (**A**) Precipitation at 1.5 m; (**B**) relative soil water content.

3.2.2. Soil Moisture

The results of the analysis of variance showed that RSWC varied highly significantly between months ($F = 146.82$, $p < 0.01$), years ($F = 163.09$, $p < 0.01$), and soil depths ($F = 285.73$, $p < 0.01$). As the month increased, the RSWC changed in a bimodal pattern. The RSWC reached a peak in April and August, and reached a maximum in August. The RSWC declined to nadir in January, June, and October, and reached a minimum in January and October (there was no significant difference between January and October). Among years, RSWC was 2011 > 2013 > 2012≈2015 > 2014≈2016 ($p < 0.01$). Among soil depths, the order of RSWC was −20 cm > −30 cm > −10 cm > −40 cm.

3.3. *Seedling Size Traits with the Increase of Age*

Seedling age had a significant effect on height ($F = 11.01$, $p < 0.01$), ground diameter ($F = 11.38$, $p < 0.01$), and biomass ($F = 18.60$, $p < 0.01$). The height increased significantly with age (Figure 4A). Ground diameter and biomass showed no significant difference among age groups I, II, and III, and increased significantly at IV (Figure 4B,C).

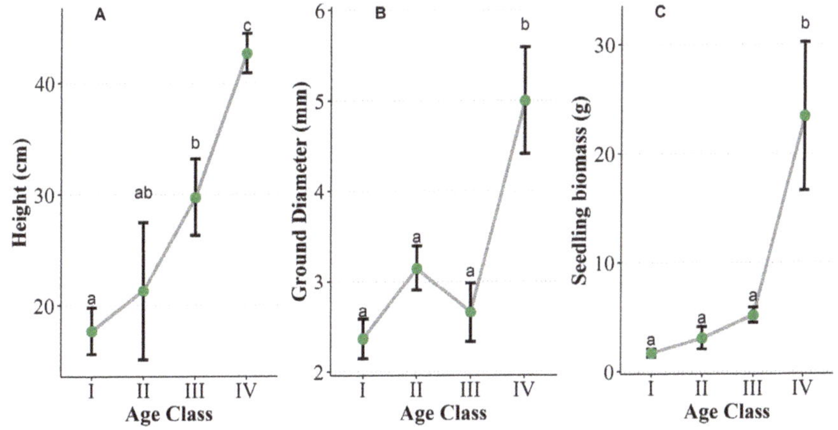

Figure 4. Changes of seedling biomass (**A**), ground diameter (**B**), and seedling biomass (**C**) with the increase of seedling age of *Quercus acutissima*. Different letters above error bars show significant differences.

3.4. Adaptation Traits of Seedlings with Increasing Age

3.4.1. Biomass Allocation

Variance analysis showed that seedling age had significant effects on LMR (F = 5.51, $p < 0.01$), RMR (F = 3.34, $p < 0.05$), LAR (F = 5.93, $p < 0.01$), and p/NP F = 5.15, $p < 0.05$), but not on SMR (F = 0.41, $p = 0.75$) nor SLA (F = 1.20, $p = 0.35$). There were no significant differences among age groups I, II, and III in LMR, LAR, and p/NP, but these indices all decreased significantly at IV (Figure 5A,E,F). As the age increased, the RMR decreased first and then increased significantly (Figure 5B).

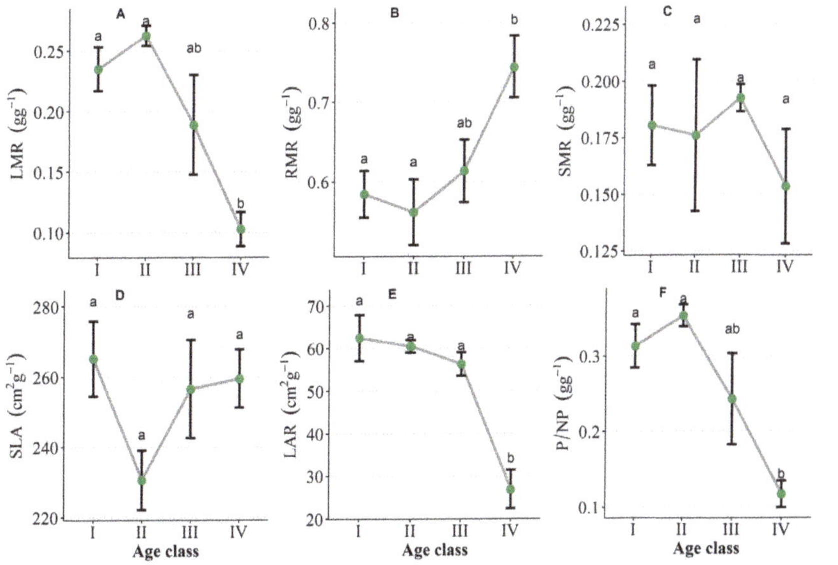

Figure 5. Changes of biomass allocation with the increase of seedling age of *Quercus acutissima*. (**A**) LMR, leaf mass ratio; (**B**) RMR, root mass ratio; (**C**) SMR, stem mass ratio; (**D**) SLA, specific leaf area; (**E**) LAR, leaf area ratio; (**F**) P/NP, photosynthetic tissues/non-photosynthetic tissues. Different letters above error bars show significant differences.

3.4.2. Root Traits

Seedling age had significant effects on total root length (F = 4.74, $p < 0.05$), root volume (F = 34.64, $p < 0.01$), average root diameter (F = 82.42, $p < 0.01$), and specific root length (F = 7.19, $p < 0.01$), but not on specific root surface area (F = 1.35, $p = 0.30$). With the increase of seedling age, root length continued to increase (Figure 6A), but specific root length and specific root surface area continued to decrease significantly (Figure 6E,F). However, RV and RAD showed no significant difference among I, II, and III, and increased significantly at IV (Figure 6C,D).

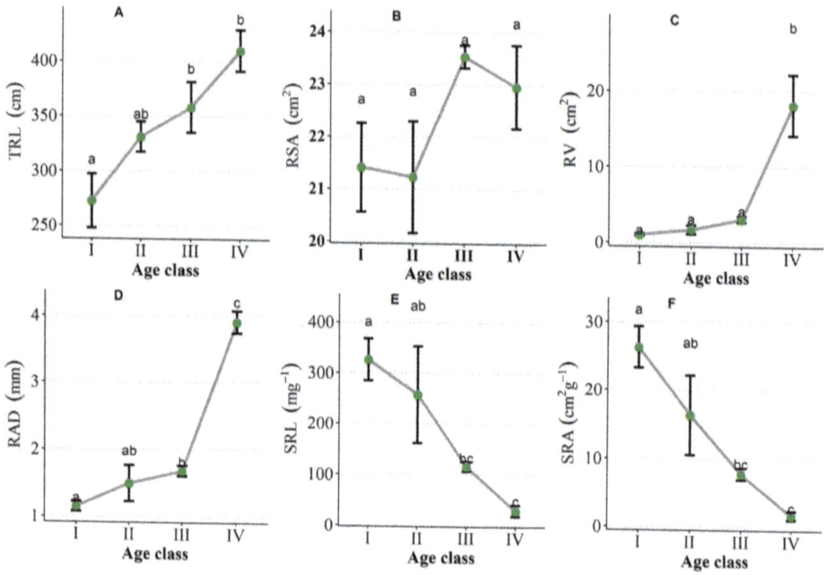

Figure 6. Changes of root traits with the increase of seedling age of *Quercus acutissima*. (**A**) TRL, total root length; (**B**) RSA, root total surface area; (**C**) RV, root volume; (**D**) RAD, root average diameter; (**E**) SRL, specific root length; (**F**) SRA, specific root area. Different letters above error bars show significant differences.

3.4.3. Non-Structural Carbohydrate (NSC)

There were significant differences in SSC (F = 312.60, $p < 0.01$) and SC (F = 329.40, $p < 0.01$) among different ages. With the increase of age, the SSC and SC of the root, stem, and leaf all increased significantly, and their orders were I < II < III < IV ($p < 0.01$, Figure 7). Organs also showed significant effects on SSC (F = 631.57, $p < 0.01$) and SC (F = 543.37, $p < 0.01$), respectively. There was no significant difference in soluble sugar between roots and stems ($p > 0.05$), which was significantly lower than that of leaves ($p < 0.01$). For starch, the root was significantly greater than that of the stem and leaf ($p < 0.01$), but there was no significant difference between the stem and leaf ($p > 0.05$).

3.5. *Correlation Analysis of Seedling Traits*

SB was negatively correlated with LMR, LAR, P/NP, SRL, and SRA, and positively correlated with RMR, TRL, RV, RAD, RSS, SSS, LSS, RS, SS, and LS (Figure 8). LMR, LAR, and P/NP were negatively correlated with RMR, TRL, RV, and RAD. RSS, SSS, LSS, RS, SS, and LS were negatively correlated with LMR, LAR, P/NP, SRL, and SRA, and positively correlated with RMR, TRL, RV, and RAD (the correlation between SS and RMR was not significant).

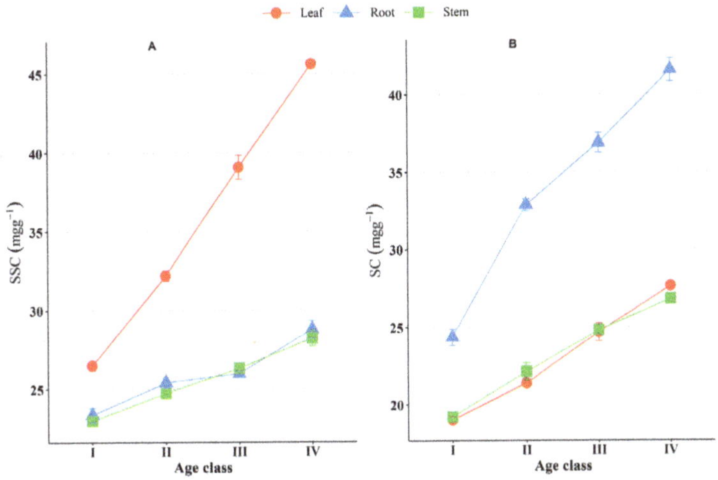

Figure 7. Changes of soluble sugar (**A**) and starch contents (**B**) in the roots, stems, and leaves with the increasing age of *Quercus acutissima* seedlings.

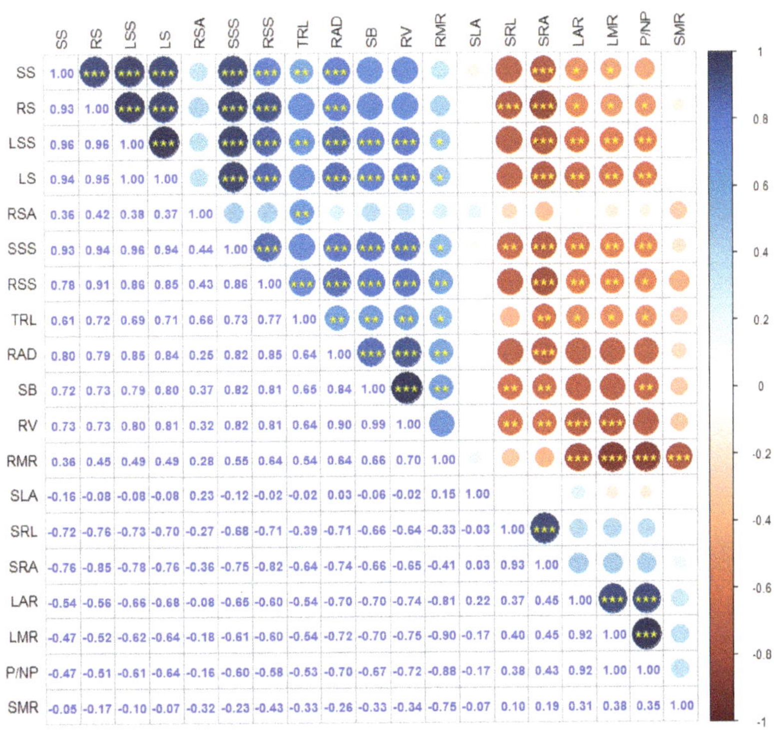

Figure 8. Correlations among seedling characteristics of *Quercus acutissima*. SB: seedling biomass; LMR: leaf mass ratio; RMR: root mass ratio; SMR: stem mass ratio; SLA: specific leaf area; LAR: leaf area ratio; TRL: total root length; RSA: root total surface area; RV: root volume; RAD: root average diameter; SRL: specific root length; SRA: specific root area; P/NP: photosynthetic tissues/non-photosynthetic tissues; RSS: root soluble sugar; SSS: stem soluble sugar; LSS: leaf soluble sugar; RS: root starch; SS: stem starch; LS: leaf starch. Significant code: *** $p < 0.001$, ** $p < 0.01$, * $p < 0.05$.

The PCA incorporating seedling traits revealed that together, the first and second principal components represented 74.15% of the total variation (Figure 9). The first principal component (62.65%) was mainly related to seedling growth, including SB, NSC (SS, SSS, LSS, RS, SS, and LS), and root traits (RV, TRL, RAD, RSA, SRL, and SRA). The second principal component (11.5%) was associated with biomass allocation, including SMR, LMR, LAR, P/NP, SLA, and RMR.

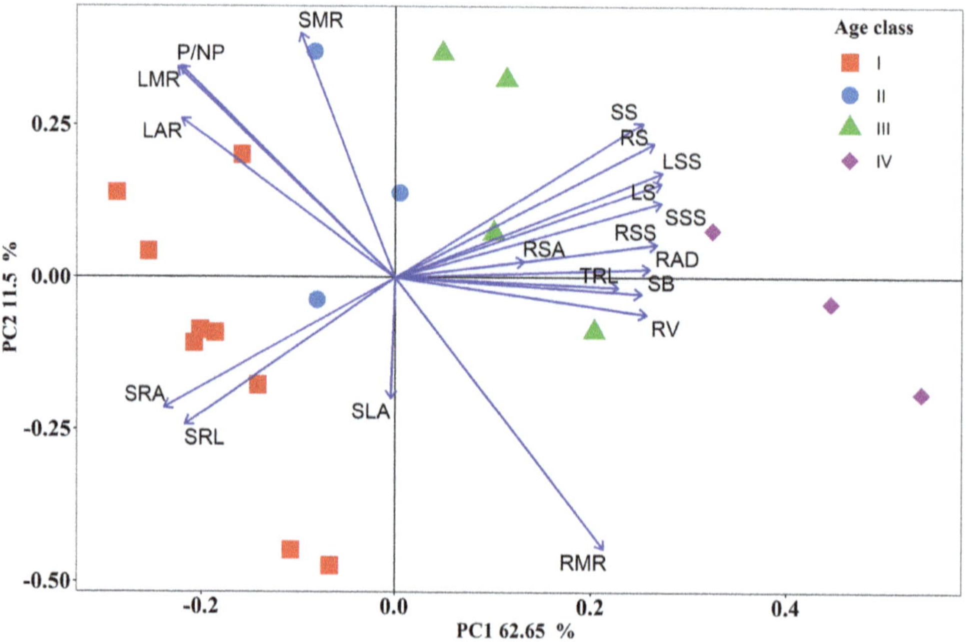

Figure 9. Principal component analysis of *Quercus acutissima* seedling trait. SB: seedling biomass; LMR: leaf mass ratio; RMR: root mass ratio; SMR: stem mass ratio; SLA: specific leaf area; LAR: leaf area ratio; TRL: total root length; RSA: total root surface area; RV: root volume; RAD: root average diameter; SRL: specific root length; SRA: specific root area; P/NP: photosynthetic tissues/ non-photosynthetic tissues; RSS: root soluble sugar; SSS: stem soluble sugar; LSS: leaf soluble sugar; RS: root starch; SS: stem starch; LS: leaf starch.

4. Discussion

Over the past decades, the predominant approach to studying successional dynamics has been to examine how community-level properties (e.g., diversity, biomass/carbon) vary over time and what factors influence their rates of change. Demographic rates of species can reveal mechanisms that underpin community assembly [49]. The seedling density of the *Q. acutissima* plantation in this paper was 14,063 N ha^{-1} in 2010 and 8667 N ha^{-1} in 2017. Thus, there were sufficient seedlings under the *Q. acutissima* plantation, which was related to the ability of shade tolerance in the seedling stage [44]. The size distribution of seedlings in 2010 and 2017 was pyramid-shaped, indicating that the seedling bank was relatively stable. Compared with 2010, the stand density changed little in 2017, but the seedling density decreased significantly, reaching only 61.63% of that in 2010. Several previous studies have suggested that low soil moisture caused by precipitation reduction is an important reason for limiting forest natural regeneration [20,50,51]. From 2014 to 2016, the relative soil water content in the *Q. acutissima* forest significantly and continuously decreased. Multiyear lags in tree drought recovery, termed 'drought legacy effects', are important for understanding the impacts of drought on forest ecosystems [52]. Thus, we thought that the decrease of seedling bank density was closely related to the drought for 3 consecutive years. However, the responses of seedlings of different sizes were significantly different. The seedling density in the height class 0–20 cm in 2010 was similar to that in 2017. Rodríguez-García et al. [22] suggested that the significant variables that best explained the total seedling and viable seedling density were the spring and autumn precipitation of the year prior to establishment. During 2014 and 2016, the higher precipitation and soil water content of the *Q. acutissima* forest in April, May, and June may have contributed to seedling establishment. However, the seedling density in the height class 20–40 cm decreased more sharply in 2017 than in 2010, which was related to the significant decrease of relative soil water content from 2014 to 2016. Our results are consistent with the conclusions drawn by Rey and Alcántara [24]. For the seedlings taller than 40 cm, the decrease of density in 2017 was more obvious than that in 2010. Their death may have been related to the increase of minimum light demand with the increase of seedling size [29]. Many studies have shown that the conversion rate from seedlings to saplings is very low due to the insufficient relative light intensity under the forest for long-term survival [14,23,24,53]. Studies on the natural regeneration of *Quercus* have suggested that higher light intensity is necessary for successful regeneration [16,54]. Therefore, the microenvironment in the forest and the adaptability of seedlings with different sizes are important factors determining the dynamics of the seedling bank.

The critical process for natural regeneration is seedling survival [24]. It was found that the decrease of *Q. acutissima* seedling density in the height class 20–40 cm was significant, and their age was less than 5 years. Ammer et al. [53] suggested that the chances of a seedling surviving intraspecific competition for *Fagus sylvatica* L. were strongly determined by its dominance ranking within the first 5 years after establishment. From 2014 to 2016, the relative soil water content in the forest was significantly lower than that in other years, and the 2–4-year seedlings experienced this stage or part of it. There were no significant differences in biomass, biomass distribution, RSA, and RV among 2–4-year seedlings, indicating that their growth was significantly inhibited. Root production would decrease at severe soil drought [55,56]. In particular, *Pinus taeda* L. ceased root growth between -0.3 and -1.2 MPa [57]. Furthermore, severe drought increased physical soil resistance, which is assumed to restrict root growth [58]. Although there was the highest soil water content at -20 cm and -30 cm, it it may have been difficult for the seedling roots of *Q. acutissima* to grow to this depth under severe drought stress. However, no significant differences in biomass allocation were found among 2–4-year seedlings. Sinz et al. [17] showed that the biomass allocation of *Fraxinus pennsylvanica* Marsh. seedlings did not change significantly. The biomass and RMR of 5–6-year seedlings increased significantly, which may be related to the fact that these seedlings experienced a more suitable soil water content in 2013, making their roots grow to a suitable soil depth. Ammer et al. [53] suggested that the older seedlings showed stronger adaptability under the forest. Therefore, the environmental variability in the *Q. acutissima* forest was a vital factor affecting the growth and survival of seedlings under the forest.

Morphological and physiological adaptation is an important way for seedlings to persist under the forest. TRL and RAD of *Q. acutissima* seedlings aged 2 to 4 years increased significantly. Correlation analysis showed that seedling biomass was significantly positively correlated with TRL and RV. The increase of root length fostered water uptake from the soil under drought stress [55,56]. The coarsening of roots was conducive to the storage of more NSC [59]. It was also found that NSC was positively correlated with RAD. Gaucher et al. [25] thought that allocating more NSC to the root system was conducive to the survival of *Acer saccharum* Marsh.seedlings under the forest. With the increase of age, the starch and soluble sugar of *Q. acutissima* seedlings increased significantly in the roots, stems, and leaves, consistent with the results of Lusk and Piper [28]. The increase of NSC did not improve the respiration of seedlings [60], and it was more sensitive than the changes of morphological indices [34]. We also found that the NSC of *Q. acutissima* seedlings aged 2 to 4 years changed more rapidly than morphological indices. Because there was a significantly positive correlation between seedling biomass and NSC indices, the increase of NSC content improved the survival of *Q acutissima* seedlings under the forest. For 5–6-year seedlings, RMR, TRL, RV, and RAD increased significantly, and the root morphological responses were more sensitive. Walters et al. [61] and Gaucher et al. [25] suggested that higher allocation to the roots favored long-term survival under the forest. Principal component analysis showed that root characteristics and NSC content were closely related to seedling growth. Therefore, the increase of root input and NSC content, as well as the regulation of biomass distribution, were the main ways for *Q. acutissima* seedlings to adapt to the variable environment under the forest.

5. Conclusions

The natural regeneration of plantations is a complex and long process, and the adaptability of seedlings under the forest is one of the key foundations for its successful regeneration. In the sample plot of the *Q. acutissima* plantation, the seedling bank was sufficient and stable, which was closely related to its shade tolerance. Compared with 2010, the seedling density decreased significantly in 2017. The most significant reduction in seedling density was observed for seedlings below 40 cm in height (seedling age $\leq$ 4 years), which is a critical stage for the natural regeneration of *Q. acutissima*.

During 2014 and 2016, precipitation decreased significantly, and SWC continued to decrease. The growth of 2–4-year seedlings was almost stagnant, and the regulation function of biomass distribution for themselves disappeared. Furthermore, their root growth was slow, which affected their root distribution in the 20–30 cm soil layer with the highest RSWC. Therefore, drought stress was the direct cause of the disappearance of 2–4-year seedlings. For those with seedling heights higher than 40 cm (seedling age $\geq$ 5a), which had root systems distributed in the appropriate soil depth, their growth was less affected. With the increase of age, the regulation ability of root morphology, NSC, and biomass allocation increased, which was the main way for *Q. acutissima* seedlings to adapt to drought stress in the understory. Therefore, as an important factor limiting the establishment of seedlings, drought stress together with seedling adaptation influenced the dynamics of the seedling bank in the understory of *Q. acutissima* plantations.

In future research, to provide theoretical support for the sustainable management of *Q. acutissima* plantations, we should strengthen the monitoring of environmental changes in the plantation, explore limiting mechanisms of seedling growth, and put forward reasonable management measures for natural regeneration.

Author Contributions: Conceptualization, P.M. and B.C.; formal analysis, R.N.; investigation, X.K. and Q.L.; resources, W.M. and B.Z.; data curation, P.G. and J.D.; writing—original draft preparation, L.G.; writing—review and editing, P.M. and B.C.; visualization, R.N.; supervision, B.C.; funding acquisition, B.C. All authors have read and agreed to the published version of the manuscript.

Funding: This research was funded by the Key R & D project of Shandong Province, China (2017CXGC0316), Funds of the Shandong "Double Tops Program" (SYL2017XTTD09) and Central Finance Forestry Reform and Development Fund (2020TG08).

Data Availability Statement: Not applicable.

Conflicts of Interest: The authors declare no conflict of interest.

References

1. Payn, T.; Carnus, J.M.; Freer-Smith, P.; Kimberley, M.; Kollert, W.; Liu, S.; Orazio, C.; Rodriguez, L.; Silva, L.N.; Wingfield, M.J. Changes in planted forests and future global implications. *For. Ecol. Manag.* **2015**, *352*, 57–67. [CrossRef]
2. Liu, S.; Yang, Y.; Wang, H. Development strategy and management countermeasures of planted forests in China: Transforming from timber-centered single objective management towards multi-purpose management for enhancing quality and benefits of ecosystem services. *Acta Ecol. Sin.* **2018**, *3*, 1–10. [CrossRef]
3. Jaworski, A.; Podlaski, R. Structure and dynamics of selected stands of primeval character in the Pieniny National Park. *Dendrobiology* **2007**, *58*, 25–41.
4. Szymura, T.H.; Dunajski, A.; Aman, I.; Makowski, M.; Szymura, M. The spatial pattern and microsites requirements of Abies alba natural regeneration in the Karkonosze Mountains. *Dendrobiology* **2007**, *58*, 51–57.

5. Madsen, P.; Hahn, K. Natural regeneration in a beech-dominated forest managed by close-to-nature principles—A gap cutting based experiment. *Can. J. For. Res.* **2008**, *38*, 1716–1729. [CrossRef]
6. Agestam, E.; Ekö, P.M.; Nilsson, U.; Welander, N.T. The effects of shelterwood density and site preparation on natural regeneration of Fagus sylvatica in southern Sweden. *For. Ecol. Manag.* **2003**, *176*, 61–73. [CrossRef]
7. Petritan, A.M.; Biris, I.A.; Merce, O.; Turcu, D.O.; Petritan, I.C. Structure and diversity of a natural temperate sessile oak (*Quercus petraea* L.)—European Beech (*Fagus sylvatica* L.) forest. *For. Ecol. Manag.* **2012**, *280*, 140–149. [CrossRef]
8. Pröll, G.; Darabant, A.; Gratzer, G.; Katzensteiner, K. Unfavourable microsites, competing vegetation and browsing restrict post-disturbance tree regeneration on extreme sites in the Northern Calcareous Alps. *Eur. J. For. Res.* **2015**, *134*, 293–308. [CrossRef]
9. Valladares, F.; Niinemets, Ü. Shade tolerance, a key plant feature of complex nature and consequences. *Annu. Rev. Ecol. Evol. Syst.* **2008**, *39*, 237–257. [CrossRef]
10. Jarčuška, B. Growth, survival, density, biomass partitioning and morphological adaptations of natural regeneration in Fagus sylvatica. A review. *Dendrobiology* **2009**, *61*, 3–11.
11. Valladares, F.; Laanisto, L.; Niinemets, Ü.; Zavala, M.A. Shedding light on shade: Ecological perspectives of understorey plant life. *Plant Ecol. Divers* **2016**, *9*, 237–251. [CrossRef]
12. Guariguata, M.R. Seed and seedling ecology of tree species in neotropical secondary forests: Management implications. *Ecol. Appl.* **2000**, *10*, 145–154. [CrossRef]
13. Barna, M.; Bosela, M. Tree species diversity change in natural regeneration of a beech forest under different management. *For. Ecol. Manag.* **2015**, *342*, 93–102. [CrossRef]
14. Pardos, M.; Montes, F.; Aranda, I.; Cañellas, I. Influence of environmental conditions on germinant survival and diversity of Scots pine (*Pinus sylvestris* L.) in central Spain. *Eur. J. For. Res.* **2007**, *126*, 37–47. [CrossRef]
15. Szewczyk, J.; Szwagrzyk, J. Spatial and temporal variability of natural regeneration in a temperate old-growth forest. *Ann. For. Sci.* **2010**, *67*, 202. [CrossRef]
16. Dobrowolska, D. Effect of stand density on oak regeneration in flood plain forests in Lower Silesia, Poland. *Forestry* **2008**, *81*, 511–523. [CrossRef]
17. Sinz, A.; Gardiner, E.S.; Lockhart, B.R.; Souter, R.A. Morphological acclimation and growth of ash (*Fraxinus pennsylvanica* Marsh.) advance regeneration following overstory harvesting in a Mississippi River floodplain forest. *For. Ecol. Manag.* **2011**, *261*, 246–254. [CrossRef]
18. Nagel, T.A.; Svoboda, M.; Rugani, T.; Diaci, J. Gap regeneration and replacement patterns in an old-growth Fagus-Abies forest of Bosnia-Herzegovina. *Plant Ecol.* **2010**, *208*, 307–318. [CrossRef]
19. Príncipe, A.; Nunes, A.; Pinho, P.; do Rosário, L.; Correia, O.; Branquinho, C. Modeling the long-term natural regeneration potential of woodlands in semi-arid regions to guide restoration efforts. *Eur. J. For. Res.* **2014**, *133*, 757–767. [CrossRef]
20. Rodriguez-Garcia, E.; Gratzer, G.; Bravo, F. Climatic variability and other site factor influences on natural regeneration of Pinus pinaster Ait. in Mediterranean forests. *Ann. For. Sci.* **2011**, *68*, 811–823. [CrossRef]
21. Dobrowolska, D.; Bolibok, L. Is climate the key factor limiting the natural regeneration of silver fir beyond the northeastern border of its distribution range? *For. Ecol. Manag.* **2019**, *439*, 105–121. [CrossRef]
22. Rodríguez-García, E.; Juez, L.; Bravo, F. Environmental influences on post-harvest natural regeneration of Pinus pinaster Ait. in Mediterranean forest stands submitted to the seed-tree selection method. *Eur. J. For. Res.* **2010**, *129*, 1119–1128. [CrossRef]
23. Szwagrzyk, J.; Szewczyk, J.; Bodziarczyk, J. Dynamics of seedling banks in beech forest: Results of a 10-year study on germination, growth and survival. *For. Ecol. Manag.* **2001**, *141*, 237–250. [CrossRef]
24. Rey, P.J.; Alcántara, J.M. Recruitment dynamics of a fleshy-fruited plant (*Olea europaea*): Connecting patterns of seed dispersal to seedling establishment. *J. Ecol.* **2000**, *88*, 622–633. [CrossRef]
25. Gaucher, C.; Gougeon, S.; Mauffette, Y.; Messier, C. Seasonal variation in biomass and carbohydrate partitioning of understory sugar maple (*Acer saccharum*) and yellow birch (*Betula alleghaniensis*) seedlings. *Tree Physiol.* **2005**, *25*, 93–100. [CrossRef]
26. Balandier, P.; Sinoquet, H.; Frak, E.; Giuliani, R.; Vandame, M.; Descamps, S.; Coll, L.; Adam, B.; Prevosto, B.; Curt, T. Six-year time course of light-use efficiency, carbon gain and growth of beech saplings (*Fagus sylvatica*) planted under a Scots pine (*Pinus sylvestris*) shelterwood. *Tree Physiol.* **2007**, *27*, 1073–1082. [CrossRef] [PubMed]
27. Claveau, Y.; Messier, C.; Comeau, P.G. Interacting influence of light and size on aboveground biomass distribution in sub-boreal conifer saplings with contrasting shade tolerance. *Tree Physiol.* **2005**, *25*, 373–384. [CrossRef]
28. Lusk, C.H.; Piper, F.I. Seedling size influences relationships of shade tolerance with carbohydrate-storage patterns in a temperate rainforest. *Funct. Ecol.* **2007**, *21*, 78–86. [CrossRef]
29. Sendall, K.M.; Reich, P.B.; Lusk, C.H. Size-related shifts in carbon gain and growth responses to light differ among rainforest evergreens of contrasting shade tolerance. *Oecologia* **2018**, *187*, 609–623. [CrossRef]
30. Gasser, D.; Messier, C.; Beaudet, M.; Lechowicz, M.J. Sugar maple and yellow birch regeneration in response to canopy opening, liming and vegetation control in a temperate deciduous forest of Quebec. *For. Ecol. Manag.* **2010**, *259*, 2006–2014. [CrossRef]
31. Soto, D.P.; Jacobs, D.F.; Salas, C.; Donoso, P.J.; Fuentes, C.; Puettmann, K.J. Light and nitrogen interact to influence regeneration in old-growth Nothofagus-dominated forests in south-central Chile. *For. Ecol. Manag.* **2017**, *384*, 303–313. [CrossRef]
32. Myers, J.A.; Kitajima, K. Carbohydrate storage enhances seedling shade and stress tolerance in a neotropical forest. *J. Ecol.* **2007**, *95*, 383–395. [CrossRef]

33. Lusk, C.H.; Falster, D.S.; Jara-Vergara, C.K.; Jimenez-Castillo, M.; Saldaña-Mendoza, A. Ontogenetic variation in light requirements of juvenile rainforest evergreens. *Funct. Ecol.* **2008**, *22*, 454–459. [CrossRef]
34. Deng, X.; Xiao, W.; Shi, Z.; Zeng, L.; Lei, L. Combined effects of drought and shading on growth and non-structural carbohydrates in Pinus massoniana lamb. Seedlings. *Forests* **2020**, *11*, 18. [CrossRef]
35. Duque, L.O.; Setter, T.L. Partitioning index and non-structural carbohydrate dynamics among contrasting cassava genotypes under early terminal water stress. *Environ. Exp. Bot.* **2019**, *163*, 24–35. [CrossRef]
36. Gleason, S.M.; Ares, A. Photosynthesis, carbohydrate storage and survival of a native and an introduced tree species in relation to light and defoliation. *Tree Physiol.* **2004**, *24*, 1087–1097. [CrossRef]
37. Veneklaas, E.J.; Den Ouden, F. Dynamics of non-structural carbohydrates in two Ficus species after transfer to deep shade. *Environ. Exp. Bot.* **2005**, *54*, 148–154. [CrossRef]
38. Sevanto, S.; Mcdowell, N.G.; Dickman, L.T.; Pangle, R.; Pockman, W.T. How do trees die? A test of the hydraulic failure and carbon starvation hypotheses. *Plant Cell Environ.* **2014**, *37*, 153–161. [CrossRef]
39. Galvez, D.A.; Landhäusser, S.M.; Tyree, M.T. Low root reserve accumulation during drought may lead to winter mortality in poplar seedlings. *New Phytol.* **2013**, *198*, 139–148. [CrossRef]
40. Ivanov, Y.V.; Kartashov, A.V.; Zlobin, I.E.; Sarvin, B.; Stavrianidi, A.N.; Kuznetsov, V.V. Water deficit-dependent changes in non-structural carbohydrate profiles, growth and mortality of pine and spruce seedlings in hydroculture. *Environ. Exp. Bot.* **2019**, *157*, 151–160. [CrossRef]
41. Tomasella, M.; Casolo, V.; Aichner, N.; Petruzzellis, F.; Savi, T.; Trifilò, P.; Nardini, A. Non-structural carbohydrate and hydraulic dynamics during drought and recovery in Fraxinus ornus and Ostrya carpinifolia saplings. *Plant Physiol. Biochem.* **2019**, *145*, 1–9. [CrossRef] [PubMed]
42. Imaji, A.; Seiwa, K. Carbon allocation to defense, storage, and growth in seedlings of two temperate broad-leaved tree species. *Oecologia* **2010**, *162*, 273–281. [CrossRef] [PubMed]
43. Piper, F.I.; Reyes-Díaz, M.; Corcuera, L.J.; Lusk, C.H. Carbohydrate storage, survival, and growth of two evergreen Nothofagus species in two contrasting light environments. *Ecol. Res.* **2009**, *24*, 1233–1241. [CrossRef]
44. Yang, Y.; Wang, C.; Liu, Y. The effect of low irradiance on growth, photosynthetic characteristics, and biomass allocation in two deciduous broad-leaved tree seedlings in southeast of Hubei Province. *Acta Ecol. Sin.* **2010**, *30*, 6082–6090.
45. Xue, W.Y.; Yang, B.; Zhang, W.H.; Yu, S.C. Spatial pattern and spatial association of Quercus acutissima at different developmental stages in the Qiaoshan mountains. *Acta Ecol. Sin.* **2017**, *37*, 3375–3384. [CrossRef]
46. Meng, Y.; Cao, B.; Dong, C.; Dong, X. Mount Taishan Forest ecosystem health assessment based on forest inventory data. *Forests* **2019**, *10*, 657. [CrossRef]
47. Meng, Y.; Cao, B.; Mao, P.; Dong, C.; Cao, X.; Qi, L.; Wang, M.; Wu, Y. Tree species distribution change study in Mount Tai based on Landsat remote sensing image data. *Forests* **2020**, *11*, 130. [CrossRef]
48. Zhou, Y.; Wu, D.; Yu, D.; Sui, C. Variations of nonstructural carbohydrate content in Betula ermanii at different elevations of Changbai Mountain, China. *Chin. J. Plant Ecol.* **2009**, *33*, 118–124. [CrossRef]
49. Lai, H.R.; Craven, D.; Hall, J.S.; Hui, F.K.C.; van Breugel, M. Successional syndromes of saplings in tropical secondary forests emerge from environment-dependent trait–demography relationships. *Ecol. Lett.* **2021**, *24*, 1776–1787. [CrossRef]
50. Dodson, E.K.; Root, H.T. Conifer regeneration following stand-replacing wildfire varies along an elevation gradient in a ponderosa pine forest, Oregon, USA. *For. Ecol. Manag.* **2013**, *302*, 163–170. [CrossRef]
51. Silva, D.; Rezende Mazzella, P.; Legay, M.; Corcket, E.; Dupouey, J.L. Does natural regeneration determine the limit of European beech distribution under climatic stress? *For. Ecol. Manag.* **2012**, *266*, 263–272. [CrossRef]
52. Kannenberg, S.A.; Schwalm, C.R.; Anderegg, W.R.L. Ghosts of the past: How drought legacy effects shape forest functioning and carbon cycling. *Ecol. Lett.* **2020**, *23*, 891–901. [CrossRef] [PubMed]
53. Ammer, C.; Stimm, B.; Mosandl, R. Ontogenetic variation in the relative influence of light and belowground resources on European beech seedling growth. *Tree Physiol.* **2008**, *28*, 721–728. [CrossRef] [PubMed]
54. Collet, C.; Lanter, O.; Pardos, M. Effects of canopy opening on the morphology and anatomy of naturally regenerated beech seedlings. *Trees Struct. Funct.* **2002**, *16*, 291–298. [CrossRef]
55. Di Iorio, A.; Montagnoli, A.; Scippa, G.S.; Chiatante, D. Fine root growth of Quercus pubescens seedlings after drought stress and fire disturbance. *Environ. Exp. Bot.* **2011**, *74*, 272–279. [CrossRef]
56. Zang, U.; Goisser, M.; Häberle, K.H.; Matyssek, R.; Matzner, E.; Borken, W. Effects of drought stress on photosynthesis, rhizosphere respiration, and fine-root characteristics of beech saplings: A rhizotron field study. *J. Plant Nutr. Soil Sci.* **2014**, *177*, 168–177. [CrossRef]
57. Torreano, S.J.; Morris, L.A. Loblolly Pine Root Growth and Distribution under Water Stress. *Soil Sci. Soc. Am. J.* **1998**, *62*, 818–827. [CrossRef]
58. Bengough, A.G.; Bransby, M.F.; Hans, J.; McKenna, S.J.; Roberts, T.J.; Valentine, T.A. Root responses to soil physical conditions; growth dynamics from field to cell. *J. Exp. Bot.* **2006**, *57*, 437–447. [CrossRef]
59. Gholz, H.L.; Cropper, W.P.J. Carbohydrate dynamics in mature pinus elliottii var. elliottii trees. *Can. J. For. Res.* **1991**, *21*, 1742–1747. [CrossRef]

60. Kobe, R.K. Carbohydrate Allocation to Storage as a Basis of Interspecific Variation in Sapling Survivorship and Growth. *Oikos* **1997**, *80*, 226. [CrossRef]
61. Walters, M.B.; Kruger, E.L.; Reich, P.B. Growth, biomass distribution and CO_2 exchange of northern hardwood seedlings in high and low light: Relationships with successional status and shade tolerance. *Oecologia* **1993**, *94*, 7–16. [CrossRef] [PubMed]

forests

MDPI

Article

Effects of Forest Gap and Seed Size on Germination and Early Seedling Growth in *Quercus acutissima* Plantation in Mount Tai, China

Peili Mao, Xiaoli Kan, Yuanxiang Pang, Ruiqiang Ni, Banghua Cao *, Kexin Wang, Jinhao Zhang, Chunxia Tan, Ying Geng, Xiaonan Cao, Shumei Wang, Peng Gao and Jinwei Dong

Taishan Mountain Forest Ecosystem Research Station, Key Laboratory of State Forestry Administration for Silviculture of the Lower Yellow River, Shandong Agricultural University, Tai'an 271018, China; maopl1979@163.com (P.M.); kanxiaoli1024@163.com (X.K.); pangyuanxiang2020@126.com (Y.P.); wind0309@163.com (R.N.); wkxsdau@163.com (K.W.); weibo18853883625@sina.com (J.Z.); tancx163@163.com (C.T.); aimee_gy@163.com (Y.G.); caoxiaonan_05@163.com (X.C.); wsm@sdau.edu.cn (S.W.); gaopengy@163.com (P.G.); jwdong@sdau.edu.cn (J.D.)

* Correspondence: caobanghua@126.com

Abstract: Elucidating the influence mechanisms of seed germination and seedling growth is important for revealing the natural regeneration of forest plantations. We collected the seeds from 58-year-old *Quercus acutissima* Carruth. forest, and the seeds were further divided into three classes: large, medium, and small, and sown under the forest gaps (I, 197.82 m^2; II, 91.85 m^2, III, understory) to observe seed germination and early seedling growth. Precipitation in the study area and soil moisture content in the forest gaps were also observed during the trial period. The results showed that the precipitation in 2019 was similar to that in 2020; both were significantly lower than the precipitation in 2021. The difference in soil water content between gaps I and II was not significant, and both were significantly lower than III. The order of seedling emergence rate in gaps was II > III > I, but the minimum was almost close to zero in I. Large and medium seeds showed significantly greater emergence rate than small seeds. The seedlings of II had higher seedling height, ground diameter, ground diameter relative growth rate, seedling biomass, root surface area, and root volume than those of III. Large seeds had the highest ground diameter, ground diameter relative growth rate, biomass, root mass ratio, root shoot ratio, and root surface area. Correlation analysis showed that seedling biomass was significantly and positively correlated with root surface area and root volume, and significantly and negatively correlated with specific root length and specific root surface area. The regulation of soil moisture in the gap and the adaptability related to seed size were two key factors influencing the seed germination and early seedling growth of *Q. acutissima*.

Keywords: precipitation; soil water content; biomass allocation; root system characteristics; natural regeneration

Citation: Mao, P.; Kan, X.; Pang, Y.; Ni, R.; Cao, B.; Wang, K.; Zhang, J.; Tan, C.; Geng, Y.; Cao, X.; et al. Effects of Forest Gap and Seed Size on Germination and Early Seedling Growth in *Quercus acutissima* Plantation in Mount Tai, China. *Forests* **2022**, *13*, 1025. https://doi.org/10.3390/f13071025

Academic Editor: Adele Muscolo

Received: 25 May 2022
Accepted: 28 June 2022
Published: 29 June 2022

1. Introduction

In recent decades, facing the pattern of global natural forest reduction and total forest resources decline, countries all over the world are rapidly developing forest plantations [1]. At present, China has the largest area of planted forests in the world, which plays an important role in greening the national territory, improving forest quality, improving the environment, and promoting the sustainable development of society. However, for a long time, the main purpose was to increase the speed of afforestation and the manufacture of timber products, ignoring the law of natural succession of forests. Current studies have shown that the natural regeneration of planted forests is extremely difficult to achieve [2,3]. In Austria, more than 2/3 of the forests required artificial measures to assist in natural regeneration. Seed germination and seedling establishment are critical stages of forest natural regeneration [4,5]. Many studies suggest that these stages are closely related to seed quality [5], seed dispersal [6], climatic factors [7–9], forest environment [8,10,11], and management modes [11–13], etc. Some scholars found that there could be a large density of seedlings under the forest canopy, but saplings were very scarce [3,10]. Therefore, the limiting factors for seed germination and seedling growth are not identical. An in-depth study of the influence mechanisms of limiting seed germination and seedling establishment is of great significance to solve the difficulties of natural regeneration in forest plantations.

Forest gap, an important factor affecting seed germination and seedling establishment of tree species, causes significant changes in environmental conditions in the forest, such as light intensity, air temperature, humidity, and soil characteristics [14]. With the increase in gap size, light intensity and air temperature increase significantly, and soil water content changes in a complex way [14]. Some have suggested that small and medium forest gaps have a small effect on seed germination, but large forest gaps are not conducive to seed germination [11], which is related to low soil water content and high temperature [3]. However, Bílek et al. [15] believed that seed germination was not a factor limiting seedling establishment as long as there was appropriate soil water content. Rozas [16] also thoughts that seed germination is not the limiting factor of oak seed germination, and this process is not affected by light. Forest gaps show a significant promotion effect on seedling growth under the forest canopy [17,18], but large forest gaps can inhibit their growth [11]. In addition, the promotion of understory seedling growth in forest gaps is also related to the shade tolerance of the tree species [12,19] and the age of the seedlings [20]. Forest gap promotes the growth of early seedlings in the forest understory [21], but few seedlings emerge in newly formed gaps [22,23]. The formation of new forest gaps provides more favorable conditions for the growth of light-demanding species [24]. Studies on oak species have shown that their natural regeneration was good in large forest gaps [16,25,26]. However, Grogan et al. [23] and Holladay et al. [27] concluded that seedling regeneration was best in medium-sized forest gaps, and excessive forest gaps can inhibit seedling regeneration. Therefore, the effect of forest gaps on seedling establishment is very complex.

Seed size is another key factor affecting seedling establishment. Numerous studies have shown that seed size is closely related to their germination and seedling growth. Large seeds have faster germination rates and seedling survival than small seeds, which has been consistently found in both inter-species and intra-species correlations [28–31]. This may be related to the fact that large seeds themselves store more nutrients available for early seedling growth [5]. Large seeds produce large seedlings with a highly competitive survival capacity [23,31,32]. Seedlings from large seeds show a lower relative growth rate than seedlings from small seeds [31,33]. However, Quero et al. [34] partially support this view. It has been found that large seeds improved the adaptive capacity of plants under adversities [30,33]. However, some studies have suggested that this advantage was not always available [31,35]. One of the studies on the germination of *Pinus thunbergii*. Parl. seeds under sand burial reported that the advantage of large seeds was shown under non-stressful stress [5]. Moreover, the advantage of large seeds gradually disappear with seedling growth [35,36]. Therefore, revealing the influence of seed size and its interaction with habitat on the seedling establishment is of great significance to the natural regeneration of the forest.

Quercus acutissima Carruth. is one of the main species of forest vegetation in warm temperate and subtropical regions of China. It is a pioneer tree species in barren hills and land and an excellent soil and water conservation species. It has high ecological, economic, and landscape value [37]. Field investigations found that, in highly shady forests, *Q. acutissima* seedlings died in large numbers and were poorly regenerated. Field investigation found that a *Q. acutissima* plantation has a sufficient and stable seedling bank [38]. The number of *Q. acutissima* seedlings in the gap was significantly higher than that in the forest understory, but the number of seedlings under the too-large gap decreased. The research on the growth of *Q. acutissima* seedlings under different light intensities found that it exhibited some shade tolerance, but it grew best under full light [39]. Therefore,

Q. acutissima has a natural regeneration ability, but its regeneration requires higher light intensity. However, there are few studies on the natural regeneration limitation mechanisms of *Q. acutissima* plantations. In this study, the effects of forest gap and seed size on seed germination and early seedling growth of *Q. acutissima* plantations, and the key restriction mechanism in the early stage of the seedling establishment, was revealed.

2. Materials and Methods

2.1. Study Area

The study site is located in Yaoxiang Forest Farm, Shandong Province (117°06′48.2″ N, 36°20′05.3″ E), with an average altitude of more than 700 m. It has a warm temperate continental monsoon climate. The average annual temperature is 13.5 °C, and the maximal and minimal temperatures are 41.5 °C and −20 °C, respectively. The frost-free period is 196 days. The average annual precipitation is 758 mm, with precipitation mainly concentrated in summer. Most of the forests at this farm were planted between the 1950s and the 1960s, with a total forest area of 9490 hm^2 and a forest coverage rate of 81.5%. The main tree species are *Q. acutissima* Carruth., *Juglans regia* L., *Robinia pseudoacacia* L., *Pinus armandii* Franch., *Castanea mollissima* Bl., *Platycladus orientalis* (L.) Franco, *Pinus thunbergii* Parl, and *Pinus densiflora* Sieb. et Zucc. The main shrubs are *Lespedeza bicolor* Turcz, *Spiraea fritschiana* Schneid., and *Vitis amurensis* Rupr. The vines mainly include *Celastrus orbiculatus* Thunb., *Rubus corchorifolius* L.f., and *Ziziphus mauritiana* Lam. The herbs mainly include *Artemisia lavandulaefolia* D.C., *Oplismenus undulatifolius* (Arduino) Beauv., *Bothriochloa ischaemum* (L.) Keng., and *Zoysia japonica* Steud.

2.2. Seed Collection

Seeds with full appearance and no obvious moth were collected by hand in the *Q. acutissima* plantation on 15 October 2018. The forest was planted in 1960 using one-year-old seedlings. Its density is 675 trees per hectare, the average tree height is 16.9 m, and the average diameter at breast height (DBH) is 21.2 cm. The collected seeds were first soaked in water to remove the floating seeds, then the remaining seeds were sorted by hand, and healthy seeds without pests and diseases were selected. According to the 1000-seed weight of *Q. acutissima* seeds, the seeds were divided into three seed size categories: large (1951.67 ± 66.90 g), medium (1634.00 ± 27.85 g), and small (826.23 ± 10.80 g).

2.3. Seed Germination and Seedling Growth Test

On 21 December 2018, according to the investigation results of the gap size of a *Q. acutissima* plantation in the forest farm, two forest gaps (I, with an area of 197.82 m^2; II, with an area of 91.85 m^2) and understory (III, 45 m^2) with the same slope, slope position, and slope aspect in the seed-harvesting *Q. acutissima* planted forest were selected for different sizes of seed germination tests. The large, medium, and small seeds (LS, MS, SS) were sown in each forest gap in four 50 cm × 50 cm plots with 40 seeds in each plot; thus, there were a total of 36 small sample plots. We removed the litter, loosened the topsoil slightly, then planted the seeds one by one, and finally re-covered the seeds with the litter to an approximately 3 cm depth. In August 2020, the number of existing seedlings was counted and the emergence rate (ER, number of seedlings/number of seeds sown × 100%) was calculated. One seedling was carefully excavated from each plot, and a total of 36 biennial seedlings were taken back to the laboratory. Then, another three to five seedlings of similar size were selected and tagged from each plot. The height and diameter of the selected seedlings were measured monthly from September 2020 to October 2021. Among these tagged seedlings, one three-year-old seedling was selected from each plot and carefully dug up. A total of 36 seedlings were taken back to the laboratory. Because the seedling emergence rate in gap I was too low and only a few seedlings from large seeds were found, they were not analyzed in the seedling growth test. The relative height growth rate (RGR_H, cm cm^{-1} d^{-1}) and relative ground diameter growth rate (RGR_D, mm mm^{-1} d^{-1}) for each month were calculated with Equations (1) and (2), respectively.

$$RGR_H = (\ln H2 - \ln H1)/(t2 - t1) \quad (1)$$

$$RGR_D = (\ln D2 - \ln D1)/(t2 - t1) \quad (2)$$

H is the seedling height (cm) and D is the seedling ground diameter (mm). Subscripts refer to the first month (1) and the next month (2), and t2 − t1 is the number of days between two adjacent months.

In the laboratory, the seedlings were divided into root, stem, and leaf. The area of each leaf of each seedling was determined using a CI-202 portable laser leaf area meter (CID Inc., Washington, DC, USA). The soil on the root system was first rinsed off with water, and after scanning the root system with an HP Scanjet 8200 scanner, the scanned images were analyzed with Delta-T Area Meter Type AMB2 root parameter analysis software to obtain root length (RL), root surface area (RSA), and root volume (RV). After scanning, the roots, stems and leaves were placed in an envelope and put inside an oven at 120 °C for 30 min to inactivate, and then the temperature was adjusted to dry the moisture at 80 °C to a constant weight, and the dry weight was weighed. Seedling biomass (SB, root dry weight + stem dry weight + leaf dry weight), specific leaf area (SLA, leaf area/leaf dry weight), leaf area ratio (LAR, leaf area/biomass), root mass ratio (RMR, root dry weight/seedling biomass), stem mass ratio (SMR, stem dry weight/seedling biomass), leaf mass ratio (LMR, leaf dry weight/seedling biomass), root/shoot ratio (RS, root dry weight)/(leaf dry weight + stem dry weight)), photosynthetic tissues/non-photosynthetic tissues (P/NP, leaf mass/(stem + root mass)), Specific root length (SRL, root length/root dry weight) and specific root surface area (SRA, root surface area/root dry weight) were calculated.

2.4. Measurement of Precipitation and Soil Water Content in the Forest

The precipitation at 1.5 m in the forest was measured from January 2019 to October 2021 using a CR3000 automatic weather station (Campbell Scientific, Logan, UT, USA), which automatically recorded daily data every 10 min. Data were collected every month.

Soil water content in the forest gaps and in the understory was measured monthly from September 2020 to October 2021. Three soil samples were randomly taken from each forest gap monthly with a soil auger at a depth of 0–10 cm and were then placed in an aluminum box and brought back to the laboratory to measure the fresh weight of the soil (m) with an electronic analytical balance. The weighed soil samples were placed in an oven at 105 °C, dried for 8 h, and then taken out to measure the mass of the dried soil samples (ms), and the soil water content (SWC) was measured according to the formula:

$$SWC\ (\%) = (m - ms)/ms \times 100\% \quad (3)$$

2.5. Statistical Analysis

The precipitation effect was tested by a two-way ANOVA, with year and month as the sources of variations. The soil water content was tested by a two-way ANOVA, with gap size and month as the sources of variation. The indexes of seedling growth (height, ground diameter, relative height growth rate, and relative ground diameter growth rate) were tested by a three-way ANOVA, with gap size, seed size, and month as the sources of variation. The biomass and morphological parameters of seedlings were also analyzed by three-factor ANOVA, with gap size, seed size, and age as the sources of variation. Before the analysis of variance, all the data passed the test of normality and homogeneity of variance. When ANOVA indicated a significant overall treatment effect, a multiple comparison test (post-hoc Least Significant Difference test) was carried out to compare the parameter means. The correlation analysis of seedling biomass, biomass allocation, and root morphological indicators was carried out. All the seedling indices were analyzed by principal component analysis (PCA) to identify the main factors relevant to seedling biomass. The statistically significant level was set at $p = 0.05$. All statistics were conducted with SPSS for Windows 26.0 (SPSS, Chicago, IL, USA).

3. Results

3.1. Monthly Dynamics of Precipitation and Soil Water Content

3.1.1. Precipitation

It was shown that the month ($F = 5.07$, $p < 0.01$) and the year ($F = 6.42$, $p < 0.01$) had significant effects on precipitation. As the month increases, precipitation first increased and then decreased, with the highest values appearing between July and September (Figure 1). The difference in precipitation was not significant between in 2019 and in 2020, and both were significantly lower than that in 2021. The precipitation distribution differed between the three years. The precipitation in 2019 was mainly concentrated in July and August, with high rainfall from April to July in 2020, and concentrated precipitation from June to October in 2021 (Figure 1).

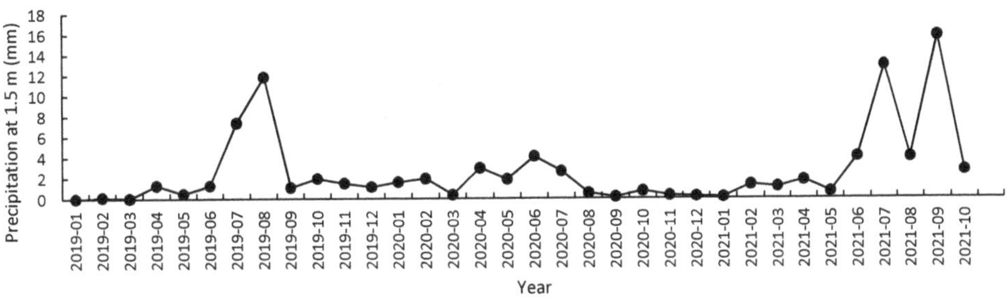

Figure 1. Change of daily average precipitation at 1.5 m in *Quercus acutissima* forest from January 2019 to October 2021.

3.1.2. Soil Water Content of Different Gap Sizes

The month ($F = 30.74$, $p < 0.01$) and the forest gap size ($F = 37.97$, $p < 0.01$) showed significant effects on soil water content. As the month increased, the soil water content fluctuated, with less variation from September 2020 to June 2021, and it varied greatly from July 2021 to September 2021, with the highest values of soil water content in September 2021, followed July 2021 (Figure 2). Among the forest gaps, the difference between the soil water content of gap I and gap II was not significant ($p > 0.05$), and both were significantly lower than that of understory III ($p < 0.01$). Significant interaction effects between the month and gap size were found ($F = 10.15$, $p < 0.01$). From November 2020 to April 2021, the rank of soil water content was gap I < II < III (Figure 2).

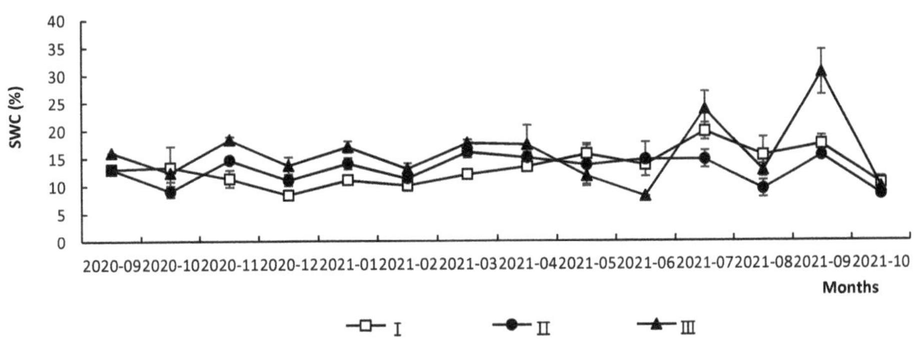

Figure 2. Monthly dynamic changes of soil water content under different gap sizes in *Quercus acutissima* forest from September 2020 to October 2021.

3.2. *Effects of Forest Gap and Seed Size on Seedling Germination Rate*

The results showed that gap size had a significant effect on ER ($F = 21.53, p < 0.01$), and seed size also had a significant effect ($F = 5.29, p < 0.05$). The interaction between the two factors was not significant ($F = 2.57, p > 0.05$). With the decreasing forest gap size, ER first increased and then decreased (Figure 3), ranked as II > III > I ($p < 0.05$). ER in gap I was extremely low, and was only 2.92 ± 4.74%. Between seed sizes, ER did not show significant differences between LS and MS ($p > 0.05$), and both were significantly higher than that of SS ($p < 0.05$).

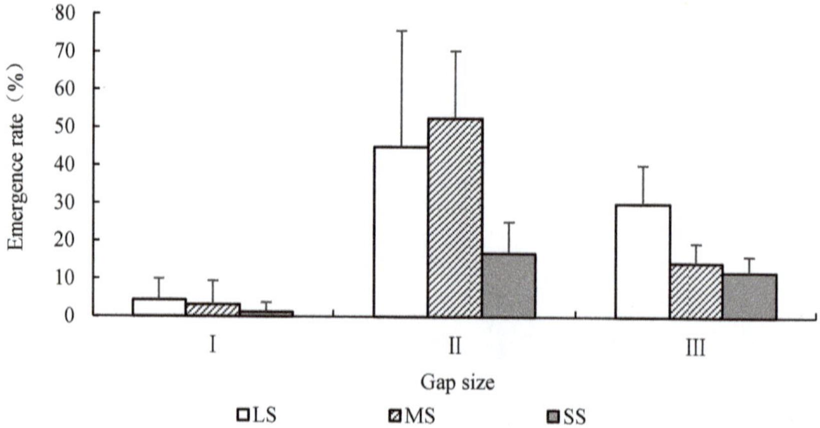

Figure 3. The emergence rate of seeds with different sizes of *Quercus acutissima* with the decrease of forest gap. LS: large seed; MS: medium seed; SS: Small seed.

3.3. *Effects of Forest Gap and Seed Size on Seedling Growth Dynamics*

3.3.1. Seedling Height and Ground Diameter

The results showed that the month had a highly significant effect on SH and GD (Table 1). SH and GD increased significantly with the month (Figure 4). The seed size did not have a significant effect on SH but had a significant effect on GD (Table 1). The difference in GD between LS and MS was not significant ($p > 0.05$) and both were significantly larger than the GD of SS ($p < 0.01$). The forest gap showed a significant effect on SH and GD (Table 1), and GD of gap II was significantly larger than that of understory III. The interaction between the month and forest gap exhibited a highly significant effect on SH and GD (Table 1). SH and GD increased significantly more between April 2021 and October 2021 than those between September 2020 and March 2021 (Figure 4). The interaction of seed size and forest gap had a highly significant effect on SH and GD (Table 1). Seedlings of different sizes in gap II were close in SH and GD, but seedlings from LS in understory III had the highest SH and GD, while the seedlings from SS were close to LS in SH and close to MS in GD (Figure 4).

Table 1. Three-factor analysis of variance of time, seed size, and gap size on seed germination and seedling indexes of *Quercus acutissima*. ER: emergence rate; SH: seedling height; GD: ground diameter; RGR_H: relative height growth rate; RGR_D: relative ground diameter growth rate.

Index	Time (A)		Seed Size (B)		Forest Gap (C)		A × B		A × C		B × C		A × B × C	
	F Value	*p* Value	F Value	*p* Value	F Value	*p* Value	F Value	*p* Value	F Value	*p* Value	F Value	*p* Value	F Value	*p* Value
ER	—	—	5.29	<0.05	21.53	<0.01	—	—	—	—	2.57	0.06	—	—
SH	85.45	<0.01	1.45	0.24	502.31	<0.01	0.23	1.00	4.68	<0.01	39.13	<0.01	0.35	1.00
GD	26.28	<0.01	27.11	<0.01	409.45	<0.01	0.74	0.83	4.12	<0.01	10.65	<0.01	0.04	1.00
RGR_H	86.81	<0.01	0.28	0.76	1.40	0.237	0.52	0.97	1.27	0.23	1.97	0.14	1.33	0.14
RGR_D	20.11	<0.01	7.70	<0.01	38.16	<0.01	1.47	0.07	3.31	<0.01	0.35	0.71	0.53	0.97

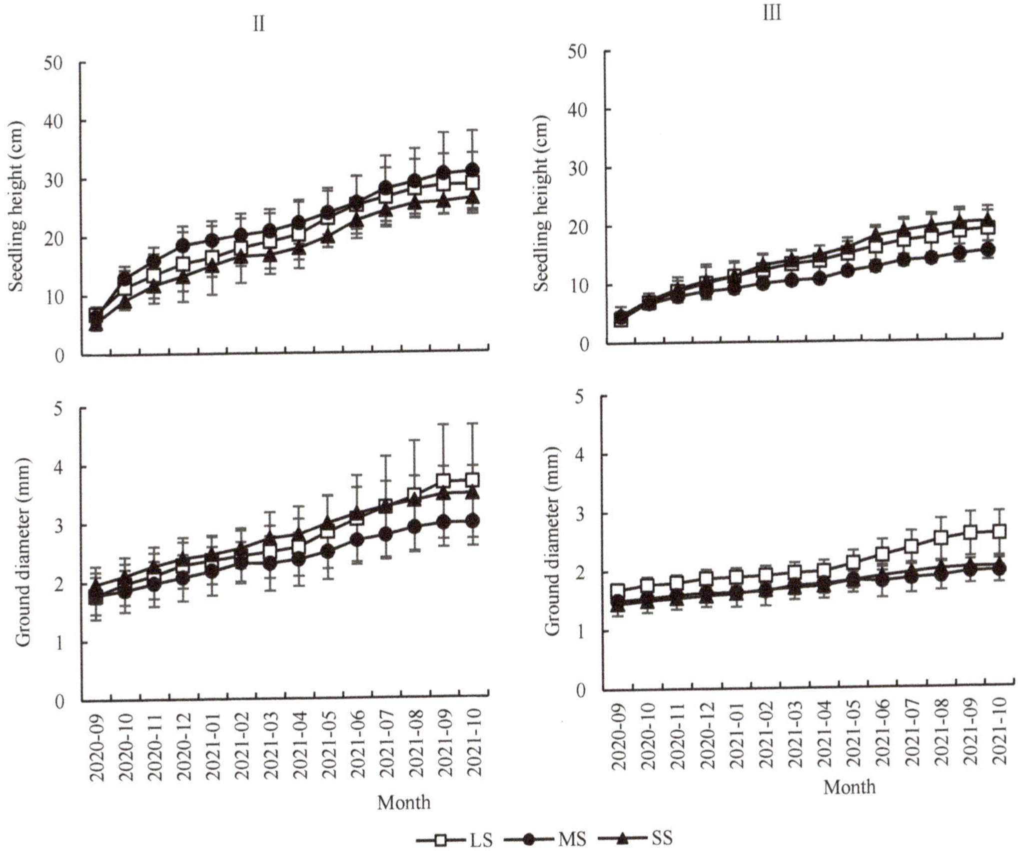

Figure 4. Monthly dynamic changes of seedling height and ground diameter of *Quercus acutissima* seedlings from different sizes of seeds under different forest gaps. LS: large seed; MS: medium seed; SS: small seed.

3.3.2. Relative Growth Rate

RGR_H was significantly affected by month but not affected by seed size or forest gap (Table 1). RGR_H decreased significantly with the increasing month in a waved pattern, with a slight increase from May to July 2021 (Figure 5). The month, seed size, and forest gap showed significantly effects on RGR_D (Table 1). RGR_D first increased and then decreased with the increasing month, with higher values between December 2020 and May 2020 to September 2020 (Figure 5). Among seed sizes, RGR_D from LS was significantly greater than that of MS and SS ($p < 0.01$), while the difference between MS and SS was not significant ($p > 0.05$). RGR_D in gap II was significantly greater than that of understory III (Table 1). The interaction of the month and forest gap showed a significant effect on RGR_D (Table 1). As the time increased, the difference in RGR_D among forest gaps became smaller (Figure 5).

3.4. Effects of Forest Gap and Seed Size on Seedling Biomass and Biomass Allocation Dynamics

The age, seed size, and forest gap showed significant effects on SB, and there was a significant interaction between seed size and forest gap (Table 2). The SB of biennial seedlings was lower than that of three-year-old seedlings ($p < 0.01$), with an average of 1.65 ± 0.88 g and 2.82 ± 1.77 g, respectively. SB was greater in gap II than in gap III (Figure 6). With different seed sizes, the difference in SB between LS and MS was not significant, and both were greater than the SB from SS. The larger the seeds, the stronger the light intensity, and then the greater the SB (Figure 6).

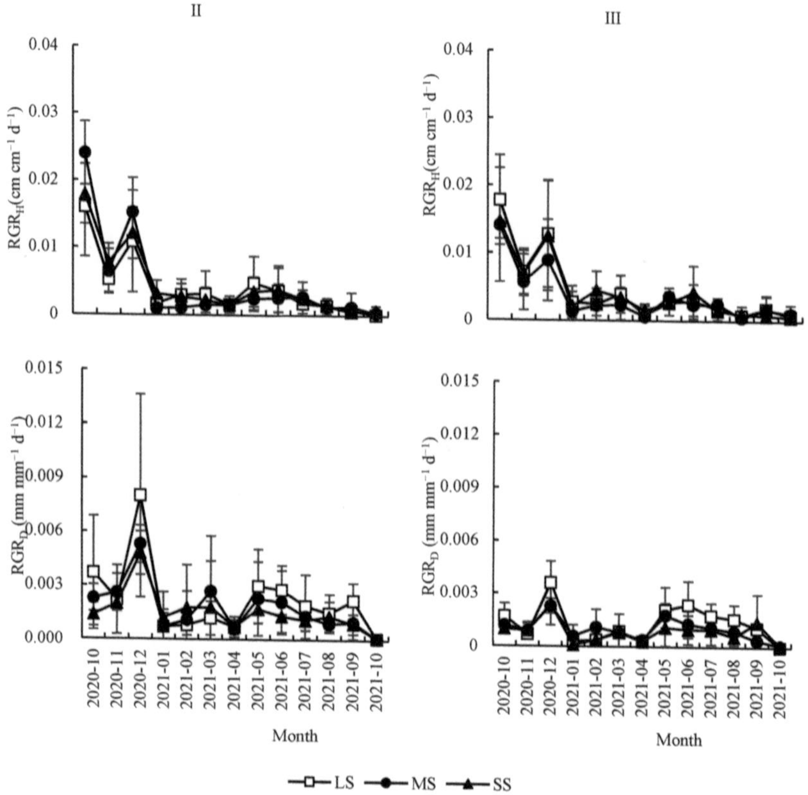

Figure 5. Monthly dynamic changes of the relative growth of height and ground diameter of *Quercus acutissima* seedlings from different sizes of seeds under different forest gaps. LS: large seed; MS: medium seed; SS: small seed.

Table 2. Three-factor analysis of variance of age, seed size, and gap size on seed germination and seedling indexes of *Quercus acutissima*. SB: seedling biomass; RMR: root mass ratio; SMR: stem mass ratio; LMR: leaf mass ratio; R/S: root/shoot ratio; P/NP: photosynthetic tissues/non-photosynthetic tissues; SLA: specific leaf area; LAR: leaf area ratio; RL: root length; RSA: root surface area; RV: root volume; SRL: specific root length; SRA: specific root surface area.

Index	Age (A)		Seed Size (B)		Forest Gap (C)		A × B		A × C		B × C		A × B × C	
	F Value	*p* Value	F Value	*p* Value	F Value	*p* Value	F Value	*p* Value	F Value	*p* Value	F Value	*p* Value	F Value	*p* Value
SB	15.59	<0.01	4.84	<0.05	22.74	<0.01	3.46	0.06	3.39	0.08	6.35	<0.01	3.04	0.08
RMR	2.19	0.16	3.76	<0.05	0.27	0.61	0.37	0.70	1.04	0.32	0.06	0.95	0.67	0.53
SMR	0.36	0.56	2.75	0.09	2.57	0.13	1.54	0.25	0.25	0.63	3.90	<0.05	0.58	0.57
LMR	3.42	0.08	0.76	0.49	0.31	0.59	0.57	0.58	0.45	0.51	1.17	0.34	1.50	0.25
R/S	3.61	0.08	4.56	<0.05	0.25	0.62	1.05	0.37	0.48	0.50	0.02	0.98	0.95	0.41
P/PN	2.52	0.13	0.78	0.48	0.06	0.81	0.62	0.55	0.78	0.39	1.00	0.39	1.35	0.29
SLA	15.04	<0.01	0.44	0.65	2.07	0.17	0.42	0.67	2.34	0.15	2.97	0.08	2.76	0.09
LAR	2.01	0.18	1.70	0.21	0.43	0.52	0.19	0.83	0.96	0.34	0.00	1.00	0.87	0.44
RL	0.34	0.57	1.14	0.34	1.31	0.27	0.58	0.57	0.09	0.77	0.35	0.71	0.99	0.39
RSA	1.98	0.18	3.90	<0.05	9.64	<0.01	2.27	0.14	0.00	0.99	0.38	0.69	1.09	0.36
RV	4.75	<0.05	3.36	0.06	12.35	<0.05	2.98	0.08	0.39	0.54	2.08	0.16	2.35	0.13
SRL	0.78	0.39	0.75	0.49	2.33	0.15	0.45	0.65	0.90	0.36	2.59	0.11	0.85	0.44
SRA	5.48	<0.05	1.40	0.28	3.80	0.07	0.11	0.89	0.46	0.51	4.00	<0.05	0.57	0.58

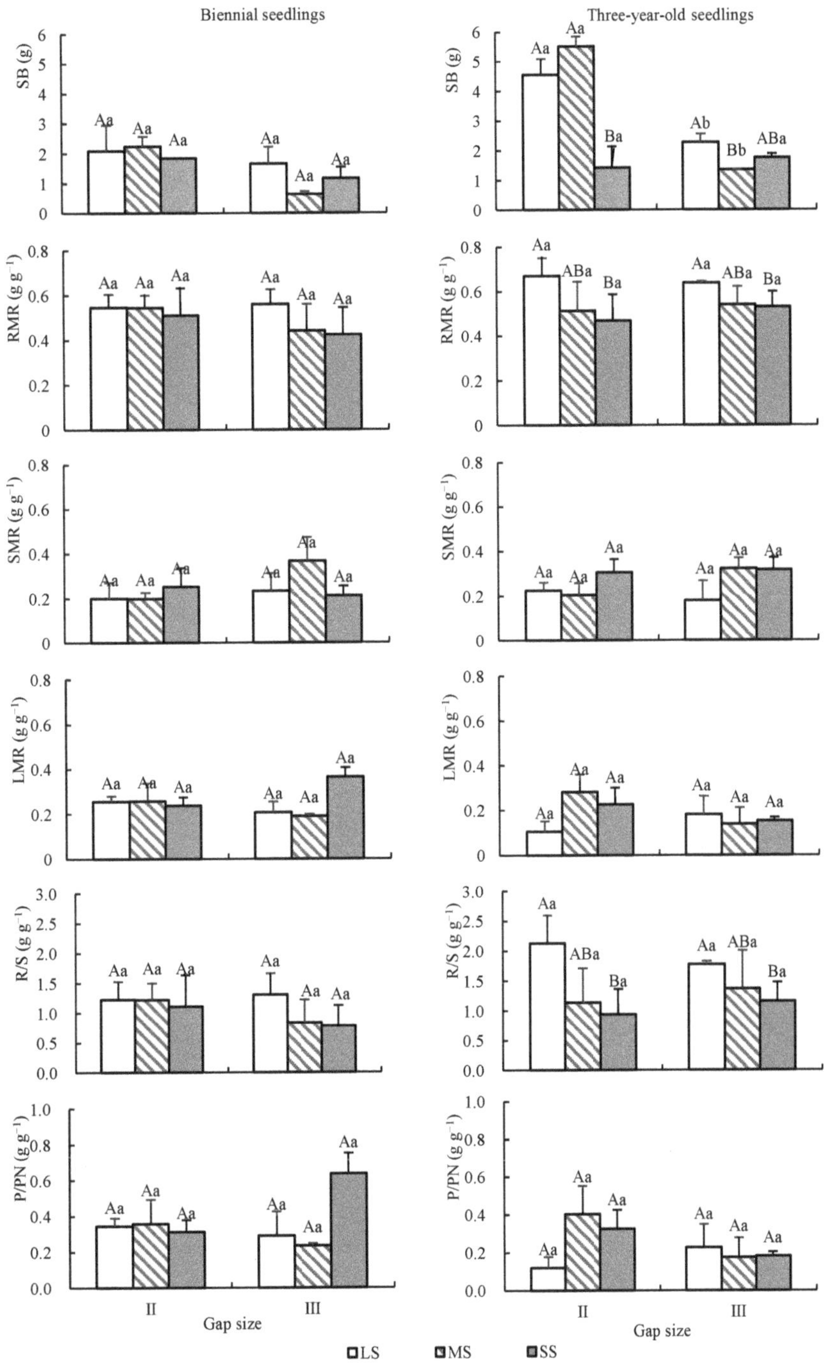

Figure 6. Dynamic changes of biomass and biomass allocation of *Quercus acutissima* seedlings from different sizes of seeds under different forest gaps. LS: large seed; MS: medium seed; SS: small seed. Different capital letters indicate that there is a significant difference between different seed size groups under the same gap ($p < 0.05$); different lower letters indicate that there is a significant difference between gaps under the same seed size ($p < 0.05$).

The age and gap size had no significant effects on RMR, SMR, LMR, R/S, or P/PN. The seed size showed significant effects on RMR and R/S, but no significant effects on SMR, LMR, and P/PN (Table 2). Among different seed sizes, LS was significantly greater than SS in RMR and R/S, but MS was not significantly different from LS and SS, respectively (Figure 6).

3.5. Effects of Forest Gap and Seed Size on Leaf Characteristics Dynamic

Results showed that age had a significant effect on SLA and no significant effect on LAR. Seed size and forest gap had no significant effect on SLA and LAR (Table 2). The SLA of three-year-old seedlings was significantly larger than that of biennial seedlings ($p < 0.01$), and the average values were 360.79 ± 214.22 $cm^2\ g^{-1}$ and 169.08 ± 29.04 $cm^2\ g^{-1}$, respectively (Figure 7).

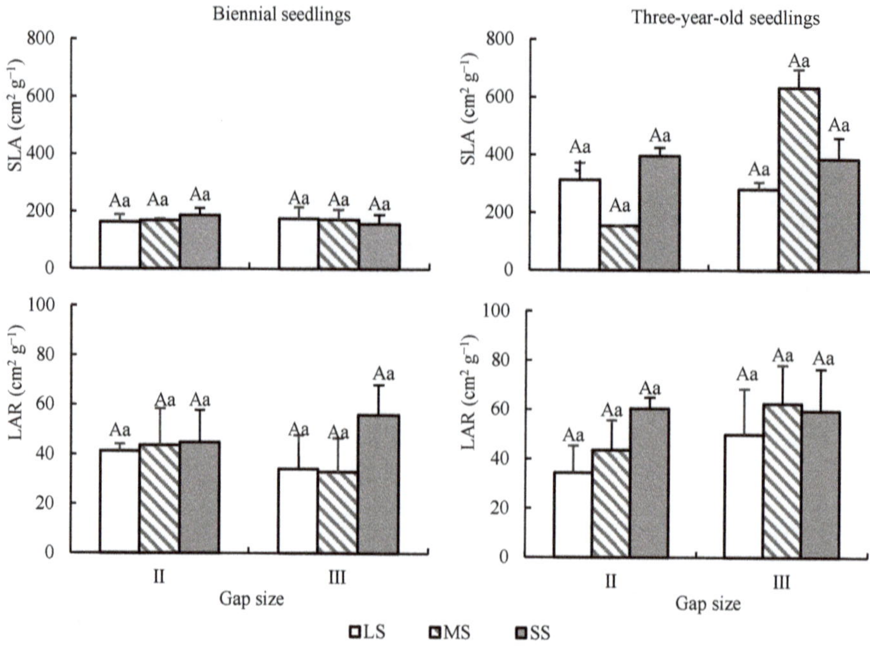

Figure 7. Dynamic changes of leaf characteristics of *Quercus acutissima* seedlings from different sizes of seeds under different forest gaps. LS: large seed; MS: medium seed; SS: small seed. Different capital letters indicate that there is a significant difference between different seed size groups under the same gap ($p < 0.05$), and different lower letters indicate that there is a significant difference between gaps under the same seed size ($p < 0.05$).

3.6. Effects of Forest Gap and Seed Size on Root Characteristics Dynamics

Among the root indicators, age had a significant effect on RV and SRA, seed size had a significant effect on RSA, forest gap had a significant effect on RSA and RV, and the effects of all other factors were not significant for root system indicators (Table 2). RV of biennial seedlings was lower than that of three-year-old seedlings ($p < 0.05$), with an average of 0.71 ± 0.35 cm^3 and 1.02 ± 0.68 cm^3, respectively. However, the SRA of biennial seedlings was larger than that of three-year-old seedlings ($p < 0.05$), and the average values were 30.13 ± 8.97 $cm^2\ g^{-1}$ and 21.87 ± 13.34 $cm^2\ g^{-1}$. Among different seed sizes, LS had higher RSA than SS, while MS was not significantly different from LS and SS. RSA and RV were higher in II than in III (Figure 8).

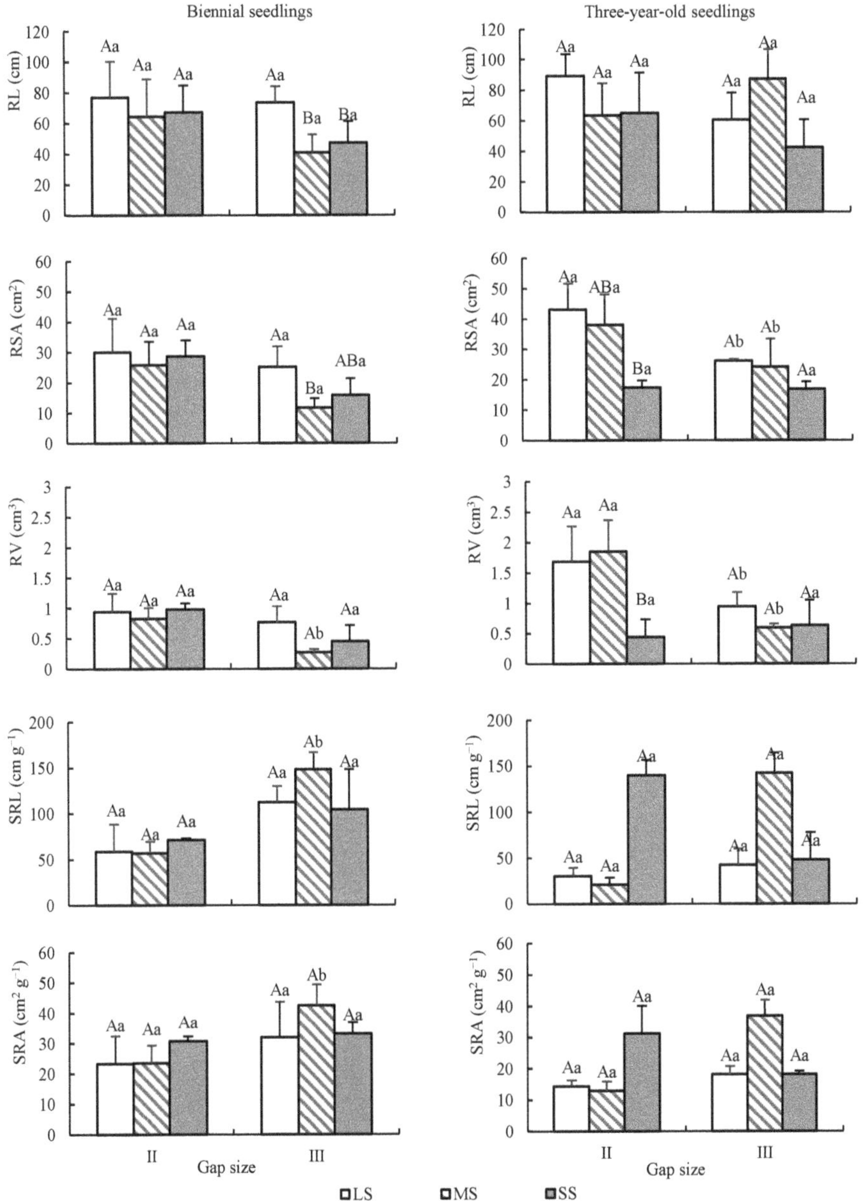

Figure 8. Dynamic changes of root characteristics of *Quercus acutissima* seedlings from different sizes of seeds under different forest gaps. LS: large seed; MS: medium seed; SS: small seed. Different capital letters indicate that there is a significant difference between different seed size groups under the same gap ($p < 0.05$), and different lower letters indicate that there is a significant difference between gaps under the same seed size ($p < 0.05$).

3.7. Correlation Analysis of Seedling Characteristics

SB was significantly and positively correlated with RSA and RV, and significantly and negatively correlated with SRL and SRA, which indicated that root characteristics were the main factors determining seedling growth. RMR was significantly negatively correlated with LMR, P/PN, SRL, and SRA, and was significantly positively correlated with R/S, RSA, and RV. RSA and RV were significantly negatively correlated with SRL and SRA (Table 3), indicating that there were differences in root resource utilization.

Table 3. Correlation analysis among seedling characteristics of *Quercus acutissima*. SB: seedling biomass; RMR: root mass ratio; SMR: stem mass ratio; LMR: leaf mass ratio; R/S: root/shoot ratio; P/NP: photosynthetic tissues/non-photosynthetic tissues; SLA: specific leaf area; LAR: leaf area ratio; RL: root length; RSA: root surface area; RV: root volume; SRL: specific root length; SRA: specific root surface area. *, $p < 0.05$; **, $p < 0.01$.

Index	SB	RMR	SMR	LMR	R/S	P/PN	SLA	LAR	RL	RSA	RV	SRL	SRA
SB	1												
RMR	0.31	1											
SMR	−0.37	−0.34	1										
LMR	−0.04	−0.72 **	−0.40 *	1									
R/S	0.37	0.96 **	−0.32	−0.70 **	1								
P/PN	−0.06	−0.72 **	−0.37	0.97 **	−0.65 **	1							
SLA	−0.09	0.34	0.35	−0.58 **	0.35	−0.49 **	1						
LAR	−0.14	−0.51 **	0.00	0.49 **	−0.52 **	0.48 *	0.19	1					
RL	0.28	−0.12	−0.01	0.16	−0.06	−0.11	0.16	−0.16	1				
RSA	0.80 **	−0.15	−0.27	0.48 *	−0.40 *	−0.17	0.50 **	−0.22	0.68 **	1			
RV	0.91 **	−0.12	−0.32	0.50 **	−0.45 *	−0.16	0.54 **	−0.19	0.33	0.91 **	1		
SRL	−0.60 **	−0.13	0.14	−0.44 *	0.46 *	0.09	−0.41 *	0.08	0.32	−0.39 *	−0.62 **	1	
SRA	−0.70 **	−0.21	0.08	−0.50 **	0.48 *	0.13	−0.48 **	0.12	0.19	−0.44 *	−0.64 **	0.94 **	1

The results of the PCA showed that the first and second principal components represented 65.73% of the total variables (Figure 9). The coefficient of each index, eigenvalue, variance contribution rate, and accumulated contribution rate are shown in Table A1. The first principal component (41.57%) reflected the root system characteristics related to biomass, including R/S, RMR, RV, RSA, SRA, SB, and SRL. The second principal component (24.16%) was related to the biomass distribution characteristics of ground parts, including LMR, P/NP, SMR, and SLA.

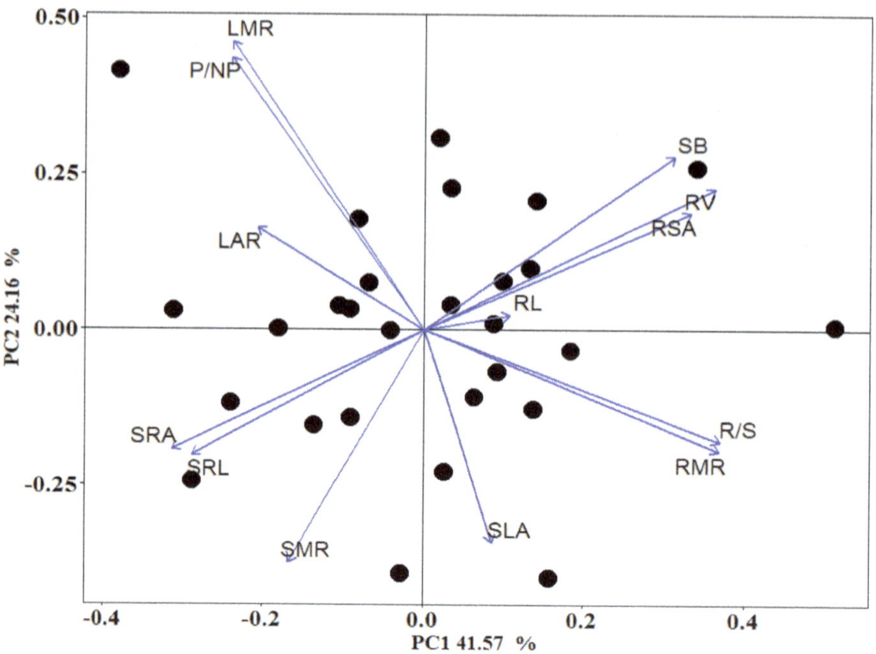

Figure 9. Principal component analysis of *Quercus acutissima* seedling traits. SB: seedling biomass; RMR: root mass ratio; SMR: stem mass ratio; LMR: leaf mass ratio; R/S: root/shoot ratio; P/NP: photosynthetic

tissues/non-photosynthetic tissues; SLA: specific leaf area; LAR: leaf area ratio; RL: root length; RSA: root surface area; RV: root volume; SRL: specific root length; SRA: specific root surface area.

4. Discussion

Precipitation was observed to be the major factor affecting soil water content in nature [11,40]. In this study, there was minimal precipitation from January to June and a significantly increased rainfall in July 2019. The seeds of *Q. acutissima* were found to germinate in July. Therefore, soil water content was a dominant factor for the seed germination of *Q. acutissima*, which is in agreement with those obtained for *Pinus sylvestris* L. [11] and *Pinus pinaster* Ait. [13]. In addition, we found that forest gaps had a significantly modulatory influence on soil water content. From November 2020 to April 2021, the soil water content of forest gap I was the lowest, which was related to less precipitation in the same period. The higher light radiation in large forest gaps than in understory contributed to the higher temperature and thus higher evaporation, leading to variable soil humidity [41]. The seeds of *Q. acutissima* are mild recalcitrance and have poor dehydration tolerance. Research on *Q. robur* L. [42] found that the lower the soil water content was, the poorer the germination capacity was. Thus, the low soil water content and long duration of drought were the major factors for the almost zero seedling emergence rate of gap I. The soil water content in gap III was the highest, and was the most important reason for the higher seedling emergence rate in the understory. However, the highest germination rate was in II, in addition to its relation to the higher soil water content, which was also associated with higher temperature. These findings clearly indicated that the smaller forest gaps were more favorable for seed germination. Moreover, we found that the seedling emergence rate of large and medium seeds was higher than that of small seeds. This finding is consistent with prior research [28–31]. The germination advantage of large seeds has been related to their high nutrient contents [5]. Large seeds had the highest germination rate and had a high-stress tolerance ability, which was in line with the finding of Paz and Martínez-Ramos [33]. Together, the above findings suggest that seed germination of *Q. acutissima* is determined mainly by precipitation and is simultaneously obviously regulated by the size of the forest gap and by seeds.

Growth is an important indicator of plant adaptation to external environments [43]. In this study, results showed that the seedlings of *Q. acutissima* in gap II had higher seedling height, ground diameter, and biomass than those in gap III. This indicated that the forest gap contributed to the growth of seedlings, which was in line with the results found by Collet et al. [17] and Diaci et al. [18]. The seedling size of *Q. acutissima* in forest gaps was higher than in understory, indicating a high demand for light, consistent with results from other research [16,25,26]. There were no significant effects of forest gaps on the relative growth rate of seedling height. Grogan et al. [23] suggested that seedling height determines growth and survival after forest gap formation. For the relative ground diameter growth rate, gap II was significantly greater than that of understory III. The high relative ground diameter growth rate reflects the increased storage capacity of the seedlings and their ability to adapt to the understorey habitat [43,44]. Therefore, the adaption ability of seedlings in forest gaps meant that they performed better than those in the understory. Additionally, we found that large seeds of *Q. acutissima* had higher biomass, ground diameter, and relative growth rate of ground diameter than small seeds, which led to the higher adaptation ability of large seeds compared to small seeds. furthermore, the reserved nutrients of large seeds affects the growth of seedlings over a longer period [35,36]. Previous studies [33,34], however, considered that the relative growth rate of small seeds was higher than that of large seeds. Quero et al. [34] found only two species supported this conclusion. These indicated that the effect of seed size on the growth of seedlings was complicated. Forest gaps and large seeds contributed to the rapid increase in seedling ground diameter, which was an important approach to improving adaptation ability to the environment within the forest.

Morphological adaptation is one of the primary ways for plants to adapt to the external environment [45]. It suggests there was no significant difference in biomass allocation of seedlings of *Q. acutissima* among forest gaps. The investigation of the natural regeneration of seedlings under *Q. acutissima* forest also showed that no significant difference existed in biomass allocation between 2–4-year-old seedlings. This was believed to be related to severe soil drought during its growth period [38]. The annual rainfall was 519.72 mm in 2020, close to the rainfall from 2014 to 2016, and thus seedlings were under severe drought stress. Therefore, the biomass allocation of seedlings was limited by severe drought stress. Seedlings of *Q. acutissima* in gap II had higher total root surface area and root volume compared to III, indicating a higher ability for the absorption of soil water and nutrients. The root is the major organ for storing non-structural carbohydrates [46,47]. The larger root volume was beneficial for *Q. acutissima* to store more non-structural carbohydrates and to improve the adaptability of seedlings in the understory [38]. Herrear [48] also argued that root development of *Quercus suber* L. was crucial for understory survival. Among seeds of different sizes, large seeds generally have a higher root mass ratio and root/shoot ratio than small seeds. In the understory, a high root/shoot ratio was beneficial to its adaptation to drought stress [47,49]. Seedlings from large seeds of *Q. acutissima* had a higher total root surface area than seedlings from small seeds, which indicated that the root development of large seeds was better than that of small seeds. Correlation analysis showed that the seedling biomass of *Q. acutissima* was significantly positively correlated with total root surface area and root volume. Principal component analysis showed that root characteristics were the key factors affecting seedling biomass. Therefore, the regulation of roots was an important way for seedlings of *Q. acutissima* to adapt to the environment within the forest, which was consistent with the results of previous studies [38].

Through research on seed germination and the early seedling growth of *Q. acutissima*, it was found that the two stages showed different requirements for forest gaps, which was consistent with the conclusions of other scholars regarding some tree species [3,10]. The low precipitation in 2019 resulted in an extremely low rate of seed germination under a large forest gap, and seedlings could not be established. Changes in precipitation patterns would have an important impact on natural forest regeneration [9]. The small gap and understory had higher soil water content, and seed germination was less affected. Rozas [16] also believed that *Quercus* seeds are large and nutrient-rich, and seed germination is not a key process that limits natural regeneration. In terms of growth, the growth of seedlings in gap II was better than that in III, which indicated that the smaller gap was suitable for the growth of early seedlings. It is easier to establish a seedling bank under the forest or under the small forest gap, which provides the foundation for rapid growth after the appearance of the larger forest gap in the future. Therefore, this paper agrees that the advanced seedling bank in the forest understory is the main source of natural regeneration [22,23]. Seed size significantly affected the germination and early seedling growth of *Q. acutissima*. Large seeds had obvious advantages in germination, seedling diameter growth, and root development, especially in the understory. Therefore, large seeds improved the adaptability of *Q. acutissima* in the understory, which was consistent with the conclusion of Quero et al. [34]. Overall, seed germination is not the key process that restricts the natural regeneration of *Q. acutissima*, but the growth and survival of seedlings are the key limiting stages of natural regeneration.

5. Conclusions

This study showed that the precipitation in this area was the key factor limiting the germination of *Q. acutissima* seeds under its plantations. The gap affected the germination of *Q. acutissima* seeds by regulating soil water content. In this paper, a suitable gap size promotes the germination of *Q. acutissima* seeds, but excessively large gaps are prone to severe drought and can cause seed germination failure. In the early growth of seedlings, the small forest gap was significantly higher than that under the forest. Therefore, seed germination and seedling growth depend on gaps differently. Large seeds have the highest germination capacity, seedling ground diameter growth rate, and root development, especially under the forest canopy, which shows that they are most conducive to the early establishment of seedlings. Therefore, forest gap and seed size are the key factors for the establishment of the understory seedling bank of the *Q. acutissima* plantation. In future research, it is necessary to strengthen the monitoring of precipitation and environmental factors in the forest. The results give insight into the relationship between seedling growth and gap size, and provide a basis for the regeneration and management of *Q. acutissima* plantation.

Author Contributions: Conceptualization, B.C. and P.M.; formal analysis, X.K. and Y.P.; investigation, J.Z., C.T., Y.G., X.C. and S.W.; resources, P.G. and J.D.; data curation, R.N.; writing—original draft preparation, X.K. and K.W.; writing—and review and editing, P.M. and B.C.; supervision, B.C.; funding acquisition, B.C. and P.M. All authors have read and agreed to the published version of the manuscript.

Funding: This research was funded by Central Finance Forestry Reform and Development Fund, grant number [2020]TG08, Germplasm resources nursery project of saline alkali tolerant tree species in the Yellow River Delta, grant number 2019-370505-05-03-035206, and Key R & D projects in Shandong Province, grant number 2017GNC10121.

Data Availability Statement: Not applicable.

Conflicts of Interest: The authors declare no conflict of interest.

Appendix A

Table A1. Coefficient, eigenvalue, variance contribution rate, and the accumulated contribution rate of principal components. SB: seedling biomass; RMR: root mass ratio; SMR: stem mass ratio; LMR: leaf mass ratio; R/S: root/shoot ratio; P/NP: photosynthetic tissues/non-photosynthetic tissues; SLA: specific leaf area; LAR: leaf area ratio; RL: root length; RSA: root surface area; RV: root volume; SRL: specific root length; SRA: specific root surface area.

Index	The First Principal Component	The Second Principal Component
R/S	0.86	−0.32
RMR	0.86	−0.34
RV	0.84	0.41
RSA	0.77	0.34
SRA	−0.73	−0.34
SB	0.73	0.50
SRL	−0.67	−0.35
LMR	−0.55	0.82
P/NP	−0.56	0.78
SMR	−0.39	−0.66
SLA	0.20	−0.60
RL	0.25	0.06
LAR	−0.48	0.30
Eigenvalue	5.40	3.15
Variance contribution rate (%)	41.57	24.16
Accumulated contribution rate (%)	41.57	65.73

References

1. Payn, T.; Carnus, J.M.; Freer-Smith, P.; Kimberley, M.; Kollert, W.; Liu, S.; Orazio, C.; Rodriguez, L.; Silva, L.N.; Wingfield, M.J. Changes in planted forests and future global implications. *For. Ecol. Manag.* **2015**, *352*, 57–67. [CrossRef]
2. Liu, S.; Yang, Y.; Wang, H. Development strategy and management countermeasures of planted forests in China Transforming from timber-centered single objective management towards multi-purpose management for enhancing quality and benefits of ecosystem services. *Acta Ecol. Sin.* **2018**, *38*, 1–10. [CrossRef]

3. Pröll, G.; Darabant, A.; Gratzer, G.; Katzensteiner, K. Unfavourable microsites, competing vegetation and browsing restrict post-disturbance tree regeneration on extreme sites in the Northern Calcareous Alps. *Eur. J. For. Res.* **2015**, *134*, 293–308. [CrossRef]
4. Rey, P.J.; Alcántara, J.M. Recruitment dynamics of a fleshy-fruited plant (*Olea europaea*): Connecting patterns of seed dispersal to seedling establishment. *J. Ecol.* **2000**, *88*, 622–633. [CrossRef]
5. Mao, P.; Guo, L.; Gao, Y.; Qi, L.; Cao, B. Effects of seed size and sand burial on germination and early growth of seedlings for coastal *Pinus thunbergii* Parl. in the Northern Shandong Peninsula, China. *Forests* **2019**, *10*, 281. [CrossRef]
6. Hutchins, H.E.; Hutchins, S.A.; Liu, B.W. The role of birds and mammals in Korean pine (*Pinus koraiensis*) regeneration dynamics. *Oecologia* **1996**, *107*, 120–130. [CrossRef]
7. Zhang, S.; Kang, H.; Yang, W. Climate change-induced water stress suppresses the regeneration of the critically endangered forest tree *Nyssa yunnanensis*. *PLoS ONE* **2017**, *12*, e0182012. [CrossRef]
8. Dodson, E.K.; Root, H.T. Conifer regeneration following stand-replacing wildfire varies along an elevation gradient in a ponderosa pine forest, Oregon, USA. *For. Ecol. Manag.* **2013**, *302*, 163–170. [CrossRef]
9. Khaine, I.; Woo, S.Y.; Kwak, M.; Lee, S.H.; Je, S.M.; You, H.; Lee, T.; Jang, J.; Lee, H.K.; Cheng, H.C.; et al. Factors affecting natural regeneration of tropical forests across a precipitation gradient in Myanmar. *Forests* **2018**, *9*, 143. [CrossRef]
10. Szwagrzyk, J.; Szewczyk, J.; Bodziarczyk, J. Dynamics of seedling banks in beech forest: Results of a 10-year study on germination, growth and survival. *For. Ecol. Manag.* **2001**, *141*, 237–250. [CrossRef]
11. Pardos, M.; Montes, F.; Aranda, I.; Cañellas, I. Influence of environmental conditions on germinant survival and diversity of Scots pine (*Pinus sylvestris* L.) in central Spain. *Eur. J. For. Res.* **2007**, *126*, 37–47. [CrossRef]
12. Gasser, D.; Messier, C.; Beaudet, M.; Lechowicz, M.J. Sugar maple and yellow birch regeneration in response to canopy opening, liming and vegetation control in a temperate deciduous forest of Quebec. *For. Ecol. Manag.* **2010**, *259*, 2006–2014. [CrossRef]
13. Rodríguez-García, E.; Juez, L.; Bravo, F. Environmental influences on post-harvest natural regeneration of Pinus pinaster Ait. in Mediterranean forest stands submitted to the seed-tree selection method. *Eur. J. For. Res.* **2010**, *129*, 1119–1128. [CrossRef]
14. Zhu, J.J.; Tan, H.; Li, F.Q.; Chen, M.; Zhang, J.X. Microclimate regimes following gap formation in a montane secondary forest of eastern Liaoning Province, China. *J. For. Res.* **2007**, *18*, 167–173. [CrossRef]
15. Bílek, L.; Remeš, J.; Zahradník, D. Natural regeneration of senescent even-aged beech (*Fagus sylvatica* L.) stands under the conditions of central Bohemia. *J. For. Sci.* **2009**, *55*, 145–155. [CrossRef]
16. Rozas, V. Regeneration patterns, dendroecology, and forest-use history in an old-growth beech-oak lowland forest in Northern Spain. *For. Ecol. Manag.* **2003**, *182*, 175–194. [CrossRef]
17. Collet, C.; Lanter, O.; Pardos, M. Effects of canopy opening on the morphology and anatomy of naturally regenerated beech seedlings. *Trees—Struct. Funct.* **2002**, *16*, 291–298. [CrossRef]
18. Diaci, J.; Adamic, T.; Rozman, A. Gap recruitment and partitioning in an old-growth beech forest of the Dinaric Mountains: Influences of light regime, herb competition and browsing. *For. Ecol. Manag.* **2012**, *285*, 20–28. [CrossRef]
19. Čater, M.; Diaci, J. Divergent response of European beech, silver fir and Norway spruce advance regeneration to increased light levels following natural disturbance. *For. Ecol. Manag.* **2017**, *399*, 206–212. [CrossRef]
20. Soto, D.P.; Jacobs, D.F.; Salas, C.; Donoso, P.J.; Fuentes, C.; Puettmann, K.J. Light and nitrogen interact to influence regeneration in old-growth Nothofagus-dominated forests in south-central Chile. *For. Ecol. Manag.* **2017**, *384*, 303–313. [CrossRef]
21. Nagel, T.A.; Svoboda, M.; Rugani, T.; Diaci, J. Gap regeneration and replacement patterns in an old-growth Fagus-Abies forest of Bosnia-Herzegovina. *Plant Ecol.* **2010**, *208*, 307–318. [CrossRef]
22. Madsen, P.; Hahn, K. Natural regeneration in a beech-dominated forest managed by close-to-nature principles—A gap cutting based experiment. *Can. J. For. Res.* **2008**, *38*, 1716–1729. [CrossRef]
23. Grogan, J.; Landis, R.M.; Ashton, M.S.; Galvão, J. Growth response by big-leaf mahogany (*Swietenia macrophylla*) advance seedling regeneration to overhead canopy release in southeast Pará, Brazil. *For. Ecol. Manag.* **2005**, *204*, 399–412. [CrossRef]
24. d'Oliveira, M.V.N.; Ribas, L.A. Forest regeneration in artificial gaps twelve years after canopy opening in Acre State Western Amazon. *For. Ecol. Manag.* **2011**, *261*, 1722–1731. [CrossRef]
25. Petritan, A.M.; Nuske, R.S.; Petritan, I.C.; Tudose, N.C. Gap disturbance patterns in an old-growth sessile oak (*Quercus petraea* L.)-European beech (*Fagus sylvatica* L.) forest remnant in the Carpathian Mountains, Romania. *For. Ecol. Manag.* **2013**, *308*, 67–75. [CrossRef]
26. Kunstler, G.; Curt, T.; Bouchaud, M.; Lepart, J. Growth, mortality, and morphological response of European beech and downy oak along a light gradient in sub-Mediterranean forest. *Can. J. For. Res.* **2005**, *35*, 1657–1668. [CrossRef]
27. Holladay, C.A.; Kwit, C.; Collins, B. Woody regeneration in and around aging southern bottomland hardwood forest gaps: Effects of herbivory and gap size. *For. Ecol. Manag.* **2006**, *223*, 218–225. [CrossRef]
28. Baskin, J.M.; Lu, J.J.; Baskin, C.C.; Tan, D.Y.; Wang, L. Diaspore dispersal ability and degree of dormancy in heteromorphic species of cold deserts of northwest China: A review. *Perspect. Plant Ecol. Evol. Syst.* **2014**, *16*, 93–99. [CrossRef]
29. Wang, T.T.; Chu, G.M.; Jiang, P.; Niu, P.X.; Wang, M. Effects of sand burial and seed size on seed germination, seedling emergence and seedling biomass of Anabasis Aphylla. *Pak. J. Bot.* **2017**, *49*, 391–396.
30. Larios, E.; Búrquez, A.; Becerra, J.X.; Venable, D.L. Natural selection on seed size through the life cycle of a desert annual plant. *Ecology* **2014**, *95*, 3213–3220. [CrossRef]

31. Baraloto, C.; Forget, P.M.; Goldberg, D.E. Seed mass, seedling size and neotropical tree seedling establishment. *J. Ecol.* **2005**, *93*, 1156–1166. [CrossRef]
32. Moles, A.T.; Ackerly, D.D.; Webb, C.O.; Twiddle, J.C.; Dickie, J.B.; Westoby, M. A brief history of seed size. *Science* **2005**, *307*, 576–580. [CrossRef]
33. Paz, H.; Martínez-Ramos, M. Seed mass and seedling performance within eight species of *Psychotria* (Rubiaceae). *Ecology* **2003**, *84*, 439–450. [CrossRef]
34. Quero, J.L.; Villar, R.; Marañón, T.; Zamora, R.; Poorter, L. Seed-mass effects in four *Mediterranean quercus* species (Fagaceae) growing in contrasting light environments. *Am. J. Bot.* **2007**, *94*, 1795–1803. [CrossRef]
35. Suárez-Vidal, E.; Sampedro, L.; Zas, R. Is the benefit of larger seed provisioning on seedling performance greater under abiotic stress? *Environ. Exp. Bot.* **2017**, *134*, 45–53. [CrossRef]
36. Bladé, C.; Vallejo, V.R. Seed mass effects on performance of *Pinus halepensis* Mill. seedlings sown after fire. *For. Ecol. Manag.* **2008**, *255*, 2362–2372. [CrossRef]
37. Liu, S.; Wu, S.; Wang, H. Managing planted forests for multiple uses under a changing environment in China. *N. Z. J. For. Sci.* **2014**, *44*, S3. [CrossRef]
38. Guo, L.; Ni, R.; Kan, X.; Lin, Q.; Mao, P.; Cao, B.; Gao, P.; Dong, J.; Mi, W.; Zhao, B. Effects of Precipitation and Soil Moisture on the Characteristics of the Seedling Bank under *Quercus acutissima* Forest Plantation in Mount Tai, China. *Forests* **2022**, *13*, 545. [CrossRef]
39. Yang, Y.; Wang, C.; Liu, Y. The effect of low irradiance on growth, photosynthetic characteristics, and biomass allocation in two deciduous broad-leaved tree seedlings in southeast of Hubei Province. *Acta Ecol. Sin.* **2010**, *30*, 6082–6090.
40. Mao, P.; Mu, H.; Cao, B.; Qin, Y.; Shao, H.; Wang, S.; Tai, X. Dynamic characteristics of soil properties in a *Robinia pseudoacacia* vegetation and coastal eco-restoration. *Ecol. Eng.* **2016**, *92*, 132–137. [CrossRef]
41. Diaci, J.; Pisek, R.; Boncina, A. Regeneration in experimental gaps of subalpine *Picea abies* forest in the Slovenian Alps. *Eur. J. For. Res.* **2005**, *124*, 29–36. [CrossRef]
42. Finch-Savage, W.E.; Blake, P.S. Indeterminate development in desiccation-sensitive seeds of *Quercus robur* L. *Seed Sci. Res.* **1994**, *4*, 127–133. [CrossRef]
43. Brüllhardt, M.; Rotach, P.; Bigler, C.; Nötzli, M.; Bugmann, H. Growth and resource allocation of juvenile European beech and sycamore maple along light availability gradients in uneven-aged forests. *For. Ecol. Manag.* **2020**, *474*, 118314. [CrossRef]
44. Delagrange, S.; Messier, C.; Lechowicz, M.J.; Dizengremel, P. Physiological, morphological and allocational plasticity in understory deciduous trees: Importance of plant size and light availability. *Tree Physiol.* **2004**, *24*, 775–784. [CrossRef]
45. Puglielli, G.; Laanisto, L.; Poorter, H.; Niinemets, Ü. Global patterns of biomass allocation in woody species with different tolerances of shade and drought: Evidence for multiple strategies. *New Phytol.* **2021**, *229*, 308–322. [CrossRef]
46. Myers, J.A.; Kitajima, K. Carbohydrate storage enhances seedling shade and stress tolerance in a neotropical forest. *J. Ecol.* **2007**, *95*, 383–395. [CrossRef]
47. Pedroso, F.K.J.V.; Prudente, D.A.; Bueno, A.C.R.; Machado, E.C.; Ribeiro, R.V. Drought tolerance in citrus trees is enhanced by rootstock-dependent changes in root growth and carbohydrate availability. *Environ. Exp. Bot.* **2014**, *101*, 26–35. [CrossRef]
48. Herrera, J. Acorn predation and seedling production in a low-density population of cork oak (*Quercus suber* L.). *For. Ecol. Manag.* **1995**, *76*, 197–201. [CrossRef]
49. Cavender-Bares, J.; Bazzaz, F.A. Changes in drought response strategies with ontogeny in *Quercus rubra*: Implications for scaling from seedlings to mature trees. *Oecologia* **2000**, *124*, 8–18. [CrossRef]

forests

MDPI

Article

Effects of Plant Fine Root Functional Traits and Soil Nutrients on the Diversity of Rhizosphere Microbial Communities in Tropical Cloud Forests in a Dry Season

Zhiyan Deng [1,2], Yichen Wang [1,2], Chuchu Xiao [1,2], Dexu Zhang [1,2], Guang Feng [1,2] and Wenxing Long [1,2,*]

1 Hainan Wuzhishan Forest Ecosystem Observation and Research Station, College of Forestry, Hainan University, Haikou 570228, China; 19095400210003@hainanu.edu.cn (Z.D.); ggg5724@163.com (Y.W.); 993951@hainanu.edu.cn (C.X.); 20095400210038@hainanu.edu.cn (D.Z.); fengguang@hainanu.edu.cn (G.F.)

2 Key Laboratory of Tropical Forest Flower Genetics and Germplasm Innovation, Ministry of Education, Haikou 570228, China

* Correspondence: oklong@hainanu.edu.cn; Tel.: +86-189-7656-5945

Abstract: The composition and diversity of rhizosphere microbial communities may be due to root–soil–microbial interactions. The fine root functional traits and rhizosphere soil environmental factors of 13 representative plants in the Bawangling tropical cloud forest of Hainan Island were measured, to assess the key factors driving plant rhizosphere microbial communities. Illumina MiSeq sequencing technology was used to sequence the v3-V4 region of the 16SrDNA gene of 13 plant rhizosphere soil bacteria and the ITS1 region of the fungal ITSrDNA gene. Results showed that there were 355 families, 638 genera, and 719 species of rhizosphere soil bacteria as well as 29 families, 31 genera, and 31 species of rhizosphere soil fungi in the tropical cloud forests. The fine root traits, such as root phosphorus content, the specific root length and specific root area, were significantly negatively correlated with the Faith-pd indices of the bacterial community but were not correlated with the diversity of fungi communities. The soil pH was significantly and positively correlated with the Chao1 index, OTUs, Faith-pd and Simpson indices of the bacteria and fungi communities. The soil available phosphorus content was significantly and negatively correlated with the bacteria Simpson and the fungus Faith-pd indices. ABT analysis showed that soil pH and soil available phosphorus were the most important environmental conditions contributing to the rhizosphere bacterial and fungi communities, respectively. Our findings demonstrate that the soil environments had more influence on rhizosphere soil microbial diversity than the fine root functional traits.

Keywords: Hainan island; soil pH; soil available phosphorus; root–soil–microbial interactions

Citation: Deng, Z.; Wang, Y.; Xiao, C.; Zhang, D.; Feng, G.; Long, W. Effects of Plant Fine Root Functional Traits and Soil Nutrients on the Diversity of Rhizosphere Microbial Communities in Tropical Cloud Forests in a Dry Season. *Forests* **2022**, *13*, 421. https://doi.org/10.3390/f13030421

Academic Editor: Timo Domisch

Received: 9 January 2022
Accepted: 28 February 2022
Published: 8 March 2022

1. Introduction

The rhizosphere is the soil region directly affected by plant roots and is the place where roots, soil microorganisms, and soil interact [1,2]. Due to the effects of the rhizosphere, there are more microorganisms in rhizosphere soil than in bulk soil [3]. The community structure, abundance, and diversity of the rhizosphere microorganisms are affected by root–soil–microbe interactions [4]. Plant roots and soil are two important components of the root–soil–microbe interface and are the main factors that influence the composition and structure of the rhizosphere microbial community [5,6].

Plant roots continuously secrete various substances to promote plants' absorption of mineral elements and provide the rhizosphere soil microbes with sugars, sugar alcohols, amino acids and phenolics as nutrient and energy supplies [7]. The type and quantity of the root exudates determine the type and quantity of rhizosphere microorganisms and affect the rhizosphere soil microbial community structure and carbon source utilization [8]. That is to say, the plants "actively" select rhizosphere soil microorganisms through root exudates [9].

As the main plant organs for the absorption, storage, and transport of nutrients and water of the plant root system, fine roots (diameter ≤ 2 mm) are crucial for the growth and distribution of plants [10]. They are sensitive to the changes in the soil environment and adapt to the environmental changes by changing their shapes and other characteristics [11]. Studies have found that the functional traits of fine roots are significant for predicting soil microbial groups and functional communities [12,13]. For example, Sweeney et al. (2021) found that temperate grassland plant functional traits, especially root traits, affect the composition of the rhizosphere fungal community and can be used to predict the fungal community. Similarly, Spitzer et al. (2021) found that the spectrum of subarctic tundra meadow plants' fine root economic trait chemical axis is positively correlated with the rhizosphere fungus/bacteria ratio. Although many studies have focused on the relationship between root traits and rhizosphere soil microbes [12,14–20], no studies we are aware of have explored these relationships in a high-altitude tropical forest ecosystem.

Soil is one of the ecosystems with the most abundant microbes on earth and is regarded as a reservoir for rhizosphere microbial communities [21]. The complex properties of soil affect the initial microbial community during the assembly process of the rhizosphere soil microorganisms by directly changing the soil microbial community composition [21]. Furthermore, soil can indirectly change the composition and relative abundance of the rhizosphere soil microbiome by influencing plant physiological activities [22]. For example, Zhao et al. [23] found that the nutrient content of soil is the main factor affecting the soil microbial structure of subtropical mountain forests. Glassman et al. [24] showed that pH and soil nutrient locally drive the assembly of a fungal community; global meta-analyses also indicate that soil physical and chemical properties, especially soil pH, are the dominant factors affecting the characteristics and diversity of soil microbial communities [25–28].

Few studies were conducted to examine the combined effect of rhizosphere soil factors and fine root functional traits on rhizosphere soil microorganisms [21], especially in high-altitude forests (such as tropical cloud forests). The tropical cloud forest is a typical ecosystem sensitive to climate change, and is the most threatened and least studied forest in the world [29]. Compared with low-altitude tropical forests, tropical cloud forests have unique community structures and rich species diversity and are mainly distributed on high-altitude mountain tops or ridges, with frequent occurrence of clouds and fog, low temperatures, strong winds, low tree heights, small tree diameter and frequently water-saturated soil [30].

In this study, our objective was to explore the effects of the rhizosphere environment and fine root functional traits on rhizosphere microbial communities in tropical cloud forest. Our hypotheses are: (1) plant fine root functional traits and rhizosphere soil nutrients together affect the rhizosphere microbial community diversity; (2) the soil abiotic environment plays a predominant determinant role in the assembly of the rhizosphere microbial community.

2. Materials and Methods

2.1. Study Site

The study site was located in the Bawangling area of Hainan Tropical Rainforest National Park (18°50′–19°05′N, 109°05′–109°25′E), with an altitude range of 100–1654 m (Figure 1). This region has a tropical monsoon climate with obvious dry and wet seasons, with a wet season from May to October and the dry season from November to April [31]. The typical type of soil is latosol developed from granite and sandstone as parent materials, which gradually transitions into mountain red soil, yellow soil and meadow soil with an increased altitude [32]. The main vegetation types include lowland rainforest, mountain rainforest and cloud forest [33]. The tropical cloud forest in Bawangling is mainly distributed in the shape of islands on ridges or mountaintops above 1250 m above sea level. There are two forest types in the tropical cloud forest in Bawangling, which are generally categorized as a tropical montane evergreen forest (TMEF) and a tropical dwarf forest (TDF) [32]. They are primary old-growth forests. The dominant species in TMEF include

Cryptocarya chinensis (Hance) Hemsl., *Cyclobalanopsis championii* (Benth.) Oerst., *Ternstroemia gymnanthera* (Wright et Arn.) Sprague, *Exbucklandia tonkinensis* (Lec.) Steen., *Cinnamomum tsoi* Allen. and *Syzygium araiocladum* Merr. et Perry. In TDF, dominant species include *Distylium racemosum* Sieb. and Zucc., *Syzygium buxifolium* Hook. et Arn., *Engelhardtia roxburghiana* Wall., *Symplocos poilanei* Guill., *Rhododendron moulmainense* Hook. F and C. tsoi Allen [32]. The species in tropical cloud forest systems are assembled with multiple ecological processes [34]. The tree and shrub species within the tropical cloud forest were non-random trait-based assembled [35].

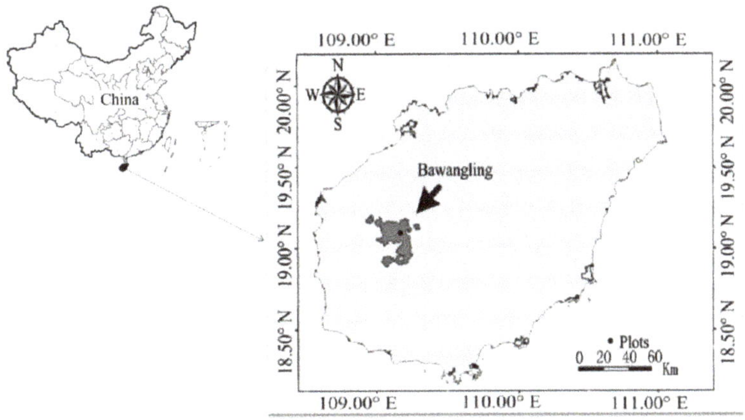

Figure 1. Location map of the study area.

2.2. Samples Collection

The experimental samples were collected in January 2021. According to the early plant species diversity, the data of 21 (20 m × 20 m) plots with diameter at breast height (DBH) larger than 1 cm in Bawangling tropical cloud forest and considered as community dominant species, rare species, phylogenetic relationships of plants, plant classification groups (gymnosperms and angiosperms), and plant life forms (shrub and macrophanerophytes) were obtained. In the present study, we chose 13 representative tree species (Table 1). DBH was measured by DBH ruler at the height of 1.3 m above the ground, and plant height of all individual trees appearing in the study plot was measured using a clinometer.

After clearing debris about 1 m around the target tree trunk (inside the crown), fine roots were collected by the root-tracking method. After digging out the fine roots, large pieces of soil were shaken off and soil within 5 mm on the root surface was collected [36]. For the soil samples, after a soil sample was taken from each of the three different directions of the plant with 10–30 cm depth (there were 3 soil samples of a plant taproot), then the three soil samples were mixed together. Finally, we obtained a mixed soil sample for each individual plant. For the root samples, a root sample was taken from each of the three different directions, and then we obtained three root samples for each individual plant. Altogether, 56 soil samples and 168 root samples were collected. The rhizosphere soil samples were brought to the laboratory for air drying and determination of physical and chemical properties. Furthermore, the soil samples were stored in −80 °C liquid nitrogen for high-throughput rhizosphere soil microbial sequencing. The fine root samples were put into moisturizing sample bags and were placed in a cryogenic storage box. After the cryogenic storage box was transported back to the laboratory, the root samples were put into refrigerator at 4 °C.

Table 1. Plant life forms, categories, stem diameter and height of plant species involved in the study.

Plant Species	Plant Life Forms	Plant Categories	DBH [1] (cm)	Height [1] (m)
Michelia mediocris Dandy	Tree	Angiosperm	6.22 ± 5.31	5.76 ± 2.43
Podocarpus neriifolius D. Don	Tree	Gymnosperm	5.62 ± 5.19	4.600 ± 1.39
Syzygium buxifolium Hook. et Arn.	Shrub or small tree	Angiosperm	8.02 ± 3.29	5.560 ± 1.13
Cyclobalanopsis disciformis (Chun et Tsiang) Y. C. Hsu et H. W. Jen	Tree	Angiosperm	7.62 ± 11.45	4.18 ± 2.20
Manglietia fordiana var. hainanensis (Dandy) N. H. Xia	Tree	Angiosperm	1.76 ± 0.53	2.90 ± 1.25
Pinus fenzeliana Hand.-Mzt.	Tree	Gymnosperm	36.24 ± 16.41	12.14 ± 5.21
Castanopsis faberi Hance	Tree	Angiosperm	2.63 ± 1.79	2.50 ± 0.87
Osmanthus didymopetalus P. S. Green	Tree	Angiosperm	2.92 ± 2.56	4.12 ± 1.89
Distylium racemosum Sieb. et Zucc.	Shrub or small tree	Angiosperm	16.46 ± 4.17	9.00 ± 1.00
Allomorphia balansae Cogn.	Shrub	Angiosperm	2.44 ± 1.45	2.75 ± 0.60
Olea dioica Roxb.	Shrub or small tree	Angiosperm	2.17 ± 1.10	2.93 ± 0.66
Syzygium championii (Benth.) Merr. et Perry	Shrub to small tree	Angiosperm	2.54 ± 1.59	3.820 ± 2.27
Melastoma penicillatum Naud.	Shrub	Angiosperm	2.13 ± 0.79	3.00 ± 0.84

[1] The values presented are "mean ± standard error"; DBH: diameter at breast height.

2.3. Selection and Measurement of Fine Root Functional Traits

We selected fine root morphology and chemical traits that reflect plant ecological strategies which are closely related to rhizosphere microorganisms, such as specific root length, root tissue density, specific root area, root carbon content, root nitrogen content and root phosphorus content (Table 2). The values of three root samples from the same plant were averaged and taken for the fine root functional traits of each plant.

Table 2. Fine root functional traits selected and ecological strategies.

	Traits	Abbreviation	Unit	Ecological Strategies
Morphology traits	Specific root length	SRL	cm/g	Resource acquisition.
	Root tissue density	RTD	g/cm^3	Transport, support and defense.
	Specific root area	SRA	cm^2/g	Resource acquisition and defense.
Chemical traits	Root carbon content	RC	g/kg	Microbial carbon source.
	Root nitrogen content	RN	g/kg	Microbial nutrient source.
	Root phosphorus content	RP	g/kg	Microbial nutrient source.

The retrieved roots were put into a 0.15 mm mesh bag and the impurities on the root surface were cleaned with low-temperature deionized water. After cleaning, the collected root samples were scanned by digital scanner (ESPON Chops V700 PHOTO). The WinRHIZO Pro 2011B (Regent Instruments, Canada) root image analysis software was used to analyze the scanned fine root images with diameter <2 mm to obtain the information of fine root length (cm), surface area (cm^2) and volume (cm^3). The scanned and analyzed fine roots were put into marked envelope bags, and then dried in an oven at 65 °C until constant weight was taken out and weighed by an electronic balance (AR2140, Ohaus, USA) to obtain the dry weight. The specific root length, root tissue density and specific root area were calculated by using the following formula [37,38]:

$$\text{Specific root length (cm/g)} = \text{Root length (cm)}/\text{ Dry weight of fine root (g)};$$

$$\text{Root tissue density (g/cm}^3) = \text{Dry weight of fine root (g)}/\text{Fine root volume (cm}^3);$$

$$\text{Specific root area (cm}^2/\text{g}) = \text{Fine root surface area (cm}^2)/\text{Dry weight of fine root (g)}.$$

The carbon content of fine roots was determined by potassium dichromate oxidation-external heating method after being crushed through a 40-mesh sieve [39]. After extracting root nitrogen by semi-trace Kelvin method, automatic flow analyzer (ProxiMA1022/1/1, Allians Scientific Instruments Co., LTD., Paris, France) was used to measure the root nitrogen content. Determination of phosphorus content in fine roots was carried out by molybdenum-antimony resistance colorimetry after concentrated sulfuric acid-perchloric acid cooking [39].

2.4. Determination of Soil Physical and Chemical Properties

The collected fresh soil was dried naturally in the laboratory and then crushed by a grinder after 100 mesh sieve was sent for testing. Soil organic matter content was determined by potassium dichromic oxidation-external heating method, soil total nitrogen was extracted by semi-trace Kelvin method and then measured by automatic flow analyzer (ProxiMA1022/1/1, Allians Scientific Instruments Co., LTD., Paris, France), soil available nitrogen content was determined by alkaline hydrolysis diffusion method [40]. The content of total phosphorus in soil was measured by molybdenum-antimony resistance colorimetric method after soil was boiled with concentrated sulfuric acid-perchloric acid [40]. The content of available phosphorus in soil was also measured by molybdenum-antimony resistance colorimetric method [40]. The pH value of soil was measured by potentiometric method [40].

2.5. Sequencing of Rhizosphere Microorganisms

About 5 g of fresh soil samples (kept on dry ice) of each sample were weighed and sent to the Shenzhen Weishengtai Technology Co., Ltd. for fungal ITS and bacterial 16S high-throughput sequencing. Before sequencing, the total DNA was extracted according to the instructions of the E.Z.N.A.® soil kit (Omega Bio-Tek, Norcross, GA, U.S.). The DNA concentration and purity were examined by using the NanoDrop 2000 UV-vis spectrophotometer (Thermo Scientific, Wilmington, DC, USA), and the DNA extraction quality was tested by 1% agarose gel electrophoresis. After qualitative analysis, the primers ITS1F (5′-CTTGGTCATTTAGAGGAAGTAA-3′) and ITS2R (5′-GCTGCGTTCTTCATCGATGC-3′) were used for polymerase chain reaction (PCR) amplification of the hypervariable region of the ribosomal ITSI; the primers 338F (5′-ACTCCTACGGGAGGCAGCAG-3′) and 806R (5′-GGACTACHVGGGTWTCTAAT-3′) were used for PCR amplification of the variable region of ribosomal 16SV3-V4. The PCR products were recovered using 2% agarose gel, then purified using the AxyPrep DNA Gel Extraction Kit (Axygen Biosciences, Union City, CA, USA), eluted with Tris-HCl, examined by 2% agarose electrophoresis, and quantified using the QuantiFluor™- ST (Promega, Madison, WI, USA). Based on the Illumina MiSeq sequencing platform, a PE 2*300 library was constructed using the purified amplified fragments following standard operating procedures.

The original sequence fastq file was imported using the import plug-in (part of the QIIME tools) into a file format that can be processed by QIIME2. Then, the QIIME2 dada2 plug-in was used for quality control, trimming, denoising, splicing, and removal of chimeras, and the final characteristic sequence table was obtained. Then, the QIIME2 feature-classifier plug-in was used to compare the ASV representative sequence to the pre-trained database (version 13.8) with a 99% similarity to obtain the classification information table of the species. After that, the QIIME2 feature-table plug-in was used to remove all contaminating mitochondria and chloroplast sequence.

From the sequencing, a barcode tag sequence representing the source information of the sample was obtained, and a valid sequence was identified. Based on the statistics of the effective sequences of the 56 samples collected, a total of 3,276,455 effective sequences were obtained from the 16S sequencing of the 56 tropical cloud forest plant rhizosphere soil samples, of which the lowest contained 44,197 effective sequences and the highest had 73,658 effective sequences. The ITS method gave a total of 3,802,639 effective sequences, with the lowest number of effective sequences of 46,633 and the highest effective sequences of 145,148.

2.6. Data Analysis

Species annotation and abundance analysis were performed after splicing and filtering Reads and clustering of the Operational Taxonomic Units (OTUs) (97%). The species composition information of rhizosphere soil microorganisms was obtained. The alpha diversity indices, such as OTUs, Chao1, Faith-pd, Shannon, Simpson index, etc., of the rhizosphere soil bacteria and fungi were calculated. The correlation between environmental factors (including soil environmental factors and fine root functional traits) around the rhizosphere microorganisms, as well as the correlation between soil, fine root functional traits, and rhizosphere soil microbes, were analyzed using the Spearman correlation. The OTU abundance of the microbial community was used as the dependent variable, and the functional traits of soil and fine roots were used as independent variables. The relative influence of the functional traits of soil and fine roots on rhizosphere microbial diversity was analyzed using Aggregated Boosted Tree (ABT). All calculations and plots were conducted through R language (R x64 version 4.0.2); ABT analysis was performed using the R language "dismo" package.

3. Results

3.1. Rhizosphere Microbial Diversity

In the present study, 33 phyla, 90 classes, 204 orders, 355 families, 638 genera, and 719 species of rhizosphere soil bacteria were found. The 3 phyla, 11 classes, 22 orders, 29 families, 31 genera, and 31 species of rhizosphere soil fungi were found in the plants collected from the Bawangling tropical cloud forest. The top three ranking phyla in terms of relative abundance among the 33 bacterial phyla are Proteobacteria (25.811–48.381%), Actinobacteria (14.754–48.243%), and Acidobacteria (7.452–37.415%) (Figure 2). One of the three fungal phyla was not annotated, and the other two phyla are Ascomycota (32.453–98.077%) and Basidiomycota (0.395–15.562%) (Figure 3). The Chao1 index of bacteria was found in the range of 547.111–1247.667, the Shannon index 7.867–9.245, and the Simpson index was in the range of 0.988–0.998098189. Similarly, the Chao1 index of fungi was found in the range of 165–896, the Shannon index 1.815–7.144, and the Simpson index ranged from 0.310 to 0.987.

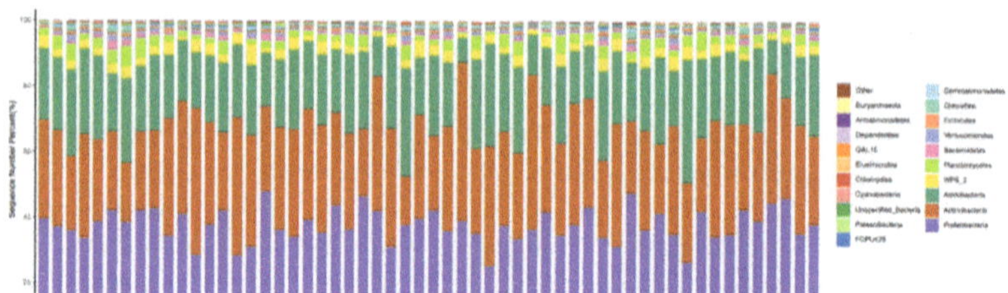

Figure 2. Relative abundance of the first 20 phylum levels of bacterial communities in the samples.

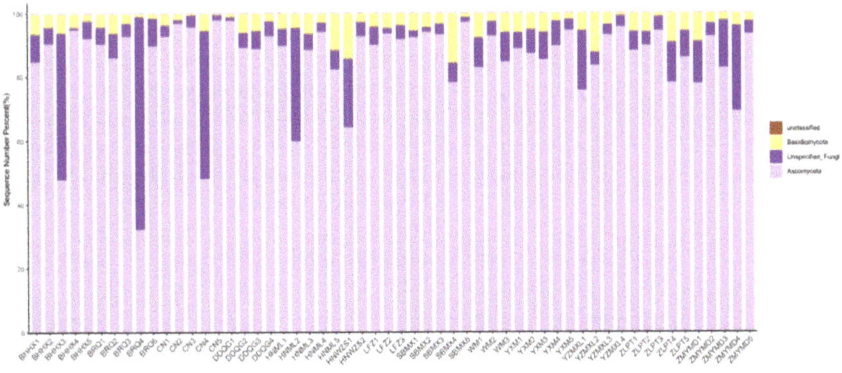

Figure 3. Relative abundance at phylum level of fungal communities in the samples.

3.2. Correlation between Fine Root Functional Traits and Rhizosphere Soil

The nitrogen contents of fine roots were significantly and negatively correlated with soil organic matter ($r = -0.54$, $p < 0.001$) and total nitrogen content ($r = -0.51$, $p < 0.001$). The root tissue density was significantly and negatively correlated with soil total phosphorus ($r = -0.35$, $p < 0.001$). The specific root length (SRL) was significantly and positively correlated with soil available nitrogen ($r = 0.28$, $p = 0.03$), but significantly negatively correlated with soil pH ($r = -0.32$, $p = 0.02$). The root area had a significant positive correlation with soil total phosphorus ($r = 0.37$, $p = 0.005$), and available nitrogen ($r = 0.30$, $p = 0.02$), and a significant negative correlation with soil pH ($r = -0.33$, $p = 0.01$) (Figure 4).

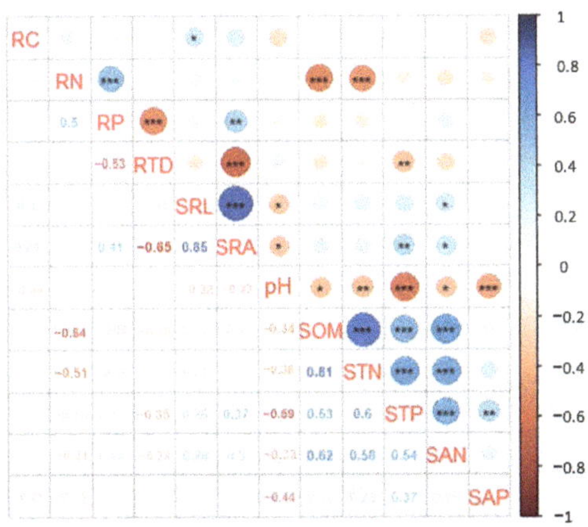

Figure 4. Correlation between fine root functional traits and soil factors (specific root length: SRL; root tissue density: RTD; specific root area: SRA; root carbon content: RC; root nitrogen content: RN; root phosphorus content: RP; soil pH: pH; soil total phosphorus: STP; soil available phosphorus: SAP; soil total nitrogen: STN; soil available nitrogen: SAN; soil organic matter: SOM; * $p < 0.05$, ** $p < 0.01$, *** $p < 0.001$).

3.3. Effects of Fine Root Functional Traits and Rhizosphere Soil on Microbial Diversity

There was a significant negative correlation between the fine root phosphorus content and the OTUs ($r = -0.27$, $p = 0.04$), Chao1 ($r = -0.28$, $p = 0.04$), and Faith-pd ($r = -0.27$, $p = 0.04$) indices of the bacterial community. The specific root length and specific root area of fine roots were significantly and negatively correlated with the Faith-pd index of the bacterial community (specific root length, $r = -0.28$, $p = 0.04$; specific root area, $r = -0.27$, $p = 0.04$). The soil pH was significantly and positively correlated with the Chao1 index of the bacteria (r = 0.34, p = 0.01), and had a significant positive correlation with the bacteria OTUs (r = 0.35, p = 0.009), Faith-pd (r = 0.48, p < 0.001), Shannon (r = 0.40, p = 0.002), and Simpson indices (r = 0.37, p = 0.004). The soil pH was significantly and positively correlated with the Shannon index of fungi (r = 0.31, p = 0.02), and positively correlated with the OTUs (r = 0.44, p = 0.001), Chao1 (r = 0.44, p = 0.001), and Faith-pd indices of fungi (r = 0.48, p = <0.001). The soil available phosphorus content was significantly and negatively correlated with the bacteria Simpson (r = −0.29, p = 0.03) and the fungus Faith-pd indices (r = −0.30, p = 0.03) (Table 3).

Table 3. Fine root functional traits and the correlation between rhizosphere soil and soil microbial community diversity.

		RC	RN	RP	RTD	SRL	SRA	pH	SOM	STN	STP	SAN	SAP
Bacteria	OTUs	0.55	0.59	0.04 *	0.23	0.26	0.15	0.009 **	0.85	0.94	0.67	0.16	0.28
	Chao1	0.56	0.57	0.04 *	0.22	0.28	0.16	0.01 *	0.86	0.96	0.66	0.18	0.31
	Faith-pd	0.89	0.33	0.04 *	0.35	0.04 *	0.04 *	0.0002 ***	0.79	0.44	0.43	0.06	0.12
	Shannon	0.34	0.69	0.16	0.32	0.25	0.16	0.002 **	0.85	0.82	0.72	0.31	0.08
	Simpson	0.40	0.60	0.31	0.60	0.22	0.19	0.004 **	0.89	0.93	0.84	0.83	0.03 *
Fungi	OTUs	1.00	0.23	0.47	0.97	0.57	0.56	0.001 **	0.62	0.44	0.37	0.58	0.05
	Chao1	1.00	0.23	0.47	0.97	0.57	0.56	0.001 **	0.62	0.44	0.37	0.58	0.05
	Faith-pd	0.90	0.39	0.47	0.88	0.43	0.47	0.0002 ***	0.40	0.19	0.17	0.99	0.03 *
	Shannon	0.69	0.18	0.85	0.31	0.56	0.33	0.02 *	0.87	0.53	0.79	0.10	0.19
	Simpson	0.89	0.21	0.60	0.25	0.44	0.21	0.06	0.75	0.46	0.86	0.18	0.20

Note: * p < 0.05, ** p < 0.01, *** p < 0.001; specific root length: SRL; root tissue density: RTD; specific root area: SRA; root carbon content: RC; root nitrogen content: RN; root phosphorus content: RP; soil pH: pH; soil total phosphorus: STP; soil available phosphorus: SAP; soil total nitrogen: STN; soil available nitrogen: SAN; soil organic matter: SOM.

ABT analysis showed that the main environmental factors contributing to the changes in the rhizosphere bacterial community were soil pH (14.1%), fine root carbon content (9.8%), soil organic matter content (9.5%), and soil available phosphorus content (9.3%). The main environmental factors contributing to the diversity of rhizosphere fungi were soil available phosphorus content (10.9%), soil organic matter content (9.9%), soil pH (9.4%), and soil total phosphorus content (8.7%) (Figure 5).

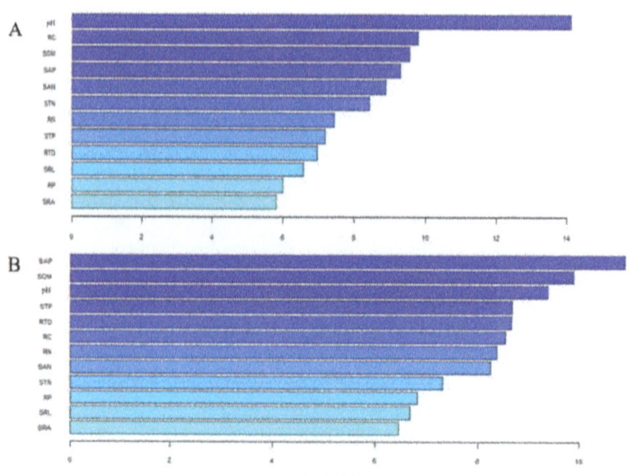

Figure 5. Fine root functional traits and the relative influence of rhizosphere soil on the diversity of rhizosphere microbial community ((**A**) represents bacteria, (**B**) represents fungi; specific root length: SRL; root tissue density: RTD; specific root area: SRA; root carbon content: RC; root nitrogen content: RN; root phosphorus content: RP; soil pH: pH; soil total phosphorus: STP; soil available phosphorus: SAP; soil total nitrogen: STN; soil available nitrogen: SAN; soil organic matter: SOM.).

4. Discussion

4.1. Diversity of Rhizosphere Microbial Community

In this study, the Proteobacteria, Actinomycota, and Acidobacteria were the top three dominant phyla in rhizosphere soil bacteria (Figure 2). According to the Global Soil Bacteria Atlas, Proteobacteria and Actinomycota are the two most abundant bacteria in global soils [41]. The Proteobacteria is the largest division among the bacteria, and its members are all Gram-negative bacteria and include multiple metabolic species [42]. Bacteria of the Proteobacteria and Actinomycetes types are the main degraders of organic matter, even complex organic compounds [43]. Another dominant division, Acidobacteria, found in the rhizosphere soil bacteria in this study, belongs to acidophilus, which is consistent with the characteristics of acidic soil in the study area. In tropical regions around the world, forest soils are generally acidic, and their soil microbial compositions are similar. For example, the dominant bacterial phyla of Sarawakian forests in tropical Southeast Asia include Acidobacteria and Proteobacteria [44]. Similarly, the main bacterial phyla of the microbial community of the virgin tropical forest soil in southern Vietnam are Proteobacteria and Acidobacteria [45]. Furthermore, Lan et al. [46] also found that the Acidobacteria are one of the main groups of soil bacterial communities in tropical rainforests when studying the soil microbial community in the tropical rainforest of Hainan Island. Therefore, it may be hypothesized that the composition of rhizosphere bacteria in tropical cloud forest plants might be affected by the combination of the soil bacteria pool and local soil characteristics.

Ascomycota and Basidiomycota are the dominant fungal phyla in the rhizosphere soil of tropical cloud forest plants (Figure 3). Among them, the Ascomycota has the highest abundance and is primarily of the saprophyte type [47]. The main functional microorganisms of saprophytes decompose refractory organic substances and improve the organic matter and soil nutrient content. Saprophytic fungi play a significant role in the soil and atmospheric carbon and nitrogen cycles [48]. Basidiomycota mainly grows in relatively moist soil [49], which is a consequence of the high humidity in the soil environment of tropical cloud forests. Globally, in tropical rainforests with a hot climate and abundant rainfall, the most abundant fungal phyla in the soil include the Ascomycota and Basidiomycota, for example, soil in the tropical rainforest of Queensland, Australia [50]. The soil fungal communities in tropical forests in Puerto Rico are mainly Ascomycota and Basidiomycota [51]. Similarly, Lan et al. [46] found that Ascomycota and Basidiomycota are the major phyla of soil fungi communities in Hainan Island. Therefore, the results of this study indicate that the microbial composition of the rhizosphere soil of high-altitude tropical forests is similar to that of low-altitude tropical forests.

4.2. Impact of the Soil Environment on the Microbial Diversity of the Rhizosphere Soil Relative to the Functional Traits of Fine Roots

In this study, the physical and chemical properties of the rhizosphere soil and the functional characteristics of fine roots significantly influence the rhizosphere soil microbe behavior (Table 3). The studies of Zhao [23], Glassman [24] and other scholars [25–28] mentioned above also showed similar results. Plant roots and rhizosphere soil microorganisms are closely connected, and fine roots and their functional traits affect rhizosphere microorganisms through root exudates and litter quality [13]. The function of specific metabolites in root exudates is indicative of the characteristics of the root economic spectrum, which includes the plant root functional traits. These traits could be used to explain the composition of plant root exudates. Therefore, to a certain extent, root functional traits could also be used to explain the composition and diversity of rhizosphere microorganisms [52]. The results of this study are consistent with the previous studies [12,13,53].

In the tropical cloud forest ecosystem, soil environmental factors have a more significant impact on the rhizosphere soil microbial community than the functional traits of fine roots (Figure 5). Root litter and exudates eventually enter the rhizosphere soil and become important nutrients for rhizosphere microorganisms. Root exudates change the chemical composition of the soil by increasing or decreasing the availability of soil nutrients [54].

Furthermore, the complex physical and chemical properties of the soil cause differences in the initial microbial community during the assembly process of the rhizosphere microbiome, affecting the composition of the rhizosphere microbiome [55]. In addition, the complex interaction between the physical and chemical properties of the soil also affects plant growth and plant physiology, which results in a change in the composition and relative abundance of the rhizosphere microbiome [56,57]. In addition, the mycorrhizal fungi, which by their common mycelial networks interconnect different plants in an ecosystem, further complicate the factors that affect the rhizosphere microbiome. Common mycelial networks (CMNs) can influence plant and microorganism community compositions, induce an efficient nutrient exchange, and improve interplant nutrition and growth through plant–plant facilitation [58]. Mycorrhizas are widespread and abundant, and they are ubiquitous in most temperate and tropical ecosystems [59]. In this study, there are many plant species that have mycorrhizas. For example, *Pinus* is a typical ectomycorrhizal tree genus [60], and *Pinus fenzeliana* Hand.-Mzt belongs to the *Pinus* genus.

4.3. Soil pH Effects on Rhizosphere Soil Microbes in the Soil Environment

Soil pH is the main driving factor for changes in the diversity of bacterial and fungal communities in the rhizosphere soil of tropical cloud forests on Hainan Island (Table 3). ABT analysis shows that soil pH changes the composition of the bacterial and fungal communities most in the tropical cloud forest rhizosphere in Bawangling (Figure 5). Other soil factors, such as SOM, STN, STP, SAN, and SAP, were significantly correlated with soil pH (Figure 4). Changes in soil pH result in changes in the distribution of various nutrient elements in the soil and changes in ion activity, thereby leading to changes in soil fertility [61]. Soil pH also significantly affects the availability of soil nutrients. The solubility of cationic nutrients in strongly alkaline soils decreases, and the sensitivity to loss by leaching or by erosion strongly increases in acidic soils, decreasing the availability of cationic salt nutrients. The availability of anionic nutrients usually shows an opposite trend to that of the cationic nutrients under varying soil pH [62]. Both the cationic and anionic salt nutrients in the soil are essential nutrients for the growth of soil microorganisms and affect the diversity of the microorganisms in the soil. The soil pH is also an important factor influencing cell metabolic activity. Plants select and adapt to rhizosphere bacteria [63]. Plant growth and fine root traits also change accordingly due to the influence of soil pH on soil fertility [64], which also results in changes in the composition and diversity of rhizosphere soil microorganisms. On a large spatial scale, the diversity of soil microbes increases with increasing soil pH [20], and the microbial diversity of acidic soils is usually significantly lower than the neutral soils [65]. The significant positive correlation between soil pH and rhizosphere soil bacterial diversity and fungal diversity found in this study provides new insights into the correlation of the soil pH and the diversity of the rhizosphere soil bacteria in the tropical high-altitude forest ecosystems.

Our results of the Spearman correlation analysis and Aggregated Boosted Tree analysis are also consistent with the meta-analysis results of Zhou et al. [27]. They integrated the results of 1235 global change factors of eight ecosystems including agricultural land, tundra, temperate forests, tropical and subtropical forests, Mediterranean vegetation, grasslands, deserts and wetlands, and concluded that soil pH is the most important factor that can be used to predict the impact of global change factors on microbial alpha diversity [27]. In particular, the results are also consistent with the results of Flores-Rentería et al. [66] and Lan et al. [46]. The composition of soil microorganisms in tropical rainforests is related to soil pH. Tropical cloud forests are high-altitude tropical forests. Therefore, our research provides new insights into the impact of soil factors on rhizosphere microbes. Soil pH in tropical forest communities at varied altitudes is an important factor affecting soil microbial diversity. At the same time, soil temperature is an important riding factor affecting many soil parameters at altitudes >1000 m ASL. Soil acidity at high altitudes is often found due to the low temperatures and accumulation of litter [67,68]. Soil temperature, moreover, affects soil pH mainly by affecting rock weathering rate [69] Therefore, effects of soil temperature, an important environmental factor, on rhizosphere microorganisms should also be considered in the future.

4.4. Effects of Soil Available Phosphorus on Rhizosphere Microorganisms

Soil available phosphorus content has a significant impact on the rhizosphere soil bacteria and the rhizosphere soil fungi diversity of tropical cloud forest plants (Table 3). It is also the environmental factor with the greatest impact on the rhizosphere soil fungal community (Figure 5). Phosphorus limitation is more prominent in tropical and subtropical forests, which are characterized by high temperatures, heavy rainfall, and strong weathering and leaching [69]. In tropical cloud forests, the ratio of nitrogen to phosphorus in plant leaves is greater than 17, and plant growth is affected by low soil phosphorus content stress [32]. Among rhizosphere microorganisms, arbuscular mycorrhizal (AM) fungi and phosphate solubilizing bacteria (PSB) are specific groups of microorganisms that play an important role in the process of soil phosphorus conversion and plant phosphorus absorption [70]. Therefore, soil available phosphorus may be a critical limiting factor for the composition of the rhizosphere soil microbial community in tropical cloud forests. First, the soil available phosphorus content might change the nutritional status of fungi, causing competition between bacteria, soil animals, and also changing fungal groups [71]. In addition, as an important nutrient factor for plant growth, soil available phosphorus changes can result in changes in plant biomass and indirectly affect soil fungi [71]. In addition, soil available phosphorus affects soil microorganisms by influencing other physical and chemical properties of the soil (such as pH) [72]. Cai et al. [73] studied soil microbes in tropical rainforests in Xishuangbanna and also found that soil available phosphorus limits the composition of tropical forest soil microbes.

Our results show that soil pH has a greater impact on the composition of rhizosphere soil bacteria, while soil available phosphorus has a greater impact on the composition of rhizosphere soil fungi. This might be due to the fact that soil fungi can grow in a wide range of pH conditions [74], whereas soil bacteria are more sensitive to soil pH [75]. Another study has shown that fungi solubilize phosphorus better than bacteria [76] and are more sensitive to changes in soil available phosphorus content than bacteria [77,78]. Therefore, fungi are more likely than bacteria to enhance absorption of soil available phosphorus by plant roots by solubilizing phosphorus when there is a demand for available phosphorus in the plant rhizosphere.

5. Conclusions

In this study, we measured the fine root functional traits and rhizosphere soil environmental factors of 13 representative plants in the Bawangling tropical cloud forest of Hainan Island, and assessed the effects of soil conditions and fine root functional traits on rhizosphere microbial communities. We found that both soil conditions and fine root functional traits had important effects on rhizosphere bacteria diversity, but we did not detect the correlations between fine root traits and fungi diversity. The rhizosphere soil environment is more important than fine root functional traits when it comes to affecting rhizosphere soil microbial composition.

Author Contributions: Z.D. and W.L. conceived and designed the experiments; Z.D., C.X., D.Z., and Y.W. performed the experiments; Z.D. and G.F. analyzed the data; Z.D. wrote the manuscript; W.L. reviewed and revised the manuscript. All authors have read and agreed to the published version of the manuscript.

Funding: This research was funded by the National Natural Science Foundation of China (Grant No: 31870508, 32171772) and the National Key Research and Development Project (Grant No: 2021YFD2200403).

Institutional Review Board Statement: Not applicable.

Informed Consent Statement: Not applicable.

Data Availability Statement: The data and results of this study are available upon reasonable request. Please contact the main author of this publication.

Acknowledgments: We are thankful to all students and teachers of Hainan University Tropical Forest Plant Diversity and Ecosystem Function Innovation Team. In particular, I would like to thank Liu Guoyin, Liu's students, and Feng Suoyong for their help in the experiment, Saraj Bahadur for his help with grammar revision of manuscript, and Liu lan and Luo Wenqi for their suggestions for experimental design. We would also like to thank the staves of the Bawangling Area of Hainan Tropical Rainforest National Park for their help during the trial.

Conflicts of Interest: We all authors have no conflict of interest to declare.

References

1. Hiltner, L. Ber neuere erfahrungen und probleme auf dem gebiete der bodenbakteriologie unter besonderer bercksichtigung der grndngung und brache. *Arb. Dtsch. Landwirtsch. Ges.* **1904**, *98*, 59–78.
2. Banerjee, S.; Kaushik, S.; Tomar, R.S. Rhizobacters as Remedy of Stress Tolerance in Potato. In *Antioxidants in Plant-Microbe Interaction*; Bahadur Singh, H., Vaishnav, A., Eds.; Springer: Singapore, 2021; pp. 395–412.
3. Zhang, P.; Cui, Z.; Guo, M.; Xi, R. Characteristics of the soil microbial community in the forestland of *Camellia oleifera*. *PeerJ* **2020**, *8*, e9117. [CrossRef] [PubMed]
4. Wang, P.; Marsh, E.L.; Ainsworth, E.A.; Leakey, A.D.; Sheflin, A.M.; Schachtman, D.P. Shifts in microbial communities in soil, rhizosphere and roots of two major crop systems under elevated CO_2 and O_3. *Sci. Rep.* **2017**, *7*, 15019. [CrossRef] [PubMed]
5. Philippot, L.; Raaijmakers, J.M.; Lemanceau, P.; van der Putten, W.H. Going back to the roots: The microbial ecology of the rhizosphere. *Nat. Rev. Microbiol.* **2013**, *11*, 789–799. [CrossRef]
6. Wei, Z.; Hu, X.; Li, X.; Zhang, Y.; Jiang, L.; Li, J.; Guan, Z.; Cai, Y.; Liao, X. The rhizospheric microbial community structure and diversity of deciduous and evergreen forests in Taihu Lake area, China. *PLoS ONE* **2017**, *12*, e0174411. [CrossRef]
7. Chaparro, J.M.; Badri, D.V.; Bakker, M.G.; Sugiyama, A.; Manter, D.K.; Vivanco, J.M. Root exudation of phytochemicals in *Arabidopsis* follows specific patterns that are developmentally programmed and correlate with soil microbial functions. *PLoS ONE* **2013**, *8*, e55731.
8. Dennis, P.G.; Miller, A.J.; Hirsch, P.R. Are root exudates more important than other sources of rhizodeposits in structuring rhizosphere bacterial communities? *FEMS Microbiol. Ecol.* **2010**, *72*, 313–327. [CrossRef]
9. Hannula, S.; Morrien, E.; van der Putten, W.; de Boer, W. Rhizosphere fungi actively assimilating plant-derived carbon in a grassland soil. *Fungal Ecol.* **2020**, *48*, 100988. [CrossRef]
10. Freschet, G.T.; Roumet, C. Sampling roots to capture plant and soil functions. *Funct. Ecol.* **2017**, *31*, 1506–1518. [CrossRef]
11. Hendricks, J.J.; Nadelhoffer, K.J.; Aber, J.D. Assessing the role of fine roots in carbon and nutrient cycling. *Trends Ecol. Evol.* **1993**, *8*, 174–178. [CrossRef]
12. Sweeney, C.J.; de Vries, F.T.; van Dongen, B.E.; Bardgett, R.D. Root traits explain rhizosphere fungal community composition among temperate grassland plant species. *New Phytol.* **2021**, *229*, 1492–1507. [CrossRef] [PubMed]
13. Spitzer, C.M.; Lindahl, B.; Wardle, D.A.; Sundqvist, M.K.; Gundale, M.J.; Fanin, N.; Kardol, P. Root trait-microbial relationships across tundra plant species. *New Phytol.* **2021**, *229*, 1508–1520. [CrossRef] [PubMed]
14. Merino-Martín, L.; Griffiths, R.I.; Gweon, H.S.; Furget-Bretagnon, C.; Oliver, A.; Mao, Z.; Le Bissonnais, Y.; Stokes, A. Rhizosphere bacteria are more strongly related to plant root traits than fungi in temperate montane forests: Insights from closed and open forest patches along an elevational gradient. *Plant* **2020**, *450*, 183–200. [CrossRef]
15. Elhaissoufi, W.; Khourchi, S.; Ibnyasser, A.; Ghoulam, C.; Rchiad, Z.; Zeroual, Y.; Lyamlouli, K.; Bargaz, A. Phosphate solubilizing rhizobacteria could have a stronger influence on wheat root traits and aboveground physiology than rhizosphere P solubilization. *Front. Plant Sci.* **2020**, *11*, 979. [CrossRef] [PubMed]
16. Ren, C.; Chen, J.; Deng, J.; Zhao, F.; Han, X.; Yang, G.; Tong, X.; Feng, Y.; Shelton, S.; Ren, G. Response of microbial diversity to C:N:P stoichiometry in fine root and microbial biomass following afforestation. *Biol. Fertil. Soils* **2017**, *53*, 457–468. [CrossRef]
17. Ding, X.; Liu, G.; Fu, S.; Chen, H.Y.H. Tree species composition and nutrient availability affect soil microbial diversity and composition across forest types in subtropical China. *CATENA* **2021**, *201*, 105224. [CrossRef]
18. Barberán, A.; McGuire, K.L.; Wolf, J.A.; Jones, F.A.; Wright, S.J.; Turner, B.L.; Essene, A.; Hubbell, S.P.; Faircloth, B.C.; Fierer, N. Relating belowground microbial composition to the taxonomic, phylogenetic, and functional trait distributions of trees in a tropical forest. *Ecol. Lett.* **2015**, *18*, 1397–1405. [CrossRef]
19. Rivest, M.; Whalen, J.K.; Rivest, D. Tree diversity is not always a strong driver of soil microbial diversity: A 7-yr-old diversity experiment with trees. *Ecosphere* **2019**, *10*, e02685. [CrossRef]
20. Oh, Y.M.; Kim, M.; Lee-Cruz, L.; Lai-Hoe, A.; Go, R.; Ainuddin, N.; Rahim, R.A.; Shukor, N.; Adams, J.M. Distinctive Bacterial Communities in the Rhizoplane of Four Tropical Tree Species. *Microb. Ecol.* **2012**, *64*, 1018–1027. [CrossRef]
21. Liu, L.; Huang, X.; Zhang, J.; Cai, Z.; Jiang, K.; Chang, Y. Deciphering the relative importance of soil and plant traits on the development of rhizosphere microbial communities. *Soil Biol. Biochem.* **2020**, *148*, 107909. [CrossRef]
22. Lareen, A.; Burton, F.; Schäfer, P. Plant root-microbe communication in shaping root microbiomes. *Plant Mol. Biol.* **2016**, *90*, 575–587. [CrossRef] [PubMed]

23. Zhao, R.; Yang, X.; Tian, Q.; Wang, X.; Liao, C.; Li, C.-L.; Liu, F. Soil microbial community structure, metabolic potentials and influencing factors in a subtropical mountain forest ecosystem of China. *Environ. Pollut. Bioavailab.* **2020**, *32*, 69–78. [CrossRef]
24. Glassman, I.S.; Wang, I.J.; Bruns, T.D. Environmental filtering by pH and soil nutrients drives community assembly in fungi at fine spatial scales. *Mol. Ecol.* **2017**, *26*, 6960–6973. [CrossRef] [PubMed]
25. Xun, W.; Huang, T.; Zhao, J.; Ran, W.; Wang, B.; Shen, Q.; Zhang, R. Environmental conditions rather than microbial inoculum composition determine the bacterial composition, microbial biomass and enzymatic activity of reconstructed soil microbial communities. *Soil Biol. Biochem.* **2015**, *90*, 10–18. [CrossRef]
26. Wan, W.; Hao, X.; Xing, Y.; Liu, S.; Zhang, X.; Li, X.; Chen, W.; Huang, Q. Spatial differences in soil microbial diversity caused by pH-driven organic phosphorus mineralization. *Land Degrad. Dev.* **2021**, *32*, 766–776. [CrossRef]
27. Zhou, Z.; Wang, C.; Luo, Y. Meta-analysis of the impacts of global change factors on soil microbial diversity and functionality. *Nat. Commun.* **2020**, *11*, 3072. [CrossRef]
28. Malik, A.A.; Puissant, J.; Buckeridge, K.M.; Goodall, T.; Jehmlich, N.; Chowdhury, S.; Gweon, H.S.; Peyton, J.M.; Mason, K.E.; van Agtmaal, M.; et al. Land use driven change in soil pH affects microbial carbon cycling processes. *Nat. Commun.* **2018**, *9*, 3591. [CrossRef]
29. Loope, L.L.; Giambelluca, T.W. Vulnerability of island tropical montane cloud forests to climate change, with special reference to East Maui, Hawaii. In *Potential Impacts of Climate Change on Tropical Forest Ecosystems*; Markham, A., Ed.; Springer: Dordrecht, The Netherlands, 1998; pp. 363–377.
30. Long, W.; Zhou, Y.; Schamp, B.S.; Zang, R.; Yang, X.; Poorter, L.; Xiao, C.; Xiong, M. Scaling relationships among functional traits are similar across individuals, species, and communities. *J. Veg. Sci.* **2020**, *31*, 571–580. [CrossRef]
31. Cheng, Y.; Zhang, H.; Zang, R.; Wang, X.; Long, W.; Wang, X.; Xiong, M.; John, R. The effects of soil phosphorus on aboveground biomass are mediated by functional diversity in a tropical cloud forest. *Plant Soil* **2020**, *449*, 51–63. [CrossRef]
32. Long, W.; Zang, R.; Ding, Y. Air temperature and soil phosphorus availability correlate with trait differences between two types of tropical cloud forests. *Flora-Morphol. Distrib. Funct. Ecol. Plants* **2011**, *206*, 896–903. [CrossRef]
33. Jiang, Y.; Zang, R.; Lu, X.; Huang, Y.; Ding, Y.; Liu, W.; Long, W.; Zhang, J.; Zhang, Z. Effects of soil and microclimatic conditions on the community-level plant functional traits across different tropical forest types. *Plant Soil* **2015**, *390*, 351–367. [CrossRef]
34. Long, W.; Zang, R.; Ding, Y.; Huang, Y. Effects of Competition and Facilitation on Species Assemblage in Two Types of Tropical Cloud Forest. *PLoS ONE* **2013**, *8*, e60252. [CrossRef] [PubMed]
35. Long, W.; Schamp, B.S.; Zang, R.; Ding, Y.; Huang, Y.; Xiang, Y. Community assembly in a tropical cloud forest related to specific leaf area and maximum species height. *J. Veg. Sci.* **2015**, *26*, 513–523. [CrossRef]
36. Barillot, C.C.D.; Sarde, C.-O.; Bert, V.; Tarnaud, E.; Cochet, N. A standardized method for the sampling of rhizosphere and rhizoplan soil bacteria associated to a herbaceous root system. *Ann. Microbiol.* **2013**, *63*, 471–476. [CrossRef]
37. Fujita, S.; Noguchi, K.; Tange, T. Root Responses of Five Japanese Afforestation Species to Waterlogging. *Forests* **2020**, *11*, 552. [CrossRef]
38. Petriţan, I.C.; von Lüpke, B.; Petriţan, A.M. Effects of root trenching of overstorey Norway spruce (*Picea abies*) on growth and biomass of underplanted beech (*Fagus sylvatica*) and Douglas fir (*Pseudotsuga menziesii*) saplings. *Eur. J. For. Res.* **2011**, *130*, 813–828. [CrossRef]
39. Liu, F. *Practical Manual of Agricultural Environment Monitoring*; China Standards Press: Beijing, China, 2001; pp. 93–191.
40. Lu, R.K. *Soil Agrochemical Analysis Methods*; China Agricultural Science and Technology Press: Beijing, China, 2000; p. 315.
41. Delgado-Baquerizo, M.; Oliverio, A.M.; Brewer, T.E.; Benavent-González, A.; Eldridge, D.J.; Bardgett, R.D.; Maestre, F.T.; Singh, B.K.; Fierer, N. A global atlas of the dominant bacteria found in soil. *Science* **2018**, *359*, 320–325. [CrossRef] [PubMed]
42. Pereira, R.M.; Silveira ÉLd Scaquitto, D.C.; Pedrinho, E.A.N.; Val-Moraes, S.P.; Wickert, E.; Carareto-Alves, L.M.; de Macedo Lemos, E.G. Molecular characterization of bacterial populations of different soils. *Braz. J. Microbiol.* **2006**, *37*, 439–447. [CrossRef]
43. Chen, Z.; Li, Y.; Ye, C.; He, X.; Zhang, S. Fate of antibiotics and antibiotic resistance genes during aerobic co-composting of food waste with sewage sludge. *Sci. Total Environ.* **2021**, *784*, 146950. [CrossRef]
44. Miyashita, N.T. Contrasting soil bacterial community structure between the phyla Acidobacteria and Proteobacteria in tropical Southeast Asian and temperate Japanese forests. *Genes Genet. Syst.* **2015**, *90*, 61–77. [CrossRef]
45. Chernov, T.I.; Zhelezova, A.D.; Tkhakakhova, A.K.; Bgazhba, N.A.; Zverev, A.O. Microbiomes of Virgin Soils of Southern Vietnam Tropical Forests. *Microbiology* **2019**, *88*, 489–498. [CrossRef]
46. Lan, G.; Wu, Z.; Yang, C.; Sun, R.; Chen, B.; Zhang, X. Forest conversion alters the structure and functional processes of tropical forest soil microbial communities. *Land Degrad. Dev.* **2021**, *32*, 613–627. [CrossRef]
47. Peterson, D. Categorization of Orthologous Gene Clusters in 92 Ascomycota Genomes Reveals Functions Important for Fungal Phytopathogenicity. Ph.D. Thesis, Northern Illinois University, Dekalb, IL, USA, 2020.
48. Harley, J. Fungi in ecosystems. *J. Ecol.* **1971**, *59*, 653–668. [CrossRef]
49. Warcup, J. Studies on Basidiomycetes in soil. *Trans. Br. Mycol. Soc.* **1959**, *42*, 45–52. [CrossRef]
50. Curlevski, N.J.A.; Xu, Z.; Anderson, I.C.; Cairney, J.W.G. Converting Australian tropical rainforest to native Araucariaceae plantations alters soil fungal communities. *Soil Biol. Biochem.* **2010**, *42*, 14–20. [CrossRef]
51. Urbina, H.; Scofield, D.G.; Cafaro, M.; Rosling, A. DNA-metabarcoding uncovers the diversity of soil-inhabiting fungi in the tropical island of Puerto Rico. *Mycoscience* **2016**, *57*, 217–227. [CrossRef]

52. Williams, A.; Langridge, H.; Straathof, A.L.; Muhamadali, H.; Hollywood, K.A.; Goodacre, R.; de Vries, F.T. Root functional traits explain root exudation rate and composition across a range of grassland species. *J. Ecol.* **2021**, *110*, 21–33. [CrossRef]
53. Ballauff, J.; Schneider, D.; Edy, N.; Irawan, B.; Daniel, R.; Polle, A. Shifts in root and soil chemistry drive the assembly of belowground fungal communities in tropical land-use systems. *Soil Biol. Biochem.* **2021**, *154*, 108140. [CrossRef]
54. Bais, H.P.; Weir, T.L.; Perry, L.G.; Gilroy, S.; Vivanco, J.M. The role of root exudates in rhizosphere interactions with plants and other organisms. *Annu. Rev. Plant Biol.* **2006**, *57*, 233–266. [CrossRef]
55. Hou, Q.; Wang, W.; Yang, Y.; Hu, J.; Bian, C.; Jin, L.; Li, G.; Xiong, X. Rhizosphere microbial diversity and community dynamics during potato cultivation. *Eur. J. Soil Biol.* **2020**, *98*, 103176. [CrossRef]
56. Hartman, K.; Tringe, S.G. Interactions between plants and soil shaping the root microbiome under abiotic stress. *Biochem. J.* **2019**, *476*, 2705–2724. [CrossRef] [PubMed]
57. Angers, D.A.; Caron, J. Plant-induced changes in soil structure: Processes and feedbacks. *Biogeochemistry* **1998**, *42*, 55–72. [CrossRef]
58. Wipf, D.; Krajinski, F.; van Tuinen, D.; Recorbet, G.; Courty, P.-E. Trading on the arbuscular mycorrhiza market: From arbuscules to common mycorrhizal networks. *New Phytol.* **2019**, *223*, 1127–1142. [CrossRef] [PubMed]
59. Johnson, N.C.; Gehring, C.A. Chapter 4—Mycorrhizas: Symbiotic Mediators of Rhizosphere and Ecosystem Processes. In *The Rhizosphere*; Cardon, Z.G., Whitbeck, J.L., Eds.; Academic Press: Burlington, VT, USA, 2007; pp. 73–100.
60. Smith, M.E.; Douhan, G.W.; Fremier, A.K.; Rizzo, D.M. Are True Multihost Fungi the Exception or the Rule? Dominant Ectomycorrhizal Fungi on Pinus sabiniana Differ from Those on Co-occurring Quercus Species. *New Phytol.* **2009**, *182*, 295–299. [CrossRef] [PubMed]
61. Zhao, J.; Dong, Y.; Xie, X.; Li, X.; Zhang, X.; Shen, X. Effect of annual variation in soil pH on available soil nutrients in pear orchards. *Acta Ecol. Sin.* **2011**, *31*, 212–216. [CrossRef]
62. McCauley, A.; Jones, C.; Jacobsen, J. Soil pH and organic matter. *Nutr. Manag. Modul.* **2009**, *8*, 1–12.
63. Hartmann, A.; Schmid, M.; Van Tuinen, D.; Berg, G. Plant-driven selection of microbes. *Plant Soil Biol.* **2009**, *321*, 235–257. [CrossRef]
64. Freschet, G.T.; Valverde-Barrantes, O.J.; Tucker, C.M.; Craine, J.M.; McCormack, M.L.; Violle, C.; Fort, F.; Blackwood, C.B.; Urban-Mead, K.R.; Iversen, C.M.; et al. Climate, soil and plant functional types as drivers of global fine-root trait variation. *J. Ecol.* **2017**, *105*, 1182–1196. [CrossRef]
65. Zhalnina, K.; Dias, R.; de Quadros, P.D.; Davis-Richardson, A.; Camargo, F.A.; Clark, I.M.; McGrath, S.P.; Hirsch, P.R.; Triplett, E.W. Soil pH determines microbial diversity and composition in the park grass experiment. *Microb. Ecol.* **2015**, *69*, 395–406. [CrossRef]
66. Flores-Rentería, D.; Sánchez-Gallén, I.; Morales-Rojas, D.; Larsen, J.; Álvarez-Sánchez, J. Changes in the Abundance and Composition of a Microbial Community Associated with Land Use Change in a Mexican Tropical Rain Forest. *J. Soil Sci. Plant Nutr.* **2020**, *20*, 1144–1155. [CrossRef]
67. Ji, C.-J.; Yang, Y.-H.; Han, W.-X.; He, Y.-F.; Smith, J.; Smith, P. Climatic and Edaphic Controls on Soil pH in Alpine Grasslands on the Tibetan Plateau, China: A Quantitative Analysis. *Pedosphere* **2014**, *24*, 39–44. [CrossRef]
68. Zhang, Y.Y.; Wu, W.; Liu, H. Factors affecting variations of soil pH in different horizons in hilly regions. *PLoS ONE* **2019**, *14*, e0218563. [CrossRef] [PubMed]
69. Du, E.; Terrer, C.; Pellegrini, A.F.A.; Ahlström, A.; van Lissa, C.J.; Zhao, X.; Xia, N.; Wu, X.; Jackson, R.B. Global patterns of terrestrial nitrogen and phosphorus limitation. *Nat. Geosci.* **2020**, *13*, 221–226. [CrossRef]
70. Zhang, L.; Fan, J.; Ding, X.; He, X.; Zhang, F.; Feng, G. Hyphosphere interactions between an arbuscular mycorrhizal fungus and a phosphate solubilizing bacterium promote phytate mineralization in soil. *Soil Biol.* **2014**, *74*, 177–183. [CrossRef]
71. He, D.; Xiang, X.; He, J.-S.; Wang, C.; Cao, G.; Adams, J.; Chu, H. Composition of the soil fungal community is more sensitive to phosphorus than nitrogen addition in the alpine meadow on the Qinghai-Tibetan Plateau. *Biol. Fertil. Soils* **2016**, *52*, 1059–1072. [CrossRef]
72. Huang, J.; Hu, B.; Qi, K.; Chen, W.; Pang, X.; Bao, W.; Tian, G. Effects of phosphorus addition on soil microbial biomass and community composition in a subalpine spruce plantation. *Eur. J. Soil Biol.* **2016**, *72*, 35–41. [CrossRef]
73. Cai, Z.-Q.; Zhang, Y.-H.; Yang, C.; Wang, S. Land-use type strongly shapes community composition, but not always diversity of soil microbes in tropical China. *CATENA* **2018**, *165*, 369–380. [CrossRef]
74. Magan, N. Fungi in extreme environments. *Mycota* **2007**, *4*, 85–103.
75. Chen, H.; Mothapo, N.V.; Shi, W. Soil moisture and pH control relative contributions of fungi and bacteria to N_2O production. *Microb. Ecol.* **2015**, *69*, 180–191. [CrossRef]
76. Venkateswarlu, B.; Rao, A.; Raina, P. Evaluation of phosphorus solubilisation by microorganisms isolated from Aridisols. *J. Indian Soc. Soil Sci.* **1984**, *32*, 273–277.
77. Li, J.; Li, Z.; Wang, F.; Zou, B.; Chen, Y.; Zhao, J.; Mo, Q.; Li, Y.; Li, X.; Xia, H. Effects of nitrogen and phosphorus addition on soil microbial community in a secondary tropical forest of China. *Biol. Fertil. Soils* **2015**, *51*, 207–215. [CrossRef]
78. Wang, J.; Liu, G.; Zhang, C.; Wang, G.; Fang, L.; Cui, Y. Higher temporal turnover of soil fungi than bacteria during long-term secondary succession in a semiarid abandoned farmland. *Soil Tillage Res.* **2019**, *194*, 104305. [CrossRef]

forests MDPI

Article

Effects of Forest Gap on Soil Microbial Communities in an Evergreen Broad-Leaved Secondary Forest

Shiyou Chen [1], Chunqian Jiang [1,*], Yanfeng Bai [1], Hui Wang [1], Chunwu Jiang [2], Ke Huang [3], Lina Guo [1], Suping Zeng [1] and Shuren Wang [1]

1 Research Institute of Forestry, Chinese Academy of Forestry, Beijing 100091, China
2 Anhui Academy of Forestry, Hefei 230031, China
3 Huitong Experimental Station of Forest Ecology, CAS Key Laboratory of Forest Ecology and Management, Institute of Applied Ecology, Shenyang 110016, China
* Correspondence: jiangchq@caf.ac.cn

Abstract: Forest gaps play a crucial role in community succession and assembly in forest ecosystems; therefore, they have recently been recognized and implemented as effective forest management practice all over the world. Forest gaps are commonly created as small disturbances in secondary forests to improve forest regeneration, nutrient cycling, ecosystem functioning, and biodiversity. The objective of this study was to investigate the responses of the physico-chemical and biological properties and microbial communities in soil to different sizes of forest gaps—including small gaps (60–80 m^2), medium gaps (130–160 m^2), and large gaps (270–300 m^2)—and to examine the driving factors that influence soil microbial community structure and composition. The results show that Gram-positive bacteria, Gram-negative bacteria, fungi, arbuscular mycorrhizal fungi (AMF), and actinomycetes were mainly aggregated in the gaps, and the structural diversity of soil microbial communities was related to the gap size ($p < 0.05$). The soil microbial community diversity increased and then decreased with an increase in gap size. Moreover, the effects of the available phosphorus, soil pH, soil water content, available potassium, nitrate nitrogen and ammonium nitrogen on changes in microbial biomass were significant ($p < 0.05$). The gap area and gap position and their combined interactions also had significant effects on soil nutrients, which impacts the soil microbial community. Medium gaps (130–160 m^2) always significantly improved the availability of soil nutrients, and good management practices in secondary forests can provide effective microenvironments for soil microbes.

Keywords: gap size; soil microbe; soil chemical property; forest management; PLFA analysis; spatial distribution

Citation: Chen, S.; Jiang, C.; Bai, Y.; Wang, H.; Jiang, C.; Huang, K.; Guo, L.; Zeng, S.; Wang, S. Effects of Forest Gap on Soil Microbial Communities in an Evergreen Broad-Leaved Secondary Forest. *Forests* **2022**, *13*, 2015. https://doi.org/10.3390/f13122015

Academic Editor: Bartosz Adamczyk

Received: 21 September 2022
Accepted: 28 November 2022
Published: 29 November 2022

1. Introduction

Forest gaps are important places for plant and soil community succession and assembly, which play a crucial role in forest ecology [1]. In forest ecosystems, human activities (e.g., cutting and thinning) and natural disasters accelerate the formation of forest gaps, creating a heterogeneous environment for plants, animals, and soil microorganisms to grow [2–4]. The artificial creation of forest gaps as an effective forest management practice has recently been recognized and is increasingly implemented all over the world. Artificial gaps play an important role in changing the structure of forest stands and renewing forests [3]. For example, the felling of trees to create forest gaps is considered a sustainable practice, accelerating seedling height growth for indigenous trees or accelerating forest restoration within exotic *Pinus radiata* plantations [5–7]. Gap size is an important feature, reflecting the magnitude of disturbance and degree of environmental change [8,9]. Environmental factors, such as temperature, solar radiation, and moisture, are all strongly affected by the size of the forest gaps [6,10]. Moreover, microclimatic changes induced by forest gaps significantly influence local biogeochemical cycling [11,12].

Forest soil responds dramatically to microclimatic changes caused by forest gaps because nutrients are actively recycled and many soil fauna or microorganisms live in the soil [13]. Previous studies concentrated on carbon, nitrogen, and phosphorus cycles and the release of nutrients from the surface soil at sites with different-sized gaps [14–16]. Some studies reported distinctly different soil organic matter decomposition and nutrient release at sites with gaps compared to sites with closed canopies, as well as varying rates of humification at sites with different-sized gaps [17–19]. Compared to undisturbed closed forests, the amount of labile carbon reduced, and the rate of soil respiration slowed as the gap size increased. At the same time, soil nitrogen and phosphorus fractions appeared to increase due to the gaps [15,20–22].

The soil microbial community is one of the most important components of the forest ecosystem and plays an important role in aboveground and underground biological processes related to carbon, nitrogen, and phosphorus cycles and nutrient release [23–25]. The soil microbial community quickly reacts to the development of forest gap vegetation and the changes in environmental conditions [26–28]. The effects of forest gaps on soil microbial communities are poorly understood due to their relative complexity. The formation of forest gaps may improve understory vegetation and change soil microclimates, which may be beneficial for microbial communities [29,30]. However, this may sometimes have a negative impact on microorganisms [31,32]. Several studies reported that the gap sizes dramatically affect soil microbial biomass and the soil microbial community [27,28]. It has been observed that small gaps enhance soil beta-glucosidase and L-leucineaminopeptidase, microbial biomass, and enzyme activity, indicating the beneficial impact of small gaps (40–50 m^2) on microbial communities [33]. Soil microbial biomass and soil respiration decreased with the increased size of gaps [34]. Moreover, gap location is another important factor affecting soil microbial communities [35]. It was observed that the beta diversity of the fungal community increased from the gap center to the closed canopy [36]. Gap locations had a positive effect on bacterial population abundance [2,33].

Evergreen broad-leaved forests are a zonal type of vegetation found in subtropical areas [36,37] and are crucial for climate regulation, biodiversity protection, and soil and water conservation [36,38]. Many evergreen broad-leaved forests were recently destroyed, and thus a larger area of evergreen broad-leaved secondary forests formed over time. Evergreen broad-leaved secondary forests generally have low-efficiency forest sand. How to manage evergreen broad-leaved secondary forests is a great international challenge. Previous studies on forest gaps mainly focused on the impact of the physical and chemical properties of soil, including the growth of seedlings under evergreen broad-leaved secondary forests [33,39,40]. However, few studies reported the effects of forest gaps on the soil microbial community structure in evergreen broad-leaved secondary forests. The main objective of the present study was to examine how soil microbial communities respond to the formation of forest gaps with different sizes in the short term. We hypothesized that, (1) due to an improved light environment and hydrothermal condition in the evergreen broad-leaved secondary forest, soil properties would react differently in the sites with forest gaps and in the understory, and (2) the microbial community structure shift would have a strong relationship with changes in soil chemical properties caused by different-sized gaps.

2. Materials and Methods

2.1. Site Description

This study was conducted at Huitong Forest Ecological Experimental Station of Chinese Academy of Sciences in Huaihua City, Hunan Province (109°30′ E, 26°48′ N) (Figure 1). The study areas were located in low mountains and hilly landforms with an approximate altitude range of 200~500 m [41]. The study area has a typical subtropical, warm, humid, monsoon climate with an average annual temperature of 16.5 °C, an extreme maximum temperature of 36.4 °C, and minimum temperature of −4.4 °C, respectively. The average annual rainfall was 1200–1400 mm, and the average annual relative humidity exceeded 80% [42]. Red–yellow soil is a typical soil type in this study area, and was equal to alliti-udic ferrosols according to the World Reference Base for Soil Resources [43]. Evergreen broad-leaved secondary forests developed due to artificial disturbances of the original evergreen broad-leaved forest after more than 30 years of natural succession. The dominant tree species are mainly from the following families and genera: *Lauraceae, Fagaceae, Hamamelisaceae, Luteaceae cyclobalanopsis, Castanopsis, Machilus, Liquidambar*, etc. The common tree species are *Castanopsis hystrix, Cyclobalanopsis glauca,* and *Machilus pauhoi,* and other species are present in typical subtropical evergreen broadleaved secondary forests (the pioneering community that originates after secondary succession of human disturbance).

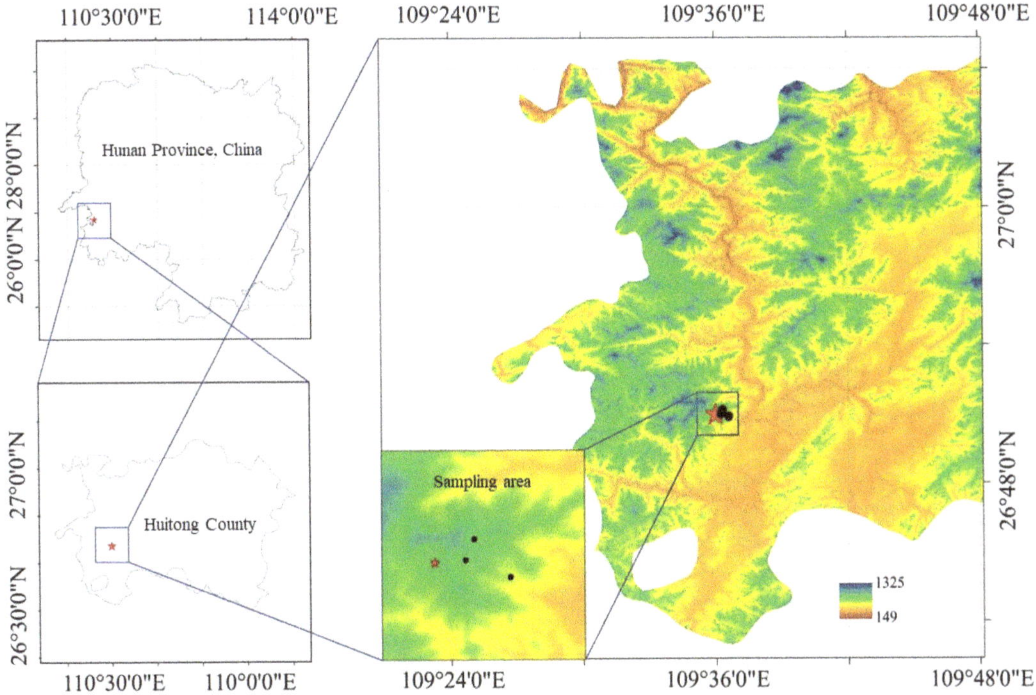

Figure 1. Study area showing location of the sample plots in an evergreen broad-leaved secondary forest of China.

2.2. Experimental Design

Study plots with a similar topography, elevation, slope, aspect, and soil type were established in a middle-age evergreen broad-leaved secondary forest, which had an average tree height of 12 m and stand density of 3200 trees /hm^2. Canopy density was about 0.9, and grass coverage was 25%. The red–yellow soil is the typical soil type in this study area. Forest gap sizes were defined using the ratios of D/H (D means gap diameter; H means stand height) [44,45]. In this study, three types of nearly circular forest gaps were created by artificial cutting, including (1) small-size forest gaps with a D/H ratio of 0.3 and an area of 60–80 m^2, (2) medium-size forest gaps with a D/H ratio of 1.3 and an area of 130–160 m^2, and (3) large size forest gaps with a D/H ratio of 2.3 and an area of 270–300 m^2. Each type of forest gap (small-, medium-, or large) had three replications, with nine individual forest gaps in total (G1–G9), which were arranged in the forest using the randomized complete-block design.). In addition, three control plots (10 m $\times$ 10 m) under the ambient closed canopy were established as controls. In this study, nine individual forest gaps and three control plots were used (Table 1). After the creation of the forest gaps, we artificially removed branches, trunks, and other residue, which remained untreated.

Table 1. General information of the forest gaps and control plots (understory) in this study.

Forest Gap	D/H	Area/m^2	Slope Direction	Geographic Position	Formation
G1	0.3	60	East	109°36′6″ E 26°51′11″ N	Cutting
G2	1.3	130	Northeast	109°36′8″ E 26°51′13″ N	Cutting
G3	2.3	270	East	109°36′7″ E 26°51′09″ N	Cutting
G4	0.3	73	Southeast	109°36′9″ E 26°51′12″ N	Cutting
G5	1.3	149	Southeast	109°36′9″ E 26°51′07″ N	Cutting
G6	2.3	286	Northeast	109°36′10″ E 26°47′10″ N	Cutting
G7	0.3	80	East	109°36′10″ E 26°50′15″ N	Cutting
G8	1.3	160	Northeast	109°36′11″ E 26°52′14″ N	Cutting
G9	2.3	300	East	109°36′15″ E 26°51′14″ N	Cutting
Understory	—	100	Southeast	109°36′16″ E 26°49′14″ N	Natural
Understory	—	100	Northeast	109°36′18″ E 26°46′14″ N	Natural
Understory	—	100	East	109°36′17″ E 26°50′14″ N	Natural

Notes: D/H represents the ratios of gap diameter to stand height; G1, G4, and G7, represent small gaps; G2, G5, and G8 represent medium gaps; and G3, G6, and G9 represent large gaps.

Soil samples were collected from a depth of 0–10 cm at different sampling locations (the center to the understory) within each of the three different-sized gaps in October 2019. In each location, four soil cores were selected by the "Circular sampling" method and dug to a depth of 10 cm by a stainless-steel cylindrical driller with a diameter of 5 cm (Figure 2). The composite soil samples were used to test soil physico-chemical properties. There are three repetitions for each location because three same-size gaps were formed. A total of 39 soil samples were collected, including the control sample. They were kept at 4 °C, transported to the laboratory, and passed through a 2 mm mesh sieve before analyses.

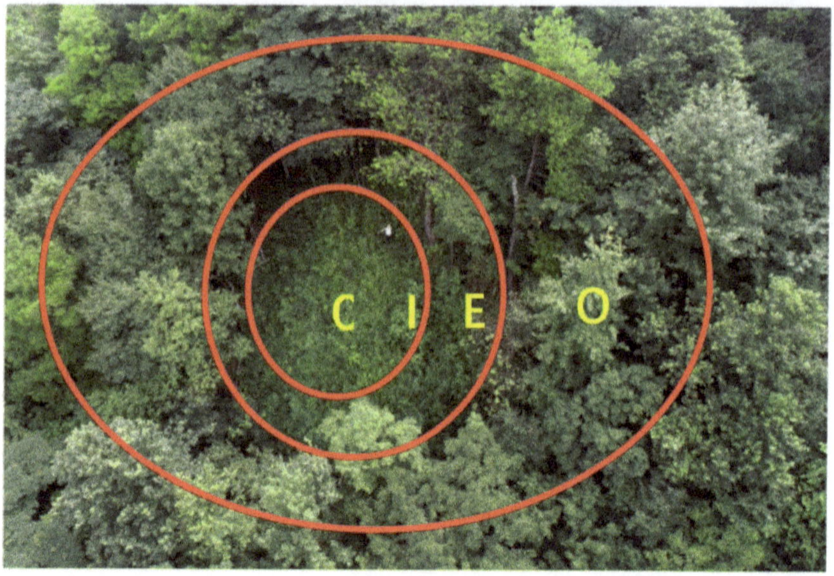

Figure 2. Soil samples from different sampling locations within the forest gaps. For sampling locations, C represents the center of forest gaps, I represents the location away from the center at 0.5 R distance, E represents the location at the edge of forest gaps, and O represents the location away from the center at 1.5 R distance (R, means gap radius).

2.3. *Analysis Methods*

2.3.1. Basic Chemical Properties of Soil

The soil pH was measured in air-dried <2 mm sieved soil with a glass electrode in a soil:water suspension (1:5) with 0.2 M KCl to determine the characteristics of the sorption complex. The total C (TC) contents were determined using an elemental analyzer (Vario Macro cube, Elementar, Germany). The total nitrogen (TN) was assayed by the Kjeldahl method and analyzed on a Kjeltec Analyzer Unit (Kjeltec 2300, FOSS, Denmark). The total P was determined using the molybdenum colorimetric method. We assayed alkaline available nitrogen (AN), available phosphorus (AP), and available potassium by extracting air-dried soils (Institute of Soil Science, Chinese Academy of Sciences 1978). The determination of AP and available potassium were analyzed on an ICP (AA-7000, Kleve, Germany). Ammonium and nitrate were extracted from fresh soil samples by shaking for 1 h with 1 mol L−1 KCl (soil: solution ratio of 1:10). The extracts were filtered, and then, the analytes were determined using a continuous-flow analyzer (AA3, SEAL, Germany). For the determination of soil-available potassium, air-dried soil samples were extracted by shaking for 30 min with 1 mol·L^{-1} CH3COONH4 (soil: solution ratio of 1:10), and then, the available potassium content of the filtrate was determined using a continuous-flow analyzer (AA3, Germany). Soil samples were sampled on site using an aluminum box and brought back to the laboratory to be dried at 105 °C in an oven until the soil moisture was measured at constant weight.

$$\text{Soil Moisture Measurement} = [(\text{Wet soil weight} - \text{dry soil weight})/\text{dry soil weight}] \times 100\% \quad (1)$$

2.3.2. Soil Microbial Community Structure

Phospholipid fatty acids (PLFA) were used to characterize the microbial community structure. PLFAs were extracted from 2 g of lyophilized soil, separated, and methylated. The resulting fatty acid methyl esters (FAMEs) [46] were separated by gas chromatography using an Agilent 7890 A GC System equipped with a HP-ULTRA 2 column and a flame ionization detector [46]. The individual FAME peaks were identified and quantified with the software Sherlock™ PLFA Method and Analysis Package. Specific PLFAs were used as biomarkers to quantify the relative abundances (mol%) of particular microbial groups (Table 2). The internal standard 19:0 phosphatidylcholine was used for the quantification of FAMEs. Although most bacterial PLFAs have acyl chain lengths between 14 and 20 carbons, there are fatty acids longer than 20 carbons that predominantly originate from bacteria or micro-eukaryotes, such as 21:0, 22:0, 22:5 ω3, 22:6 ω3, and 24:0. This software was designed to detect such fatty acids in soil samples; therefore, they were also taken into account in our study. The viable microbial biomass was calculated by summing PLFAs concentrations and reported as nanomoles of PLFA per gram of soil. Several PLFAs may have various sources. Fatty acids indicating arbuscular mycorrhizae fungi (AMF) were summed as total fungal biomass. Bacterial biomass was calculated from the residual fatty acids that could be assigned to the bacterial groups. Biomass was expressed relative to dry weight of the freeze-dried soil. The ratios of fungal/bacterial (18:2ω6 for fungi) and Gram-positive(G^+)/Gram-negative (G^-) bacterial markers were also obtained. The absolute amount of PLFA was calculated by the area normalization method using the following formula ([47,48]):

$$\text{Phospholipid Fatty Acids (PLFA)} = \frac{(\text{19:0 concentration} \times \text{total area})/\text{19:0 molarmass}}{\text{Actual weight of soil}} \times \frac{\text{Response area of a}}{\text{Response area of 19:0}} \quad (2)$$

Table 2. Phospholipid fatty acid (PLFA) profiling of soil microbial communities.

Microbial Group	Phospholipids Fatty Acids Signatures
Actinomycetes	16:0 10-methyl; 17:0 10-methyl; 17:1 w7c 10-methyl; 18:0 10-methyl; 18:1 w7c 10-methyl etc.
G+ bacteria	11:0 anteiso; 11:0 iso; 12:0 anteiso; 12:0 iso; 13:0 anteiso; 13:0 iso; 14:0 anteiso; 14:0 iso; 14:1 iso w7c; 15:0 anteiso; 15:0 iso; 15:1 anteiso w9c; 15:1 iso w6c; 15:1 iso w9c; 16:0 anteiso; 16:0 iso; 17:0 anteiso; 17:0 iso; 17:1 iso w9c; 18:0 iso; 19:0 anteiso; 19:0 iso; 20:0 iso; 22:0 iso
G− bacteria	13:1 w5c; 14:0 2OH; 14:1 w8c; 14:1 w9c; 15:1 w7c; 15:1 w8c; 5:1 w9c; 16:0 2OH; 17:0 cyclo w7c; 17:1 w3c; 21:1 w3c; 21:1 w4c; 21:1 w5c; 21:1 w6c; 21:1 w8c; 22:1 w6c; 22:1 w8c; 22:1 w9c; 24:1 w7c
Eukaryote	15:3 w3c; 15:4 w3c; 16:3 w6c; 16:4 w3c; 18:3 w6c; 19:3 w3c; 19:3 w6c; 19:4 w6c; 20:2 w6c; 20:3 w6c; 20:4 w6c; 20:5 w3c; 21:3 w3c; 21:3 w6c; 22:2 w6c; 22:4 w6c 22:5 w3c; 22:5 w6c; 22:6 w3c; 23:1 w4c; 23:1 w5c; 23:3 w3c; 23:3 w6c; 23:4 w6c; 24:1 w3c; 24:3 w3c; 24:3 w6c; 24:4 w6c
AM Fungi	18:2 w6c
Fungi	16:1 w5c

2.3.3. Saturated Fatty Acids/Monounsaturated Characteristics

SAT/MONO (Saturated fatty acids/monounsaturated) usually indicates environmental stress in soil microorganisms [49].

$$\text{SAT/MONO} = \text{Saturated fatty acids/monounsaturated} \tag{3}$$

2.4. *Statistical Analysis*

We conducted a two-way analysis of variance (ANOVA) using the General Linear Models package in SPSS 21.0 (IBM Corp., Armonk, NY, USA) to test the interaction between the forest gap size and soil position in the available phosphorus and potassium, nitrate nitrogen and ammonium nitrogen concentrations, soil water content, and soil pH. Additionally, a redundancy analysis (RDA) was performed using the Canoco 4.5 software. Six soil microbial community structures (eukaryote, fungi, actinomycetes, Gram-positive and Gram-negative bacteria, and AM fungi) and six soil chemical properties were analyzed to determine the relationship between microbial community and environmental outcomes. The experimental effects on soil microbial communities in forest gaps were assessed using a generalized linear model (GLM), applying the binomial family and default logit link function. Post hoc pairwise comparisons of significance were carried out for GLMs using a Least Significance Difference Method. All the differences were tested with a significance level of $p = 0.05$. Graphics were generated in Origin (version 2018, Origin Lab).

3. Results

3.1. *Response of Soil Microbial Community Structure to Forest Gap*

A total of 92 different PLFAs were detected from all the samples. For G− (Gram-negative) bacteria and G+ (Gram-positive) bacteria, fungi, actinomycetes, and AM fungal PLFAs, all the markers varied significantly in different-sized forest gaps. The different percentages of specific biomarkers of the PLFAS in forest gaps are shown in Figure 3.

Compared to forest gap, the contents of PLFAs (10:03OH,13:0iso,14:02OH;15:1w6c,16:1w9c, 18:1w5c,19:1w8c,20:1w8c,21:1w3c,22:5w6c,24:1w3c) were not detected. The contents of PLFAs varied with forest gap size. The PLFAS of 16:0,19:0 cyclo w7c,15:0 iso,18:1 w7c,16:0 10-methyl,18:1 w9c,17:1 iso w9c,18:0,16:0 iso,15:0 anteiso,18:0 10-methyll17:1 w7c 10-methyll17:0 isol16:1 w7c,19:0l17:0 anteiso,16:1 w5cl18:2 w6cl17:0 cyclo w7c and18:1 w5c ranked differently, as shown in Figure 2. In three gaps and the forest understory, the total amount of PLFAs was quite different, medium gap (101.73 nmol/g) > large gap (95.75 nmol/g) > small gap (94.35 nmol/g) > understory (92.36 nmol/g) in Table 3.

As shown in Figure 4, the SAT/MONO ratios were the highest in the medium gap, but the lowest in the large gap ($p < 0.001$). The SAT/MONO ratios showed no differences in different positions within the large gap, but distinct differences in the small gap, medium gap, and the understory. The SAT/MONO ratio decreased in the small gap and large gap in comparison to the understory. The medium gap was consistent with understory in the SAT/MONO ratio. Overall, the changes in environment deeply affected the soil microbial communities.

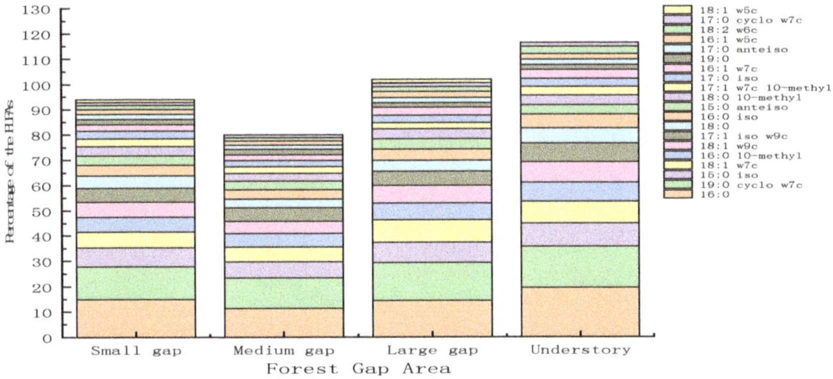

Figure 3. The percentage of PLFAS charactering microbes in forest gaps of different sizes.

Table 3. Total amount of soil microbial PLFAs under the forest gap area (mean ± SE, 95% CI).

Gap Type	Gap Position				
	Center	Inside	Edge	Outside	Total
Small gap	100.82 ± 2.21 Aa	95.71.41 ± 1.19 Bb	94.25 ± 2.4 Cb	90.08 ± 2.58 Dd	94.35 ± 3.12 a
Medium gap	108.93 ± 3.81 Ba	105.00 ± 3.34 Bb	98.48 ± 4.45 Bc	95.00 ± 3.81 Dc	101.73 ± 1.10 b
Large gap	86.75 ± 2.63 Ca	95.27 ± 1.63 Bb	116.73 ± 2.83 Cc	86.08 ± 1.41 Ba	95.75 ± 2.06 c
Understory	90.43 ± 3.62 C				90.43 ± 3.62 a

Note: A, B, C, and D indicate differences in the same gap position with different sizes; a, b, c, and d show differences in the different positions within the same gap ($p < 0.05$).

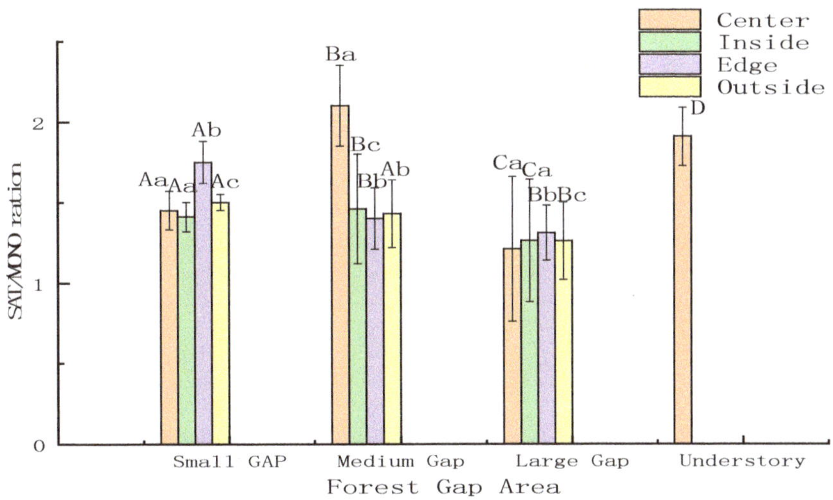

Figure 4. Comparison of SAT/MONO ratios in different gap positions (mean ± SE, 95% CI). Note: A, B, C, and D indicate differences in the same gap position with different sizes; a, b, and c show differences in different positions within the same gap ($p < 0.05$).

3.2. *Response of Soil Microbial Community Composition to Forest Gap*

3.2.1. Soil Fungi, Actinomycetes, G+ Bacteria, and G− Bacteria Response to Gap Size

The PLFA contents in fungi in different forest gaps ranged from 1.86 to 9.01 nmol g^{-1}: large gap > small gap > medium gap > understory. From the center of gap to the understory, the fungal PLFA ranged from 1.28 to 3.45 nmol g^{-1}. Across different positions within the same gap, quite different variations were shown. Soil fungi in the center and edge of the gap responded more strictly to the forest gap than the understory (Figure 5a).

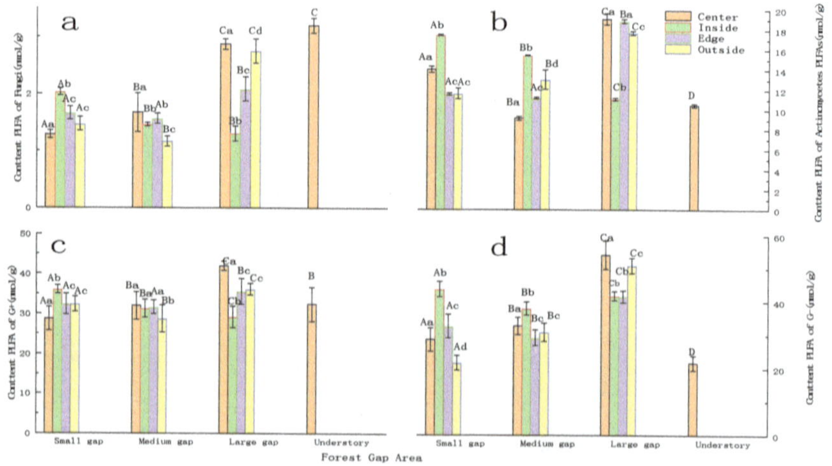

Figure 5. (**a–d**) Differences in the contents of the soil microbial community in forest gaps (mean ± SE, 95% CI). Note: A, B, C, and D indicate differences in the same gap positions with different sizes; a, b, c, and d show differences in different positions within the same gap ($p < 0.05$).

Actinomycetes were most in the forest gap in comparison with the understory. From the understory to the gaps, the order was: understory (11.06 nmol g^{-1}), small gap (12.21 nmol g^{-1}), medium gap (13.68 nmol g^{-1}), large gap (16.9 nmol g^{-1}) ($p < 0.01$). The PLFA contents of actinomycetes in the same forest gap varied with the position in the gaps. Inside of the forest gap, the PLFA contents of actinomycetes were highest; however, the PLFA contents of actinomycetes at the edge of the forest gap. Overall, the surface aggregation of actinomycetes was significantly affected by the forest gap sizes and position (Figure 5b).

The large gap (270–300 m^2) was the most active habitat of G+ bacteria, and the G+ bacterial contents at the top soil (0–10 cm) were significantly higher than those in the understory. However, there were no significant changes in the small gaps (60–80 m^2) and medium gaps (130–160 m^2). The content of G+ bacteria in the center of the gap was significantly higher than that in the same gap size at a different position. Additionally, variations in G+ bacteria differed in the same positions with different sizes ($p < 0.05$ (Figure 5c).

The gap size difference in the PLFA contents of G− bacteria was more complex. The PLFA levels in the same gap size were dramatically altered in the four sites, and changes in the gap center and outside of the gap were substantial between the small gap and large gap. Compared to the understory, the number of medium gaps (130–160 m^2) was relatively stable, and there were no significant changes in the center, inside, edge and outside areas. In the forest gaps, variations in G− bacterial PLFA contents were not clear. However, at the center and outside of the gap, a peak was observed as gap size increased (Figure 5d).

3.2.2. Soil AM Fungi and Anaerobic Response to Forest Gap

The gap size altered the surface aggregation of AM fungi in the gap and understory (Figure 6). The AM fungal PLFA content in the forest gap was similar to the PLFA contents of fungi, large gap (270–300 m^2) > small gap (60–80 m^2) > medium gap (130–160 m^2) > understory ($p < 0.01$). The PLFA contents of AM fungi ranged from 1.45 to 2.80 nmol g^{-1} (Figure 4). The PLFA contents of AM fungi were highest in the small gap; however, the large gap and medium gap reached a peak outside of the forest gap. Overall, the forest gap significantly affected the surface aggregation of AM fungi.

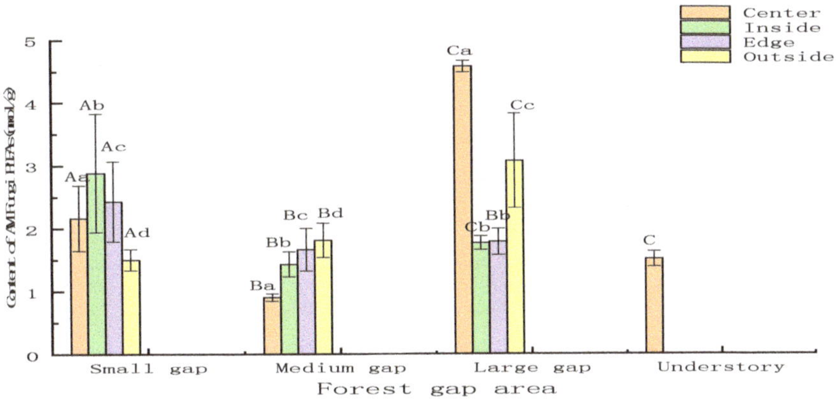

Figure 6. Contents of AM Fungi PLFAs in the different forest gap area (mean ± SE, 95% CI). Note: A, B, C, and D indicate differences in the same gap position with different sizes; a, b, c, and d show differences in a different position within the same gap ($p < 0.05$).

The anaerobic bacterial community occupies a large proportion in the structure of the anaerobic microbial community in the forest gap. From all of the samples, phospholipid fatty acids that can be used to characterize anaerobic bacteria, such as 18:1 w7c, 15:0 a, 15:0 i, 16:0 i, 17:0 i, 19:0 cy, etc., were detected. The PLFA content of the anaerobic in the small gap and the large gap accounted for 8%–11% and 4%–10%, respectively. Anaerobic bacterial PLFA content was significantly higher than those in the small gap and large gap, accounting for 22%–34%. More interestingly, no anaerobic PLFA makers were detected in the sample collected from the medium gap. This gap had a strong effect on the anaerobic microbial community.

3.3. Soil Factors Driving Soil Microbial Community Shift

The gap area of evergreen broad-leaved secondary forest influenced the physical properties of the soil (Table 4). The value of available phosphorus in the four habitats followed the order medium gap > large gap > small gap > understory, and the value of available phosphorus in the medium gap was significantly higher than that in the small gap, large gap, and understory ($p < 0.01$). The value of nitrate nitrogen in the forest gap was similar to the value of available phosphorus ($p < 0.05$). The value of available phosphorus potassium and ammonium nitrogen in the four habitats followed the order medium large gap > small gap > medium gap > understory. The gap area of evergreen broad-leaved secondary forest significantly affected the value of soil water content, as the soil water content decreased from the understory to the forest gap. Meanwhile, the differences in soil pH among the three gaps and understory were not significant ($p > 0.05$).

The results show that the differences in soil water, nitrate nitrogen, ammonium nitrogen, available potassium, and available phosphorus between center and edge and between the outside and center were significant ($p < 0.05$). Environmental outcomes evidently contribute to the size of the forest gaps. Compared to the understory, the differences in soil pH among the center, edge, and outside of the gaps were not significant ($p > 0.05$).

Furthermore, the two-way ANOVA results show that forest gaps significantly affected all the chemical properties of soil ($p < 0.05$) (Table 5). Gap area, gap position, and their combined interactions also had significant effects on available phosphorus, nitrate nitrogen, ammonium nitrogen, and soil water content.

Table 4. Effect of different sizes of forest gaps on soil chemical properties (mean ± SE, 95% CI).

Gap Size	Available Phosphorus	Available Potassium	Nitrate Nitrogen	Ammonium Nitrogen	Soil Water Content	Soil pH
Small gap	1.83 ± 0.07 A	44.10 ± 5.23 A	2.1 ± 1.09 A	8.57 ± 1.18 A	0.38 ± 0.03 A	4.28 ± 0.02 A
Medium gap	2.09 ± 0.09 B	43.88 ± 3.91 A	2.48 ± 0.71 B	8.93 ± 0.88 B	0.37 ± 0.02 A	4.37 ± 0.11 A
Large gap	2.07 ± 0.09 C	69.50 ± 7.09 B	1.58 ± 0.62 C	9.85 ± 4.43 C	0.34 ± 0.01 B	4.44 ± 0.08 A
Understory	1.98 ± 0.06 A	36.59 ± 0.3.7 C	0.37 ± 0.01 D	8.27 ± 0.04 A	0.32 ± 0.05 C	4.31 ± 0.04 A

Note: A, B, C, and D indicate differences in the forest gap with different sizes.

Table 5. Two-way analysis of variance (ANOVA) for the effects of gap size and location on soil basic chemical properties in evergreen broad-leaved secondary forest.

Factor	*df*	Available Phosphorus	Available Potassium	Nitrate Nitrogen	Ammonium Nitrogen	Soil Water Content	Soil pH
Gap area	3	<0.01	<0.05	<0.05	<0.05	<0.05	<0.05
Gap position	3	<0.05	<0.05	<0.01	<0.01	<0.01	0.679
Gap area × gap position	9	<0.05	0.269	<0.01	<0.01	<0.05	0.963

The redundancy analysis (RDA) demonstrated that soil chemical properties significantly affected the soil community structures in evergreen broad-leaved secondary forest gaps. The analysis results show that 96.2% of the soil microbial community structure information could be explained by the six selected soil chemical property indicators, among which axis 1 and axis 2 explained 78.2% and 18.0% of the variation information, respectively (Figure 7). Among all the tested environment factors, ammonium nitrogen, available potassium, and pH value had significant positive correlations with microbial communities, in which available potassium had a stronger effect (longer arrow) on microbial community on microbial community structure. Interestingly, the correlations between the fungi, eukaryote and soil water contents, and nitrate nitrogen were negative and significant. Moreover, the soil pH was not the main environmental factor affecting the soil microbial community composition.

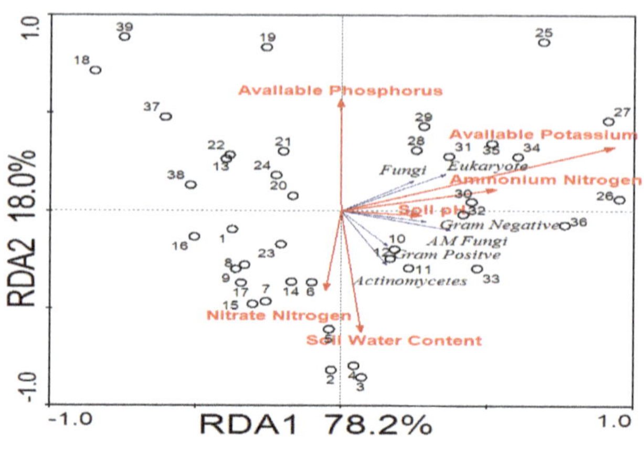

Figure 7. Redundancy analysis (RDA) of relationships between microbial community structures, physcochemical properties, and environmental factors. The variations in the cumulative interpretation of the first and second axes were 96.2%. The first axis indicates the variables in 78.2%, and the second axis explains the variables in 18.0%.

4. Discussion

Significant differences were identified in the physical properties of soil among different-sized forest gaps and the understory, indicating differences in the sensitivity of soil properties and the mineralization process in forest gaps of different sizes. The value of available phosphorus in the four habitats followed the order medium gap > large gap > small gap > understory ($p < 0.01$). The available phosphorus content in the top soil layer of middle gaps in secondary forests was significantly increased, which may be due to the accumulated rich litter and root biomass in soil [15], increasing its phosphorus content. In addition, the phosphorus in the soil is easily fixed by soil clay [19], and the vegetation and litter layer in middle forest gaps can easily intercept rainfall, resulting in a small amount of phosphorus leaching loss [50]. Wei (2021) conducted a study on the influence of forest gaps on the available nitrogen of soil in a *Pinus massoniana* plantation. It was found that the larger forest gap areas corresponded to higher concentrations of ammonium nitrogen and nitrate nitrogen, and both were more concentrated in gap areas than in the understory [51]. Interestingly, congruent changes in ammonium nitrogen were found upon analyzing the value of ammonium nitrogen in large gaps, medium gaps, small gaps, and the understory, with the following order: large gap > small gap > medium gap > understory ($p < 0.05$). In contrast, different changes in nitrate nitrogen were observed upon analyzing the value of nitrate nitrogen in large gaps, medium gaps, small gaps, and the understory. The values of nitrate nitrogen in medium gaps were significantly higher than those in the small gaps, large gaps, and understory ($p < 0.01$), which might be caused by differences in forest gap size and in gap zone, from the understory to the canopy and the expanded gaps. Additionally, we found that the value of available potassium in the four habitats followed the order large gap > small gap > medium gap > understory, which showed that the gaps could contribute to soil nutrient transformation. Forest gaps alter the soil conditions by increasing solar radiation and reducing plant water uptake; consequently, soil surface temperatures and moisture levels in the gaps are different from those of the adjacent closed-canopy forests [12,52,53]. The gap area of evergreen broad-leaved secondary forest significantly affected the value of soil water content, which decreased between the understory and the forest gap, while the differences in soil pH among three gaps and understory were not significant ($p > 0.05$). This finding is consistent with the previous findings that no significant differences in pH between the gaps and forest canopy were detected [54]. Generally, medium forest gaps may have a region of rich fertility, increasing soil nutrient availability within evergreen broad-leaved secondary forests, which is beneficial to vegetation renewal and biodiversity conservation.

The diversity of soil biota communities generally plays significant roles in mediating plant community attributes, including diversity, plant productivity, community composition, and plant–soil interactions, as well as regulating how plants respond to stress factors. [24,32,55]. Microorganisms are very active in the decomposition of matter and are important factors that influence nutrient availability in plant–soil feedback systems [33,55]. Previous studies found that gap area and gap position are the main reason for the difference in species richness between the gap and within the forest [16,56,57]. The results of our study also demonstrate that forest gap size significantly affects soil community structures and composition in evergreen broad-leaved secondary forest gaps. Among all the sizes of forest gaps, medium gaps had the most significant positive correlation with the microbial communities. A similar finding was also observed in other studies, indicating that the structure and diversity of the soil bacterial community were affected by different forest gap sizes, and the soil bacterial diversity was higher in the medium gap than the small gap [55]. The SAT/MONO ratios were highest in the medium gap, but lowest in the large gap ($p < 0.0001$). The temperature and humidity under different forest gaps are regulated by environmental factors such as illumination and rainwater, and after entering the gap, these factors are re-regulated by the difference between vegetation types under different forest gaps [58]. This is why the SAT/MONO ratio showed no differences in different positions within the large gap but were quite different in the small gap, medium gap, and understory. Within the first few years following gap creation, the physical and chemical environment of the soil changed. The litter input and the decrease in plant root exudates resulted in insufficient nutrients for microbial metabolism, which inhibited the growth and reproduction of soil microorganisms, and this inhibition effect was more evident with the increase in the gap area [27,33]. Our study had similar findings, where the total amount of PLFAs was quite different: medium gap > large gap > small gap > under-story. The comprehensive effect of gap size and sampling location on the microbial communities is quite significant [35]. The content of G+ bacteria in the center of the gap was significantly higher than that in the same gap with a different position. Interestingly, variations in G- bacterial PLFA contents were not evident in the forest gaps. Moreover, the level in the medium gaps were relatively stable, and there were no significant changes in the center, inside, edge, and outside of the gaps. In this study, due to the same site conditions, the differences between plant community structures in the different-sized gaps may be the main factor for the significant change in soil microbial community structure [52]. Gap locations caused distinct co-occurrence patterns in fungal communities, and the beta diversity of the fungal community increased from the gap center to the closed canopy [35]. Our findings also indicate that the PLFA contents in fungi in different forest gaps ranged from 1.86 to 9.01 nmol g^{-1}: large gap > small gap > medium gap > understory. In contrast, actinomycetes were most active in the forest gap in comparison with the understory, large gap (16.9 nmol g^{-1}) > medium gap (13.68 nmol g^{-1}) > small gap (12.21 nmol g^{-1}) > understory (11.06 nmol g^{-1}). Generally, the forest gap size substantially alters soil properties in the evergreen broad-leaved secondary forest, and alters the soil microbial community structure due to gap formation. The presented results may be strongly biased by the meteorological conditions in that specific moment, so we continued to track them for a long time. Forest gaps have profound effects on the biogeochemical processes of soil in evergreen broad-leaved secondary forests and, thus, the soil carbon pool and plant diversity. Our experiment was short, and forest gaps were shown to affect the availability of soil nutrients and soil microbe communities, driving underground ecological process. Gaps in evergreen broad-leaved secondary forests need to be further studied to understand locally relevant species and structural changes in the gap phase dynamics.

5. Conclusions

In this study, forest gaps significantly affected the soil chemical properties and soil microbial communities of evergreen broad-leaved secondary forests: (1) Forest gaps significantly affected all the chemical properties of soil; the gap area and position and their combined interactions determined the chemical properties and microbial communities of the soil. (2) Gap size was positively related to the community characteristics of soil microbial communities, as the structural quantity of soil microbial communities of the evergreen broad-leaved secondary first increased and then decreased with the increase in gap size. (3) The SAT/MONO ratio was highest in the medium gaps but lowest in the large gaps ($p < 0.001$). The SAT/MONO ratio showed no difference in different positions within the large gaps, but was quite different in small gaps, medium gaps, and understory. (4) The soil pH was not the main environmental factor affecting the soil microbial community composition. (5) Medium gaps (130–160 m^2) always significantly improved the soil nutrients, and provided a good microenvironment for soil bacteria, fungi, AM fungi, and actinomycetes ($p < 0.05$). In the practice of secondary forest management, medium gaps may be a good method for secondary forest conservation, biodiversity conservation, and carbon dioxide storage. Although we provide important insights into the changes in soil microbial community structure and soil nutrients driven by forest gaps, the results need to be validated in the evergreen broad-leaved secondary forest within a few years following the gap creation.

Author Contributions: Conceptualization, S.C., C.J. (Chunqian Jiang) and Y.B.; methodology, S.C. and C.J. (Chunqian Jiang); software, Y.B. and S.Z.; validation, C.J. (Chunwu Jiang); formal analysis, L.G.; investigation, S.W.; resources, K.H.; writing—original draft preparation, S.C.; writing—review and editing, S.C. and C.J. (Chunqian Jiang); visualization, H.W. and S.Z.; supervision, C.J. (Chunqian Jiang). All authors have read and agreed to the published version of the manuscript.

Funding: This research was funded by the National Key Research and Development Program of China (2022YFF1303000) (2022YFF1303005).

Data Availability Statement: Not applicable.

Acknowledgments: We feel grateful to the facilities made available by Wang Silong at the Huitong National Research Station of Forest Ecosystem. Additionally, Yu Xiaojun, Huang Ke, Zhang Xiuyong, Qi Mengjuan, Shi Shuorong and Wang Shuren spent plenty of time assisting with the experiments. We have been very appreciative of their support.

Conflicts of Interest: The authors declare that they have no known competing financial interests or personal relationships that could have appeared to influence the work reported in this paper.

References

1. Mantovani, M.T.W. Gap-phase regeneration in a tropical montane forest the effects of gap structure and bamboo species. *Plant Ecol.* **2000**, *148*, 149–155.
2. Muscolo, A.; Bagnato, S.; Sidari, M.; Mercurio, R. A review of the roles of forest canopy gaps. *J. For. Res.* **2014**, *25*, 725–736. [CrossRef]
3. Thompson, J.; Proctor, J.; Scott, D.A.; Fraser, P.J.; Marrs, R.H.; Miller, R.P.; Viana, V. Rain forest on Maracá Island Roraima Brazil artificial gaps and plant response to them. *For. Ecol. Manag.* **1998**, *102*, 305–321. [CrossRef]
4. Orman, O.; Dobrowolska, D. Gap dynamics in the Western Carpathian mixed beech old-growth forests affected by spruce bark beetle outbreak. *Eur. J. For. Res.* **2017**, *136*, 571–581. [CrossRef]
5. Forbes, A.S.; Norton, D.A.; Carswell, F.E. Artificial canopy gaps accelerate restoration within an exotic Pinus radiata plantation. *Restor. Ecol.* **2016**, *24*, 336–345. [CrossRef]
6. Zhu, J.; Matsuzaki, T.; Lee, F.-Q.; Gonda, Y. Effect of gap size created by thinning on seedling emergency, survival and establishment in a coastal pine forest. *For. Ecol. Manag.* **2003**, *182*, 339–354. [CrossRef]
7. Keram, A.; Halik, Ü.; Keyimu, M.; Aishan, T.; Mamat, Z.; Rouzi, A. Gap dynamics of natural Populus euphratica floodplain forests affected by hydrological alteration along the Tarim River: Implications for restoration of the riparian forests. *For. Ecol. Manag.* **2019**, *438*, 103–113. [CrossRef]
8. Gálhidy, L.; Mihók, B.; Hagyó, A.; Rajkai, K.; Standovár, T. Effects of gap size and associated changes in light and soil moisture on the understorey vegetation of a Hungarian beech forest. *Plant Ecol.* **2005**, *183*, 133–145. [CrossRef]
9. Zhao, Q.; Pang, X.; Bao, W.; He, Q. Effects of gap-model thinning intensity on the radial growth of gap-edge trees with distinct crown classes in a spruce plantation. *Trees* **2015**, *29*, 1861–1870. [CrossRef]
10. Schliemann, S.A.B.; James, G. Methods for studying treefall gaps: A review. *For. Ecol. Manag.* **2011**, *261*, 1143–1151. [CrossRef]
11. Vajari, K.A. The influece of forest gaps on some properties of humus in a managed beech forest, northern Iran. *Eurasian Soil Sci.* **2015**, *48*, 1131–1135. [CrossRef]
12. Ritter, E.; Dalsgaard, L.; Einhorn, K.S. Light, temperature and soil moisture regimes following gap formation in a semi-natural beech-dominated forest in Denmark. *For. Ecol. Manag.* **2005**, *206*, 15–33. [CrossRef]
13. Barreiro, A.; Martín, A.; Carballas, T.; Díaz-Raviña, M. Response of soil microbial communities to fire and fire-fighting chemicals. *Sci. Total Environ.* **2010**, *408*, 6172–6178. [CrossRef] [PubMed]
14. Baena, C.W.; Andrés-Abellán, M.; Lucas-Borja, M.E.; Martínez-García, E.; García-Morote, F.A.; Rubio, E.; López-Serrano, F.R. Thinning and recovery effects on soil properties in two sites of a Mediterranean forest, in Cuenca Mountain (South-eastern of Spain). *For. Ecol. Manag.* **2013**, *308*, 223–230. [CrossRef]
15. Hu, B.; Yang, B.; Pang, X.; Bao, W.; Tian, G. Responses of soil phosphorus fractions to gap size in a reforested spruce forest. *Geoderma* **2016**, *279*, 61–69. [CrossRef]
16. Gray, A.N.; Spies, T.A.; Easter, M.J. Microclimatic and soil moisture responses to gap formation in coastal Douglas-fir forests. *Can. J. For. Res.* **2002**, *32*, 332–343. [CrossRef]
17. Ni, X.; Yang, W.; Tan, B.; Li, H.; He, J.; Xu, L.; Wu, F. Forest gaps slow the sequestration of soil organic matter: A humification experiment with six foliar litters in an alpine forest. *Sci. Rep.* **2016**, *6*, 19744. [CrossRef]
18. Bauhus, J.; Bartsch, N. Mechanisms for carbon and nutrient release and retention in beech forest gaps. *Plant Soil* **1995**, *168*, 579–584. [CrossRef]
19. Zhang, C.; Zhao, X. Soil properties in forest gaps and under canopy in broad-leaved Pinus koraiensis forests in Changbai Mountainous Region, China. *Front. For. China* **2007**, *2*, 60–65. [CrossRef]
20. Achat, D.L.; Augusto, L.; Bakker, M.R.; Gallet-Budynek, A.; Morel, C. Microbial processes controlling P availability in forest spodosols as affected by soil depth and soil properties. *Soil Biol. Biochem.* **2012**, *44*, 39–48. [CrossRef]

21. Wu, Q.; Wu, F.; Tan, B.; Yang, W.; Ni, X.; Yang, Y. Carbon, nitrogen and phosphorus stocks in soil organic layer as affected by forest gaps in the alpine forest of the eastern Tibet Plateau. *Russ. J. Ecol.* **2015**, *46*, 246–251. [CrossRef]
22. Han, M.; Tang, M.; Shi, B.; Jin, G. Effect of canopy gap size on soil respiration in a mixed broadleaved-Korean pine forest: Evidence from biotic and abiotic factors. *Eur. J. Soil Biol.* **2020**, *99*, 103194. [CrossRef]
23. Bååth, E.; Díaz-Raviña, M.; Bakken, L.R. Microbial Biomass, Community Structure and Metal Tolerance of a Naturally Pb-Enriched Forest Soil. *Microb. Ecol.* **2005**, *50*, 496–505. [CrossRef]
24. Banning, N.C.; Murphy, D.V. Effect of heat-induced disturbance on microbial biomass and activity in forest soil and the relationship between disturbance effects and microbial community structure. *Appl. Soil Ecol.* **2008**, *40*, 109–119. [CrossRef]
25. Bhardwaj, Y.; Sharma, M.P.; Pandey, J.; Dubey, S.K. Variations in microbial community in a tropical dry deciduous forest across the season and topographical gradient assessed through signature fatty acid biomarkers. *Ecol. Res.* **2020**, *35*, 139–153. [CrossRef]
26. Berg, E.C.; Zarnoch, S.J.; McNab, W.H. Twenty-year survivorship of tree seedlings in wind-created gaps in an upland hardwood forest in the eastern US. *New For.* **2019**, *50*, 323–344. [CrossRef]
27. Shen, Y.; Yang, W.; Zhang, J.; Xu, Z.; Liu, Y.; Li, H.; You, C.; Tan, B. Forest Gap Size Alters the Functional Diversity of Soil Nematode Communities in Alpine Forest Ecosystems. *Forests* **2019**, *10*, 806. [CrossRef]
28. Mogilewsky, M.; Vasey, N.; Andriamahaihavana, M.A.; Rakotomalala, Z. Contribution of tree-fall canopy gaps to variation in nitrogen content of fruits and leaves from Varecia and Eulemur food trees in northeastern Madagascar. *Am. J. Phys. Anthropol.* **2019**, *168*, 167–168.
29. Dietz, L.; Collet, C.; Dupouey, J.; Lacombe, E.; Laurent, L.; Gégout, J. Windstorm-induced canopy openings accelerate temperate forest adaptation to global warming. *Glob. Ecol. Biogeogr.* **2020**, *29*, 2067–2077. [CrossRef]
30. Herault, B.; Ouallet, J.; Blanc, L.; Wagner, F.; Baraloto, C. Growth responses of neotropical trees to logging gaps. *J. Appl. Ecol.* **2010**, *47*, 821–831. [CrossRef]
31. Scharenbroch, B.C.; Bockheim, J.G. Impacts of forest gaps on soil properties and processes in old growth northern hardwood-hemlock forests. *Plant Soil* **2007**, *294*, 219–233. [CrossRef]
32. Kushwaha, P.; Neilson, J.W.; Barberán, A.; Chen, Y.; Fontana, C.G.; Butterfield, B.J.; Maier, R.M. Arid Ecosystem Vegetation Canopy-Gap Dichotomy: Influence on Soil Microbial Composition and Nutrient Cycling Functional Potential. *Appl. Environ. Microbiol.* **2021**, *87*, e02780-20. [CrossRef] [PubMed]
33. Yang, Y.; Geng, Y.; Zhou, H.; Zhao, G.; Wang, L. Effects of gaps in the forest canopy on soil microbial communities and enzyme activity in a Chinese pine forest. *Pedobiologia* **2017**, *61*, 51–60. [CrossRef]
34. Yang, B.; Pang, X.; Hu, B.; Bao, W.; Tian, G. Does thinning-induced gap size result in altered soil microbial community in pine plantation in eastern Tibetan Plateau? *Ecol. Evol.* **2017**, *7*, 2986–2993. [CrossRef]
35. Han, W.; Wang, G.; Liu, J.; Ni, J. Effects of vegetation type, season, and soil properties on soil microbial community in subtropical forests. *Appl. Soil Ecol.* **2021**, *158*, 103813. [CrossRef]
36. Li, D.; Li, X.; Su, Y.; Li, X.; Yin, H.; Li, X.; Guo, M.; He, Y. Forest gaps influence fungal community assembly in a weeping cypress forest. *Appl. Microbiol. Biotechnol.* **2019**, *103*, 3215–3224. [CrossRef] [PubMed]
37. Kisanuki, H.; Nakasu, M.; Nakai, A.; Yurugi, Y. Predicting the population dynamics of three understory broad-leaved evergreen species under the influence of Sika deer in primary and secondary forests of mid-western Japan. *J. For. Res.* **2008**, *13*, 52–58. [CrossRef]
38. Li, X.; Wilson, S.; Song, Y. Secondary succession in two subtropical forests. *Plant Ecol.* **1999**, *143*, 13–21. [CrossRef]
39. Pang, X.; Bao, W.; Zhu, B.; Cheng, W. Responses of soil respiration and its temperature sensitivity to thinning in a pine plantation. *Agric. For. Meteorol.* **2013**, *171–172*, 57–64. [CrossRef]
40. Bolat, I. The effect of thinning on microbial biomass C, N and basal respiration in black pine forest soils in Mudurnu, Turkey. *Eur. J. For. Res.* **2013**, *133*, 131–139. [CrossRef]
41. Xiao, F. Mensuration of respiration amount in the community of secondary evergreen broadleaved forests in Huitong County, Hu'nan Province. *J. Beijing For. Univ.* **2006**, *28*, 40–44.
42. Yan, S.; Wang, S.; Hu, Y.; Gao, H.; Zhang, X. A comparative study on soil fauna in native secondary evergreen broad-leaved forest and Chinese fir plantation forests in subtropics. *Chin. J. Appl. Ecol.* **2004**, *15*, 1792–1796.
43. Wang, J.; Zhang, Y.; Ma, Y.; Liu, Y.; Li, Y.; Duan, W. Linear character of the sunshine of the gap in the artifical forest in Xishuangbanna. *J. Plant Resour. Environ.* **2000**, *9*, 27–30.
44. IUSS Working Group WRB. World Reference Base for Soil Resources 2014, Update 2015. In *International Soil Classification System for Naming Soils and Creating Legends for Soil Maps, World Soil Resources Reports No. 106*; FAO: Rome, Italy, 2015.
45. Runkle, J.R. Patterns of Disturbance in Some Old-Growth Mesic Forests of Eastern North America. *Ecology* **1982**, *63*, 1533–1546. [CrossRef]
46. Amir, A.A.D.; Norman, C. Distinct characteristics of canopy gaps in the subtropical mangroves of Moreton Bay, Australia. *Estuar. Coast. Shelf Sci.* **2019**, *222*, 66–80. [CrossRef]
47. Jiménez, J.J.; Igual, J.M.; Villar, L.; Benito-Alonso, J.L.; Abadias-Ullod, J. Hierarchical drivers of soil microbial community structure variability in "Monte Perdido" Massif (Central Pyrenees). *Sci. Rep.* **2019**, *9*, 8768. [CrossRef]
48. Qiao, H.; Luan, Y.; Wang, B.; Dai, W.; Zhao, M. Analysis of spatiotemporal variations in the characteristics of soil microbial communities in Castanopsis fargesii forests. *J. For. Res.* **2019**, *31*, 1975–1984. [CrossRef]

49. Pollierer, M.M.; Ferlian, O.; Scheu, S. Temporal dynamics and variation with forest type of phospholipid fatty acids in litter and soil of temperate forests across regions. *Soil Biol. Biochem.* **2015**, *91*, 248–257. [CrossRef]
50. Ou, J.; Liu, Y.; Zhang, J.; Zhang, J.; Cui, N.; Deng, C.; Ji, T.; He, R. Early influence of forest gap harvesting on soil phosphorus in Pinus massoniana plantation in a hilly area of the reaches of Yangtze River. *Chin. J. Ecol.* **2014**, *33*, 592–601.
51. Wei, Q. Infuence of forest gaps on the available nitrogen of soil in a pinus massoniaan plantation. *J. For. Environ.* **2021**, *41*, 124–131.
52. Zhu, J.-J.; Tan, H.; Li, F.-Q.; Chen, M.; Zhang, J.-X. Microclimate regimes following gap formation in a montane secondary forest of eastern Liaoning Province, China. *J. For. Res.* **2007**, *18*, 167–173. [CrossRef]
53. Wang, Q.; Pieristè, M.; Liu, C.; Kenta, T.; Robson, T.M.; Kurokawa, H. The contribution of photodegradation to litter decomposition in a temperate forest gap and understorey. *New Phytol.* **2021**, *229*, 2625–2636. [CrossRef] [PubMed]
54. Zhou, Y.-G.; Hao, K.-J.; Li, X.-W.; Fan, C.; Chen, Y.-L.; Liu, Y.-K.; Wang, X. Effects of forest gap on seasonal dynamics of soil organic carbon and microbial biomass car- bon in Picea asperata forest in Miyaluo of Western Sichuan, Southwest China. *Chin. J. Appl. Ecol.* **2014**, *25*, 2469–2476.
55. Wang, X.; Liu, J.; He, Z.; Xing, C.; Zhu, D. Forest gaps mediate the structure and function of the soil microbial community in a *castanopsis kawakamii* forest. *Ecol. Indic.* **2021**, *122*, 107288. [CrossRef]
56. Chen, L.; Han, W.; Liu, D.; Liu, G. How forest gaps shaped plant diversity along an elevational gradient in Wolong National Nature Reserve? *J. Geogr. Sci.* **2019**, *29*, 1081–1097. [CrossRef]
57. Tang, C.Q.; Han, P.-B.; Li, S.; Shen, L.-Q.; Huang, D.-S.; Li, Y.-F.; Peng, M.-C.; Wang, C.-Y.; Li, X.-S.; Li, W.; et al. Species richness, forest types and regeneration of Schima in the subtropical forest ecosystem of Yunnan, southwestern China. *For. Ecosyst.* **2020**, *7*, 35. [CrossRef]
58. Modrow, T.; Kuehne, C.; Saha, S.; Bauhus, J.; Pyttel, P.L. Photosynthetic performance, height growth, and dominance of naturally regenerated sessile oak (Quercus petraea Mattuschka Liebl.) seedlings in small-scale canopy openings of varying sizes. *Eur. J. For. Res.* **2020**, *139*, 41–52. [CrossRef]

forests

MDPI

Article

AM Fungi Endow Greater Plant Biomass and Soil Nutrients under Interspecific Competition Rather Than Nutrient Releases for Litter

Bangli Wu, Yun Guo, Minhong He, Xu Han, Lipeng Zang, Qingfu Liu, Danmei Chen, Tingting Xia, Kaiping Shen, Liling Kang and Yuejun He *

Research Center of Forest Ecology, Forestry College, Guizhou University, Guiyang 550025, China; wubangli1116@163.com (B.W.); zihanyun2013@163.com (Y.G.); heminhong1994@163.com (M.H.); hanxukumo@163.com (X.H.); cafzanglp@163.com (L.Z.); qingfuliu@gzu.edu.cn (Q.L.); dmchen3@gzu.edu.cn (D.C.); xtt1268@163.com (T.X.); skp0825@163.com (K.S.); kangliling0727@163.com (L.K.)

* Correspondence: hyj1358@163.com

Abstract: Plant competition affects belowground ecological processes, such as litter decomposition and nutrient release. Arbuscular mycorrhizal (AM) fungi play an essential role in plant growth and litter decomposition potentially. However, how plant competition affects the nutrient release of litter through AM fungi remains unclear especially for juvenile plants. In this study, a competitive potting experiment was conducted using juvenile seedlings of *Broussonetia papyrifera* and *Carpinus pubescens* from a karst habitat, including the intraspecific and interspecific competition treatments. The seedlings were inoculated by AM fungus or not inoculated, and the litter mixtures of *B. papyrifera* and *C. pubescens* were added into the soil or not added. The results were as follows: Litter addition significantly increased the root mycorrhizal colonization of two species in intraspecific competition. AM fungus significantly increased the biomass of *B. papyrifera* seedings and nitrogen release and decreased nitrogen concentration and N/P ratio of litter and further improved the total nitrogen and N/P ratio of soil under litter. The interspecific competition interacting with AM fungus was beneficial to the biomass accumulation of *B. papyrifera* and improvement of soil nutrients under litter. However, intraspecific competition significantly promoted nutrient releases via AM fungus. In conclusion, we suggest that AM fungi endow greater plant biomass and soil nutrients through interspecific competition, while intraspecific competition prefers to release the nutrients of litter.

Keywords: arbuscular mycorrhizal fungi; intraspecific competition; interspecific competition; litter decomposition; nutrient release

Citation: Wu, B.; Guo, Y.; He, M.; Han, X.; Zang, L.; Liu, Q.; Chen, D.; Xia, T.; Shen, K.; Kang, L.; et al. AM Fungi Endow Greater Plant Biomass and Soil Nutrients under Interspecific Competition Rather Than Nutrient Releases for Litter. *Forests* **2021**, *12*, 1704. https://doi.org/10.3390/f12121704

Academic Editor: Douglas Godbold

Received: 11 November 2021
Accepted: 30 November 2021
Published: 5 December 2021

1. Introduction

Litter is an essential carrier for carbon storage and nutrient cycling in terrestrial ecosystems [1] and is the main source of nutrients that flow into the soil through decomposition and nutrient release [2], which is closely related to the aboveground and belowground ecological process [3]. The decomposition of litter in the soil contributes significantly to the productivity of the terrestrial ecosystem [4]. Moreover, the nutrients released from litter influences the nutrient turnover between plants and soil, and this is mainly controlled by litter chemistry [5].

The chemical properties of litter affect the litter mass-loss rate [6], thus, affecting its decomposition in the soil at different stages [7]. Generally, most plants return to the soil in the form of litter, which provides substantial nutrients and energy for the soil [3,8]. This process is affected by soil microorganisms, which are considered to be the major drivers of biogeochemical cycles [9–11].

Hence, the significance of soil microorganisms on ecosystem processes cannot be ignored. Arbuscular mycorrhizal (AM) fungi are essential to soil functional microorganisms and form a symbiotic relationship with the roots in over 80% of terrestrial plant species [12] and play a critical role in soil chemical properties and plant communities by affecting the plant nutrient uptake [13]. Research has shown that AM fungi can improve N and P acquisition for plants [14]. N and P are generally thought to be the main elements that limit plant growth and terrestrial ecosystem productivity [15], and the acquisition of these nutrients may depend on the effect of AM fungi on litter decomposition.

Several studies have suggested that AM fungi can influence litter breakdown [16,17], and consequently accelerate the release of litter nutrients [18,19]. For instance, AM fungi can promote the decomposition of complex organic materials in soil through the function of saprophytes [20]. AM fungal extraradical mycelium can penetrate into the litter to facilitate the decomposition [21].

Nutrients released from litter are mainly returned to the soil, and only small amounts are transferred to host plants via AM hyphae [22]. However, these transfers of nutrients can change the competitive relationship between plants [23]. Despite the importance of AM fungi in competitive plant relationships and litter nutrient release, how competition affects decomposition and nutrient release in the presence of AM fungi has not yet been addressed.

Belowground competition is an important force in structuring plant communities [24] and affects the distribution of plant species and the diversity of plant communities [25]. Additionally, belowground competition is more intense than aboveground competition because plant roots compete for soil nutrients [26]. AM fungi can regulate the plant competition associated with plant roots through interconnected mycorrhizal networks that exchange C, N and P among different species and change the nutrient status between plants and soil [27,28].

Aponte et al. [29] proposed that plant competition indirectly affects the decomposition rate by inducing changes in the microbial community. Specifically, interspecific differences drive litter mass changes that positively affect the decomposition process [30], but intraspecific differences are generally considered to influence plant residue breakdown more intensely due to almost complete niche overlap [31]. Meanwhile, AM fungi play an important role in mediating intra- and interspecific competitions [32].

Therefore, AM fungi and competitive plant interactions could affect plant nutrient acquisition strategies [28]. The variable nutrient acquisition strategies of plants reflect the difference in litter decomposition rates and mediate the population of decomposer organisms [33], which may have different impacts on the decomposition and release of litter nutrients, especially in nutrient-poor environments.

The southwest of China is the largest karst distribution area in the world. It is a typical fragile ecosystem characterized by nutrient deficiency that severely restricts the primary productivity of plants [34]. Furthermore, the small amount of litter reduces the nutrient turnover of the karst ecosystem [35]. AM fungi are widely distributed in karst areas and can coexist with woody plants, such as *Broussonetia papyrifera* [36]. AM fungi can also regulate plant nutrient competition in karst regions.

For instance, Shen et al. [37] reported that AM fungi confer in invasive plants a competitive advantage in nutrient acquisition compared to native species, and Xia et al. [38] showed that AM fungi change plant root phenotypic traits and resource acquisition strategies to increase host competitiveness. Additionally, AM fungi could promote litter decomposition and the transfer of N from litter to host plants by enhancing the interaction among plants in karst habitats [21].

Therefore, plant competition affects the decomposition and nutrient release of litter via AM fungi. However, the effects of AM fungi on the release of litter nutrients associated with intra- versus interspecific competitions in karst soils remains unclear. Thus, our aim is to clarify the role of AM fungi in the nutrient release of litter under intra- and interspecific competitions. We hypothesized that:

Hypothesis 1 (H1). *The interspecific competition increases more plant biomass and soil nutrients than the intraspecific competition when associated with AM fungi and litter in karst soil.*

Hypothesis 2 (H2). *AM fungi can promote the nutrient release of litter in karst soil.*

We verify the arbuscular mycorrhizal roles in soil nutrient maintenance and forest ecosystem management of karst areas.

2. Materials and Methods

2.1. Experimental Treatments

A potting experiment was conducted in a plastic pot (22 × 20 × 28 cm, caliber × bottom diameter × height) in which there is a 1 cm hole in the bottom of the pot for draining water. The experimental treatments included competition patterns, AM fungus and litter addition. The competition pattern contained the intraspecific competition treatment (Intra-) and the interspecific competition treatment (Inter-) using seedlings of *Broussonetia papyrifera* and *Carpinus pubescens*, which are common plant species from karst habitat. Two seedlings of *B. papyrifera* or *C. pubescens* were monoculture in a pot as the Intra-treatment, in contrast to mixed planting for two species as the Inter-treatment.

A 100 g of AM inoculum of the fungus *Claroideoglomus etunicatum* was taken into the growth substrate of soil in each pot as the inoculated M^+ treatment, or a sterilized inoculum of 100 g of *C. etunicatum* was applied into the substrate as the control of M^-. The litter addition wrapped with 10 g litter mixture (5 g *B. papyrifera* and 5 g *C. pubescens*) in 1 mm aperture nylon mesh bag (5.0 cm × 5.0 cm) was regarded as the L^+ treatment and was not added as the L^- treatment (Figure 1).

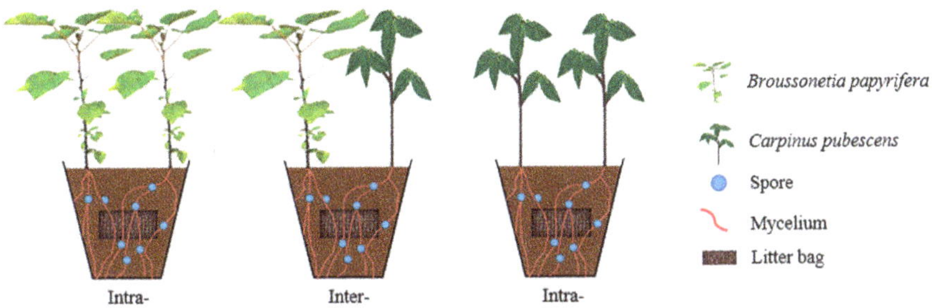

Figure 1. Diagram of the experimental design. The woody plant of *Broussonetia papyrifera* and *Carpinus pubescens* seedlings were inoculated by AM fungus of *Claroideoglomus etunicatum*. The litter bags = a 10 g of litter mixture of *B. papyrifera* and *C. pubescens*. Intra- = intraspecific competition of *B. papyrifera* or *C. pubescens*; Inter- = interspecific competition of *B. papyrifera* and *C. pubescens*.

The litters were soaked in 2% H_2O_2 for 3 min and oven-dried for 48 h at 65 °C before the experimental treatment. The initial chemistry of the litter mixtures had nitrogen 1.368 mg g^{-1} and phosphorus 0.268 mg g^{-1}. A 1300 g of growth substrate of limestone soil (Calcaric regosols, FAO) [36] was sterilized at 121 °C and 0.14 Mpa for one hour and was then placed into a pot before the start of the experiment. The growth substrate had a pH of 6.93, total nitrogen of 0.718 g kg^{-1}, total phosphorus of 0.491 g kg^{-1}, available nitrogen of 201.3 mg kg^{-1} and available phosphorus of 233.15 mg kg^{-1}.

Specifically, 550 g of growth substrate was added into the bottom of each pot. Then, the litter bag was placed on the growth substrate and covered with 500 g of sterilized substrate. Finally, five seeds of *B. papyrifera* and *C. pubescens* were sterilized with 1/1000 solution of $KMnO_4$ for 10 min and rinsed with sterile water three times, and were placed into the pot and covered with 250 g soil substrate. Two plants were retained after three weeks of seedling growth in each pot (Figure 1). There were 40 pots with five replicates for each treatment in this experiment.

In particular, the AM inoculum contained colonized root pieces, hyphae and approximately 100 spores per gram, and an extra 10 mL of filtrate was taken into M^- from 100 g of inoculum and dissolved into 1000 mL sterilized water in order to maintain a consistent microflora with the inoculated treatment except for the AM fungus. Additionally, *C. etunicatum* had been initially isolated from a local karst site in Guizhou of southwest China, purchased from the Institute of Nutrition Resources, Beijing Academy of Agricultural and Forestry Sciences (BGA0046).

The soil, plant seeds and litter were collected from a typical karst habitat in Huaxi District, Guiyang city, Guizhou Province, China. The plants were cultured in a greenhouse at Guizhou University, China (106°22′ E, 29°49′ N, 1120 m above sea level). All materials of plants and soil were harvested for determination after 24 weeks.

2.2. Measurements of the Root Mycorrhizal Colonization, Spore Density, Hyphal Length, Plant Biomass and the Concentrations of Nitrogen and Phosphorus in Litter and Soil

All materials containing the soil substrates, plants and residual litter were harvested for determination. The plant tissues, including roots, stems and leaves, were weighed after being dried at 65 °C for constant weight, and then added together for the total biomass of the plant individual. The spore density of *C. etunicatum* was determined in soil using the method described by Biermann and Linderman [39], and the hyphal length was determined using the gridline intersect method [40].

The root mycorrhizal colonization rate was used the magnified gridline intersection method from Mcgonigle et al. [41]. Nitrogen and phosphorus concentrations regarding the residual litter and soil materials were determined using the Kjeldahl method [42] and the molybdenum-antimony anti-colorimetric method [43]. The available nitrogen (AN) and the available phosphorus (AP) of soil were determined, respectively, by the alkali hydrolysis diffusion method [44] and the colorimetry method [45].

2.3. Calculations of the Release of Nitrogen and Phosphorus

Decomposing traits, including the mass-loss rate and nutrient release of litter, were calculated via the equation referenced from Bragazza et al. [46] and Wu [47] as follows:

$$\text{Litter mass-loss rate (\%)} = [(W_0 - W_1)/W_0] \times 100\%$$

$$\text{Litter nutrient release} = [(X_0W_0 - X_1W_1)/X_0W_0] \times X_0W_0$$

where the W_0 and W_1 are the initial and final litter weights, respectively; meanwhile, X_0 and X_1 are the initial and final litter nutrient concentrations of nitrogen and phosphorus, respectively. Positive or negative values mean the net mineralization or net immobilization, respectively.

2.4. Statistical Analysis

For comparing Intra- and Inter- competition when analyzing nutrients of litter and soil, these data were integrated through all monoculture treatments combined the respective *B. papyrifera* with *C. pubescens* seedlings as intraspecific competition data, except for the mycorrhizal colonization rate and plant biomass for each species. All data were tested for normality and homogeneity of variance before analysis. Two-way ANOVAs were applied for analyzing the effects of AM fungus (M^+ vs. M^-) and competition pattern (Intra- vs. Inter-) and their interactions on decomposing traits of litter and soil nutrients under L^+ or L^-. All statistical analyses were performed using the SPSS 25.0 software, and all the graphs were drawn through the Origin 2018 software.

3. Results

3.1. The Root Mycorrhizal Colonization of Two Plants and The Spore Density and Hyphal Length in Different Competition and Litter Treatments when Inoculated with AM Fungus

The significant Inter- > Intra- of mycorrhizal colonization was presented in *C. pubescens* under L^- (Table 1). However, there was no significant difference in mycorrhizal colonization of the two species in Intra- and Inter- competitions under L^+. In addition, a significant $L^+ > L^-$ was observed for mycorrhizal colonization of two species in Intra-treatment. The soil spore density of the Inter-treatment was significantly greater than the Intra-treatment under L^+ but not for under L^-.

Table 1. The root mycorrhizal colonization of *Broussonetia papyrifera* and *Carpinus pubescens*, the spore density and the hyphal length through the treatments of competition pattern and with or without litter addition under AM fungus inoculation.

Treatments		Mycorrhizal Colonization Rate (%)		Spore Density (g soil^{-1})	Hyphae Length (m soil^{-1})
		Broussonetia papyrifera	*Carpinus pubescens*		
Intra-	L^+	63.3 ± 4.6 ax	39.0 ± 7.1 ax	8.6 ± 1.3 by	15.2 ± 4.1 ax
	L^-	44.7 ± 2.1 ay	14.6 ± 0.9 by	12.5 ± 0.6 ax	9.9 ± 1.7 ax
Inter-	L^+	59.3 ± 5.3 ax	17.4 ± 4.3 ax	17.9 ± 0.6 ax	20.6 ± 8.9 ax
	L^-	46.0 ± 3.3 ax	23.0 ± 1.6 ax	4.4 ± 0.2 by	21.9 ± 7.6 ax

Abbreviations: M^+ = with AM fungus; M^- = without AM fungus; Intra- = intraspecific competition; Inter- = interspecific competition; L^+ = with litter addition; L^- = without litter addition; the different lowercase letters (a,b) means a significant difference between Intra- and Inter-treatments under L^+ or L^- ($p < 0.05$); the different lowercase letters (x,y) means a significant difference between L^+ and L^- treatments under Intra- or Inter- ($p < 0.05$). The values are the mean ± SE.

A significant $L^+ > L^-$ of spore density was presented in Inter-treatment; there was no significant difference between L^+ and L^- under Intra- and Inter-treatments for the hyphal length (Table 1). This indicates that litter addition significantly increased the root mycorrhizal colonization of two species except for interspecific competition conditions; the intraspecific competition decreased the spore density, while the interspecific competition increased the spore density in this experiment.

3.2. The Biomass of *B. papyrifera* and *C. pubescens* Seedlings in Different Competition and Litter Treatments

A significant $M^+ > M^-$ of biomass was presented in *B. papyrifera* seedlings regardless of Intra- and Inter- under litter and no litter addition treatments (Figure 2a,c). However, the biomass of *C. pubescens* seeding was not significantly different when comparing M^+ to M^- under Intra- or Inter- competition (Figure 2b), except for the biomass of M^+ was significantly less than the biomass of M^- in Intra-treatment (Figure 2d). The significant Inter- > Intra- of biomass were presented in *B. papyrifera* seedlings under M^+ of litter or no litter addition and also M^- of litter addition (Figure 2a,c) but not for *C. pubescens* seedlings even presenting a not significant Inter- > Intra- under no litter (Figure 2d).

3.3. The Concentration and Release of Residual Litter Nutrients on Nitrogen and Phosphorus in Competition Interacting with AM Fungus

The mass-loss rate was not significantly different among AM fungus (M), competition (C) and their interactions M × C (Table 2), and there was no significant difference between M^+ and M^- in the litter mass-loss rate under Intra- and Inter- competitions (Figure 3a). AM fungus (M) significantly affected the N concentration, N/P ratio and the N and P releases of residual litter but not the P concentration (Table 2). The significant $M^+ < M^-$ were presented with the N concentration, N/P ratio and P release under Intra-treatment (Figure 3b,d,f).

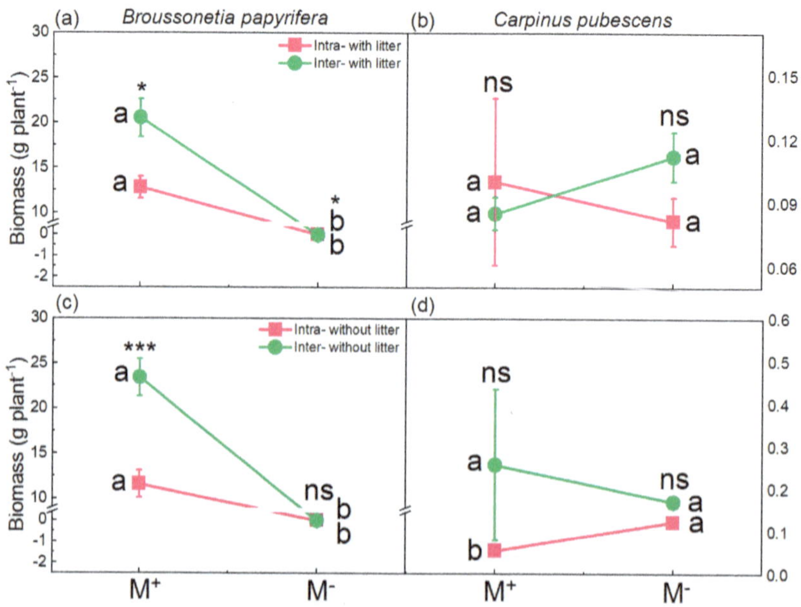

Figure 2. The biomass of *B. papyrifera* and *C. pubescens* seedlings through the treatments of competition pattern, AM fungus and with (**a**,**b**) or without (**c**,**d**) litter addition. Abbreviations: M^+ = with AM fungus; M^- = without AM fungus; Intra- = intraspecific competition; Inter- = interspecific competition; the *, ** and *** mean different difference between Intra- and Inter- treatments under M^+ or M^- at $p < 0.05$, $p < 0.01$ and $p < 0.001$, respectively; while the 'ns' means a not significant difference; the different lowercase letters (a,b) means a significant difference between M^+ and M^- treatments under Intra- or Inter- ($p < 0.05$).

Table 2. Two-Way ANOVAs for the effects of AM fungus (M^+ vs. M^-) and competition pattern (Intra- vs. Inter-) on the concentrations and releases of nitrogen and phosphorus and N/P ratio in residual litter.

Factors	*df*	Litter Mass-Loss Rate (%)		N Concentration (mg g^{-1})		P Concentration (mg g^{-1})		N/P Ratio		N Release (mg g^{-1})		P Release (mg g^{-1})	
		F	*p*	*F*	*p*	*F*	*p*	*F*	*p*	*F*	*p*	*F*	*p*
M	1	1.936	0.176	24.302	0.000	3.907	0.060	20.675	0.000	10.746	0.003	5.113	0.033
C	1	0.115	0.737	46.680	0.000	15.909	0.001	4.270	0.050	28.123	0.000	9.259	0.006
M × C	1	2.537	0.123	4.495	0.045	5.170	0.032	0.000	0.994	0.792	0.382	0.835	0.370

Abbreviations: M = AM fungus treatments; C = Competition pattern treatments. $p < 0.05$ and $p < 0.01$ indicate significant differences, and $p < 0.001$ indicates highly significant differences.

However, the significant $M^+ > M^-$ were presented for N release and P concentration under Intra-treatment (Figure 3c,e). The competition pattern (C) significantly affected the N and P concentrations, N/P ratio and the N and P releases of residual litter. The significant Inter- > Intra- of N concentration (Figure 3b) and the significant Inter- < Intra- of N release (Figure 3e) were presented under M^+ or M^- treatment.

The P concentration and P release under M^- showed Inter- > Intra- or Inter < Intra-, respectively, while there were no significant differences between Inter- and Intra-treatments under M^+ (Figure 3c,f). The N/P ratio of Inter-treatment was greater than Intra-treatment but was not significant under M^+ or M^- in the residual litter (Figure 3d). The interaction of M × C significantly affected the concentrations of N and P but not for the N/P ratio and the releases of N and P (Table 2).

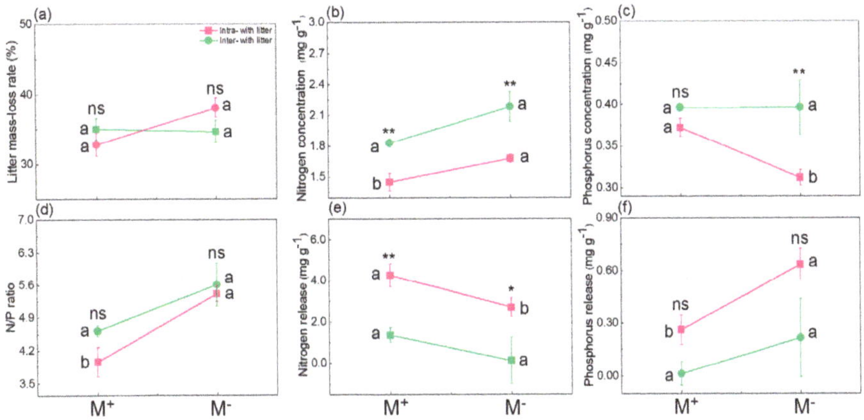

Figure 3. The residual litter on mass-loss rate (**a**), on the concentrations of nitrogen and phosphorus (**b**,**c**) and N/P ratio (**d**) and nutrient release of nitrogen and phosphorus (**e**,**f**) in two competition soils through AM fungus treatment. Abbreviations: see Figure 2 for explanations of M^+, M^-, Intra-, Inter-, * and **, the lowercase letters (a,b) and the ns.

3.4. The Nutrients of Nitrogen and Phosphorus in Two Competition Soil with AM Fungus and Litter Treatments

Under the litter addition, the AM fungus (M) significantly affected the total nitrogen (TN) and N/P ratio but not for the total phosphorus (TP), the available nitrogen (AN) and the available phosphorus (AP) (Table 3). The significant $M^+ > M^-$ of the TN and N/P ratio was found under Intra- and Inter-treatments (Figure 4a,c); however, a significant $M^+ < M^-$ of TP was found under Intra-treatment (Figure 4b). The competition pattern (C) significantly affected the TN but did not significantly affect the TP and N/P ratio; furthermore, the interaction of M × C significantly affected TN and TP via the two-way ANOVAs analysis (Table 3).

Table 3. Two-Way ANOVAs for the effects of AM fungus (M^+ vs. M^-) and competition pattern (Intra- vs. Inter-) on total nitrogen and phosphorus and the N/P ratio and the available nitrogen and phosphorus of soil under L^+ and L^- treatments.

	Factors	*df*	Total Nitrogen (mg g^{-1})		Total Phosphorus (mg g^{-1})		N/P Ratio		Available Nitrogen (mg kg^{-1})		Available Phosphorus (mg kg^{-1})	
			F	*p*	*F*	*p*	*F*	*p*	*F*	*p*	*F*	*p*
L^+	M	1	64.409	0.000	1.784	0.193	9.522	0.005	0.006	0.940	0.591	0.449
	C	1	41.446	0.000	2.578	0.120	0.401	0.532	0.254	0.619	0.101	0.753
	M × C	1	34.943	0.000	4.814	0.037	0.001	0.977	0.790	0.382	0.947	0.340
L^-	M	1	1.244	0.275	15.942	0.000	11.148	0.003	1.141	0.295	4.009	0.056
	C	1	9.046	0.006	0.002	0.969	0.013	0.909	3.871	0.060	0.129	0.723
	M × C	1	3.019	0.094	0.949	0.339	0.001	0.981	0.181	0.674	0.627	0.436

Abbreviations: L^+ = with litter addition treatment; L^- = without litter addition treatment; M = AM fungus treatments; C = Competition pattern treatments. $p < 0.05$ and $p < 0.01$ indicate significant differences, and $p < 0.001$ indicates highly significant differences.

Under M^+, the TN and TP of Inter-treatment were significantly greater than Intra-treatment (Figure 4a,b). In addition, the AM fungus (M), competition pattern (C) and their interaction did not significantly affect AN and AP, and there were no differences of AN and AP among Intra- and Inter- compared with other soil nutrients under the M^+ and M^- treatments (Figure 4d,e).

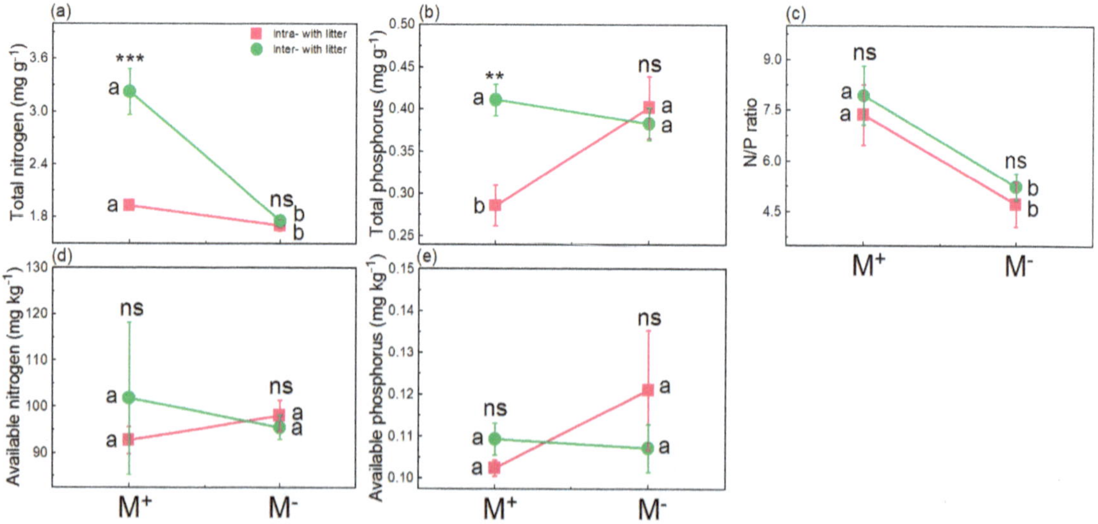

Figure 4. The soil nutrient properties on the total nitrogen and phosphorus (**a**,**b**) and the N/P ratio (**c**) and the available nitrogen and phosphorus (**d**,**e**) in two competition soils through AM fungus under litter addition treatment. Abbreviations: see Figure 2 for explanations of M^+, M^-, Intra-, Inter-, ** and ***, the lowercase letters (a,b) and the ns.

Under the no litter addition, the AM fungus (M) significantly affected TP and N/P ratio but not for TN, AN and AP (Table 3); the TN and N/P ratio were showed as $M^+ > M^-$ under Intra-, but the TP was showed as $M^+ < M^-$ under Intra- and Inter- (Figure 5a–c). The competition pattern (C) significantly affected the TN but not for the TP, the N/P ratio and the AN and AP (Table 3), by significantly increasing the TN of Intra-treatment (Figure 5a). The interaction of M × C did not significantly affect the nutrients of TP, AN, AP and N/P ratio via this experiment (Table 3).

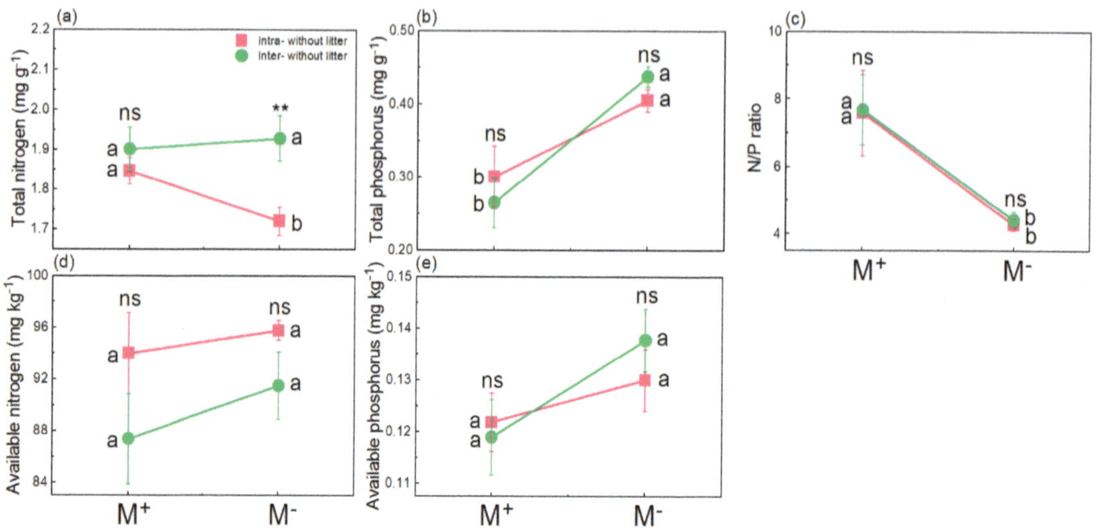

Figure 5. The soil nutrient properties on the total nitrogen and phosphorus (**a**,**b**) and the N/P ratio (**c**) and the available nitrogen and phosphorus (**d**,**e**) in two competition soils through AM fungus under none-litter addition treatment. Abbreviations: see Figure 2 for explanations of M^+, M^-, Intra-, Inter- and **, the lowercase letters (a,b) and the ns.

4. Discussion

4.1. Intraspecific and Interspecific Competition Mediated The Accumulation of Plant Biomass and Soil Nutrients and The Releases of Litter Nutrient via AM Fungi

In this study, the interspecific competition significantly increased the biomass of *B. papyrifera* than the intraspecific competition via AM fungus regardless of litter addition (Figure 2a,c), which indicated that interspecific competition endowed plant biomass accumulation an advantage over the intraspecific competition. Matos et al. [48] found that interspecific competition enhanced the biomass accumulation of *Bidens pilosa* and *Ipomoea grandifolia* than the intraspecific competition; Heuermann et al. [49] also observed that interspecific competition greatly increased the biomass of catch plants.

In addition, the interspecific competition significantly increased the soil N and P compared with intraspecific competition under AM fungus inoculation and litter addition (Figure 4a,b), indicating that interspecific competition was more beneficial to increase soil nutrients when compared with the intraspecific competition, which was similar to the previous studies. For instance, the interaction of interspecific plants can positively influence soil properties and, thus, reducing soil nutrients loss, while this was not obvious in competitive intraspecific plants [50].

Additionally, when compared to intraspecific competition, interspecific competition could enhance soil nutrient availability through altering the densities of plants intercropping [51]. Therefore, the interspecific competition confers greater soil nutrients than intraspecific competition; this exactly verified the hypothesis of H1. Overall, the interspecific competition plays an important role in plant biomass accumulation and soil nutrient improvement in nutrient-limited karst soil.

The competition significantly affected the N and P releases of litter (Table 2), and the N and P releases in the intraspecific competition were all greater than in interspecific competition (Figure 3e,f), indicating that intraspecific competition was an important factor affecting the nutrient release of litter in nutrient strategies of competitive plants.

Yang et al. [52] proposed that the intraspecific competition was the main competition mode, which greatly intensified the plant nutrient acquisitions due to niche overlap, thus, driving the decomposition of organic matter and nutrient release [53]. Meanwhile, Tedersoo et al. [54] discovered that the interspecific competition reduced the nutrient demands of plants due to niche differentiation, thus, slowing down litter decomposition and nutrient release [55]. These results supported that intraspecific competition promoted the greater nutrient releases for litter when compared with the interspecific competition in this experiment.

4.2. Arbuscular Mycorrhizal Fungi Regulate the Release of Nitrogen and Phosphorus in Decomposing Litter

AM fungus significantly increased the N release of litter under intra- and interspecific competitions (Figure 3e), indicating that N release positively responded to AM fungus, which was consistent with the study of Hodge et al. [56] and Tan et al. [19], who found that AM fungi stimulated N release during the process of litter decomposition. The N release of litter via AM fungi was mainly achieved through two pathways. One is that AM fungi accelerated N release through decomposing nitrogen compounds in the litter [57], and the other is that AM fungi could facilitate the transformation of organic N into inorganic N; thereby, the N was released from decomposed litter [58].

Previous studies have shown that AM fungi can directly or indirectly affect the decomposition of litter and then promote nutrient release. Directly, the length of AM fungal extraradical mycelium can generally exceed 13-fold of the root length [59], which penetrates into the litter to facilitate the decomposition [21]. Indirectly, AM fungi accelerated decomposition by altering the fungal community composition associated with litter breakdown [60]. In addition, AM fungi hyphal exudates could stimulate litter decomposition by increasing soil enzyme activity [61].

Therefore, AM fungi contribute to the nutrient release of litter directly or indirectly; this exactly verifies the hypothesis of H2. Nevertheless, whether most of the released nutrients are transferred to host plants or soil through AM fungi remains unclear. Therefore, isotope tracing experiments are necessary to explore the nutrient transport among competitive plants via AM mycelium in the future.

4.3. Arbuscular Mycorrhizal Fungi Differentially Affected Plant Biomass and Soil Nutrients

AM fungus significantly increased the biomass of *B. papyrifera* seedlings under intra- and interspecific competitions in this study (Figure 2a,c), which was accordant with AM fungi enhancing the biomass of *Cinnamomum camphora* [62]. Wu et al. [63] also proposed that AM fungi facilitated the biomass accumulation of *Populus cathayana* seedlings.

In this experiment, the AM fungus had no significant effect on the biomass of *C. pubescens* compared with *B. papyrifera* (Figure 2a–d), indicating that the two plants differently responded to AM fungus in biomass production, which proved that the AM fungi had the selectivity in regulating plant growth. For example, AM fungi enabled the preferential allocation of nutrient resources to high-quality host plants and aggravated the growth differences between plants [64].

Meanwhile, when two plants were mixed, one of the plants obtained higher N and P and subsequently promoted the biomass accumulation than the neighboring plant through AM fungi [65]. Smith and Read [12] and Helgason and Fitter [66] also agreed that there was a selectivity between specific AM fungi and host plants. Thereby, AM fungi are essential for plant growth, but the mutual selection is not excluded between AM fungi and host plant, precisely as the *B. papyrifera* and *C. pubescens* seedlings presented further growth under interspecific or intraspecific competition.

In this experiment, AM fungus increased soil N and P under interspecific competition with litter addition (Figure 4a,b) compared with no litter addition (Figure 5a,b), which indicated that AM fungi improved soil nutrients in the presence of litter. Several studies reported that AM fungi could increase soil N content by stimulating the secretion of soil urease [67] and P content through improving the phosphatase activity [68]. In addition, the nutrients released from the decomposing litter were transferred to the surrounding soil through AM fungi, thus, affecting the soil nutrient turnover [69].

Litter is the main source of soil nutrients, and AM fungi could significantly increase the soil N and P during decomposition [19]. Moreover, AM fungus significantly increased the soil N/P ratio under litter addition, indicating that AM fungi facilitated the soil nitrogen accumulation more than phosphorus (Figure 4c). Verbruggen et al. [70] showed that AM fungi could regulate the soil N/P ratio, and a higher N/P ratio means that the contribution of AM fungi to nitrogen was greater than phosphorus.

It is possible that AM fungi prevented soil N loss through expanding nutrient interception zone [71] or acquired mobile N by absorbing NH_4^+ and NO_3^- ions in the soil [72]. Overall, these studies showed that AM fungi could improve soil nutrients associated with litter, and the improvement in nitrogen is greater than for phosphorus. However, the specific mechanism of how AM fungi regulating soil nutrients through litter needs to be further explored in the future.

5. Conclusions

In this experiment, AM fungus affected the plant growth, litter nutrient release and soil nutrients differently under intra- and interspecific competitions via litter. The litter addition had a significant improvement on the root mycorrhizal colonization of *B. papyrifera* and *C. pubescens* in intraspecific competition. The AM fungus exerted a positive influence on the biomass of *B. papyrifera*, the litter N release and soil TN and N/P ratio while showing the opposite effects for the N concentration, P release and N/P ratio of litter in two competitive patterns under litter addition.

The interspecific competition interacting with AM fungus enhanced the biomass accumulation of *B. papyrifera* and the N concentration and N/P ratio of litter, as well as the

TN and TP of soil under litter addition; however, the intraspecific competition regulated by AM fungus significantly improved the N and P releases of litter. In conclusion, the interspecific competition conferred more significant benefits over intraspecific competition in enhancing plant biomass and soil nutrients, while the intraspecific competition instead increased litter nutrient releases when associated with AM fungi in karst soil.

Author Contributions: Conceptualization, Y.H.; Data analysis, Y.G., M.H., D.C., T.X., K.S., Q.L. and L.K.; methodology, X.H., M.H. and L.Z.; writing—original draft preparation, B.W.; writing—review and editing, B.W. and Y.H. All authors have read and agreed to the published version of the manuscript.

Funding: This research was funded by National Natural Science Foundation of China (31660156, 31360106), the Science and Technology Project of Guizhou Province ([2021] General-455, [2016] Supporting-2805), the Guizhou Hundred-level Innovative Talents Project (Qian-ke-he platform talents [2020]6004), the First-class Disciplines Program on Ecology of Guizhou Province (GNYL [2017]007) and the Talent-platform Program of Guizhou Province ([2017]5788, [2018]5781).

Data Availability Statement: Not applicable.

Conflicts of Interest: The authors declare no conflict of interest.

References

1. Camenzind, T.; Httenschwiler, S.; Treseder, K.K.; Lehmann, A.; Rillig, M.C. Nutrient limitation of soil microbial processes in tropical forests. *Ecol. Monogr.* **2018**, *88*, 4–21. [CrossRef]
2. Yue, K.; Yang, W.; Peng, C.; Peng, Y.; Zhang, C.; Huang, C.; Tan, Y.; Wu, F. Foliar litter decomposition in an alpine forest meta-ecosystem on the eastern Tibetan Plateau. *Sci. Total Environ.* **2016**, *566*, 279–287. [CrossRef] [PubMed]
3. Berg, B.; McClaugherty, C. Models that describe litter decomposition. In *Plant Litter: Decomposition, Humus Formation, Carbon Sequestration*; Springer: Berlin/Heidelberg, Germany, 2014; pp. 189–199.
4. Adair, E.C.; Parton, W.J.; Del Grosso, S.J.; Silver, W.L.; Harmon, M.E.; Hall, S.A.; Burke, I.C.; Hart, S.C. Simple three-pool model accurately describes patterns of long-term litter decomposition in diverse climates. *Glob. Chang. Biol.* **2008**, *14*, 2636–2660. [CrossRef]
5. Berger, T.W.; Duboc, O.; Djukic, I.; Tatzber, M.; Gerzabek, M.H.; Zehetner, F. Decomposition of beech (*Fagus sylvatica*) and pine (*Pinus nigra*) litter along an alpine elevation gradient: Decay and nutrient release. *Geoderma* **2015**, *251*, 92–104. [CrossRef] [PubMed]
6. Scott, N.A.; Binkley, D. Foliage litter quality and annual net N mineralization: Comparison across North American forest sites. *Oecologia* **1997**, *111*, 151–159. [CrossRef]
7. Del Giudice, R.; Lindo, Z. Short-term leaching dynamics of three peatland plant species reveals how shifts in plant communities may affect decomposition processes. *Geoderma* **2017**, *285*, 110–116. [CrossRef]
8. Wardle, D.A.; Bardgett, R.D.; Klironomos, J.N.; Setala, H.; van der Putten, W.H.; Wall, D.H. Ecological linkages between aboveground and belowground biota. *Science* **2004**, *304*, 1629–1633. [CrossRef]
9. Trogisch, S.; He, J.-S.; Hector, A.; Scherer-Lorenzen, M. Impact of species diversity, stand age and environmental factors on leaf litter decomposition in subtropical forests in China. *Plant Soil* **2016**, *400*, 337–350. [CrossRef]
10. Bradford, M.A.; Veen, G.F.; Bonis, A.; Bradford, E.M.; Classen, A.T.; Cornelissen, J.H.C.; Crowther, T.W.; De Long, J.R.; Freschet, G.T.; Kardol, P.; et al. A test of the hierarchical model of litter decomposition. *Nat. Ecol. Evol.* **2017**, *1*, 1836–1845. [CrossRef] [PubMed]
11. Zhao, C.; Long, J.; Liao, H.; Zheng, C.; Li, J.; Liu, L.; Zhang, M. Dynamics of soil microbial communities following vegetation succession in a karst mountain ecosystem, Southwest China. *Sci. Rep.* **2019**, *9*, 2160. [CrossRef]
12. Smith, S.; Read, D. Mycorrhizal symbiosis. *Q. Rev. Biol.* **2008**, 3, 273–281.
13. Korb, J.; Johnson, N.; Covington, W. Arbuscular mycorrhizal propagule densities respond rapidly to ponderosa pine restoration treatments. *J. Appl. Ecol.* **2003**, *40*, 101–110. [CrossRef]
14. Kiers, E.T.; Duhamel, M.; Beesetty, Y.; Mensah, J.A.; Franken, O.; Verbruggen, E.; Fellbaum, C.R.; Kowalchuk, G.A.; Hart, M.M.; Bago, A.; et al. Reciprocal rewards stabilize cooperation in the mycorrhizal symbiosis. *Science* **2011**, 333, 880–882. [CrossRef]
15. Güsewell, S. N: P ratios in terrestrial plants: Variation and functional significance. *New Phytol.* **2004**, *164*, 243–266. [CrossRef]
16. Cheng, L.; Booker, F.L.; Tu, C.; Burkey, K.O.; Zhou, L.; Shew, H.D.; Rufty, T.W.; Hu, S. Arbuscular mycorrhizal fungi increase organic carbon decomposition under elevated CO_2. *Science* **2012**, *337*, 1084. [CrossRef] [PubMed]
17. Xu, J.; Liu, S.; Song, S.; Guo, H.; Tang, J.; Yong, J.W.H.; Ma, Y.; Chen, X. Arbuscular mycorrhizal fungi influence decomposition and the associated soil microbial community under different soil phosphorus availability. *Soil Biol. Biochem.* **2018**, *120*, 181–190. [CrossRef]
18. Kalbitz, K.; Solinger, S.; Park, J.H.; Michalzik, B.; Matzner, E. Controls on the dynamics of dissolved organic matter in soils: A review. *Soil Sci.* **2000**, *165*, 277–304. [CrossRef]

19. Tan, Q.; Si, J.; He, Y.; Yang, Y.; Shen, K.; Xia, T.; Kang, L.; Fang, Z.; Wu, B.; Guo, Y.; et al. Improvement of karst soil nutrients by arbuscular mycorrhizal fungi through promoting nutrient release from the litter. *Int. J. Phytoremediat.* **2021**, *23*, 1244–1254. [CrossRef] [PubMed]
20. Sheldrake, M.; Rosenstock, N.P.; Revillini, D.; Olsson, P.A.; Mangan, S.A.; Sayer, E.J.; Wallander, H.; Turner, B.L.; Tanner, E.V.J. Arbuscular mycorrhizal fungal community composition is altered by long-term litter removal but not litter addition in a lowland tropical forest. *New Phytol.* **2017**, *214*, 455–467. [CrossRef]
21. He, Y.J.; Cornelissen, J.H.C.; Zhong, Z.C.; Dong, M.; Jiang, C.H. How interacting fungal species and mineral nitrogen inputs affect transfer of nitrogen from litter via arbuscular mycorrhizal mycelium. *Environ. Sci. Pollut. Res.* **2017**, *24*, 9791–9801. [CrossRef]
22. Herman, D.J.; Firestone, M.K.; Nuccio, E.; Hodge, A. Interactions between an arbuscular mycorrhizal fungus and a soil microbial community mediating litter decomposition. *FEMS Microbiol. Ecol.* **2012**, *80*, 236–247. [CrossRef]
23. Daisog, H.; Sbrana, C.; Cristani, C.; Moonen, A.-C.; Giovannetti, M.; Barberi, P. Arbuscular mycorrhizal fungi shift competitive relationships among crop and weed species. *Plant Soil* **2012**, *353*, 395–408. [CrossRef]
24. Weiss, L.; Schalow, L.; Jeltsch, F.; Geissler, K. Experimental evidence for root competition effects on community evenness in one of two phytometer species. *J. Plant Ecol.* **2019**, *12*, 281–291. [CrossRef]
25. Stanescu, S.; Maherali, H. Arbuscular mycorrhizal fungi alter the competitive hierarchy among old-field plant species. *Oecologia* **2017**, *183*, 479–491. [CrossRef] [PubMed]
26. Lamb, E.G.; Cahill, J.F., Jr. When competition does not matter: Grassland diversity and community composition. *Am. Nat.* **2008**, *171*, 777–787. [CrossRef] [PubMed]
27. Bever, J.D.; Dickie, I.A.; Facelli, E.; Facelli, J.M.; Klironomos, J.N.; Moora, M.; Rillig, M.C.; Stock, W.D.; Tibbett, M.; Zobel, M. Rooting theories of plant community ecology in microbial interactions. *Trends Ecol. Evol.* **2010**, *25*, 468–478. [CrossRef]
28. Scheublin, T.R.; Logtestijn, R.S.P.V.; Heijden, M.G.A.V.D. Presence and identity of arbuscular mycorrhizal fungi influence competitive interactions between plant species. *J. Ecol.* **2010**, *95*, 631–638. [CrossRef]
29. Aponte, C.; Maranon, T.; Garcia, L.V. Microbial C, N and P in soils of Mediterranean oak forests: Influence of season, canopy cover and soil depth. *Biogeochemistry* **2010**, *101*, 77–92. [CrossRef]
30. Hattenschwiler, S.; Tiunov, A.V.; Scheu, S. Biodiversity and litter decomposition in terrestrial ecosystems. *Annu. Rev. Ecol. Evol. Syst.* **2005**, *36*, 191–218. [CrossRef]
31. Chesson, P. Mechanisms of maintenance of species diversity. *Annu. Rev. Ecol. Syst.* **2000**, *31*, 343–366. [CrossRef]
32. Moora, M.; Zobel, M. Effect of arbuscular mycorrhiza on inter- and intraspecific competition of two grassland species. *Oecologia* **1996**, *108*, 79–84. [CrossRef]
33. Phillips, R.P.; Brzostek, E.; Midgley, M.G. The mycorrhizal-associated nutrient economy: A new framework for predicting carbon-nutrient couplings in temperate forests. *New Phytol.* **2013**, *199*, 41–51. [CrossRef] [PubMed]
34. Liu, C.; Lang, Y.; Li, S.; Hechun, P.; Tu, C.; Liu, T.; Zhang, W. Researches on biogeochemical processes and nutrient cycling in karstic ecological systems, southwest China: A review. *Front. Earth Sci.* **2009**, *16*, 1–12.
35. Yu, G.; Wang, S.; Rong, L.; Ran, J. Litter dynamics of major successional communities in Maolan karst forest of China. *Chin. J. Plant. Ecol.* **2011**, *35*, 1019–1028.
36. He, Y.; Cornelissen, J.H.C.; Wang, P.; Dong, M.; Ou, J. Nitrogen transfer from one plant to another depends on plant biomass production between conspecific and heterospecific species via a common arbuscular mycorrhizal network. *Environ. Sci. Pollut. Res.* **2019**, *26*, 8828–8837. [CrossRef]
37. Shen, K.; Cornelissen, J.H.C.; Wang, Y.; Wu, C.; He, Y.; Ou, J.; Tan, Q.; Xia, T.; Kang, L.; Guo, Y.; et al. AM fungi alleviate phosphorus limitation and enhance nutrient competitiveness of invasive plants via mycorrhizal networks in karst areas. *Front. Ecol. Evol.* **2020**, *8*, 125. [CrossRef]
38. Xia, T.; Wang, Y.J.; He, Y.; Wu, C.; Han, X. An invasive plant experiences greater benefits of root morphology from enhancing nutrient competition associated with arbuscular mycorrhizae in karst soil than a native plant. *PLoS ONE.* **2020**, *15*, e0234410. [CrossRef]
39. Biermann, B.; Linderman, R.G. Quantifying vercular-arbuscular mycorrhizas: Aproposed method towards standardization. *New Phytol.* **1981**, *87*, 63–67. [CrossRef]
40. Rillig, M.C.; Wright, S.F.; Shaw, M.R.; Field, C.B. Artificial climate warming positively affects arbuscular mycorrhizae but decreases soil aggregate water stability in an annual grassland. *Oikos* **2002**, *97*, 52–58. [CrossRef]
41. Mcgonigle, T.P.; Miller, M.H.; Evans, D.G.; Fairchild, G.L.; Swan, J.A. A new method which gives an objective measure of colonization of roots by vesicular—arbuscular mycorrhizal fungi. *New Phytol.* **1990**, *115*, 495–501. [CrossRef]
42. Hopkins, D.W.; Alef, K.; Nannipieri, P. Methods in applied soil microbiology and biochemistry. *J. Appl. Ecol.* **1996**, *33*, 178–188. [CrossRef]
43. Myers, S.W.; Claudio, G.; Wolkowski, R.P.; Hogg, D.B.; Wedberg, J.L. Effect of soil potassium availability on soybean aphid (Hemiptera: Aphididae) population dynamics and soybean yield. *J. Econ. Entomol.* **2005**, *98*, 113–120. [CrossRef]
44. Holliday, V.T.; Gartner, W.G. Methods of soil P analysis in archaeology. *J. Archaeol. Sci.* **2007**, *34*, 301–333. [CrossRef]
45. Stubbs, M.M.; Pyke, D.A. Available nitrogen: A time-based study of manipulated resource islands. *Plant Soil* **2005**, *270*, 123–133. [CrossRef]
46. Bragazza, L.; Siffi, C.; Iacumin, P.; Gerdol, R. Mass loss and nutrient release during litter decay in peatland: The role of microbial adaptability to litter chemistry. *Soil Biol. Biochem.* **2007**, *39*, 257–267. [CrossRef]

47. Wu, Q. Effects of snow depth manipulation on the releases of carbon, nitrogen and phosphorus from the foliar litter of two temperate tree species. *Sci. Total Environ.* **2018**, *643*, 1357–1365. [CrossRef]
48. Da Conceição de Matos, C.; da Silva Teixeira, R.; da Silva, I.R.; Costa, M.D.; da Silva, A.A. Interspecific competition changes nutrient: Nutrient ratios of weeds and maize. *J. Plant Nutr. Soil Sci.* **2019**, *182*, 286–295. [CrossRef]
49. Heuermann, D.; Gentsch, N.; Boy, J.; Schweneker, D.; Feuerstein, U.; Groß, J.; Bauer, B.; Guggenberger, G.; von Wirén, N. Interspecific competition among catch crops modifies vertical root biomass distribution and nitrate scavenging in soils. *Sci. Rep.* **2019**, *9*, 11531. [CrossRef]
50. Irvine, W.; Hollingsworth, A.D.; Grier, D.G.; Chaikin, P.M. Below-ground interactions in tropical agroecosystems: Concepts and models with multiple plant components. *Cabi Publ.* **2004**, *12*, 881–891.
51. Corre-Hellou, G.; Fustec, J.; Crozat, Y. Interspecific competition for soil N and its interaction with N 2 fixation, leaf expansion and crop growth in pea–barley intercrops. *Plant Soil* **2006**, *282*, 195–208. [CrossRef]
52. Yang, X.; Zhang, W.; He, Q. Effects of intraspecific competition on growth, architecture and biomass allocation of *Quercus liaotungensis*. *J. Plant Interact.* **2019**, *14*, 284–294. [CrossRef]
53. Sun, Y.; Zang, H.; Splettstosser, T.; Kumar, A.; Xu, X.; Kuzyakov, Y.; Pausch, J. Plant intraspecific competition and growth stage alter carbon and nitrogen mineralization in the rhizosphere. *Plant Cell Environ.* **2021**, *44*, 1231–1242. [CrossRef]
54. Tedersoo, L.; Bahram, M.; Zobel, M. How mycorrhizal associations drive plant population and community biology. *Science.* **2020**, *367*, eaba1223. [CrossRef]
55. Boberg, J.B.; Finlay, R.D.; Stenlid, J.; Ekblad, A.; Lindahl, B.D. Nitrogen and carbon reallocation in fungal mycelia during decomposition of Boreal forest litter. *PLoS ONE* **2014**, *9*, e92897. [CrossRef]
56. Hodge, A.; Campbell, C.D.; Fitter, A.H. An arbuscular mycorrhizal fungus accelerates decomposition and acquires nitrogen directly from organic material. *Nature* **2001**, *413*, 297–299. [CrossRef]
57. Kahkola, A.; Nygren, P.; Leblanc, H.A.; Pennanen, T.; Pietikainen, J. Leaf and root litter of a legume tree as nitrogen sources for cacaos with different root colonisation by arbuscular mycorrhizae. *Nutr. Cycl. Agroecosystems* **2012**, *92*, 51–65. [CrossRef]
58. Hodge, A.; Fitter, A.H. Substantial nitrogen acquisition by arbuscular mycorrhizal fungi from organic material has implications for N cycling. *Proc. Natl. Acad. Sci. USA* **2010**, *107*, 13754–13759. [CrossRef]
59. Camenzind, T.; Rillig, M.C. Extraradical arbuscular mycorrhizal fungal hyphae in an organic tropical montane forest soil. *Soil Biol. Biochem.* **2013**, *64*, 96–102. [CrossRef]
60. Lin, D.; Pang, M.; Fanin, N.; Wang, H.; Qian, S.; Zhao, L.; Yang, Y.; Mi, X.; Ma, K. Fungi participate in driving home-field advantage of litter decomposition in a subtropical forest. *Plant Soil* **2019**, *434*, 467–480. [CrossRef]
61. Alguacil, M.M.; Caravaca, F.; Azcon, R.; Roldan, A. Changes in biological activity of a degraded Mediterranean soil after using microbially-treated dry olive cake as a biosolid amendment and arbuscular mycorrhizal fungi. *Eur. J. Soil Biol.* **2008**, *44*, 347–354. [CrossRef]
62. Kang, L.; He, Y.; Zang, L.; Si, J.; Yang, Y.; Shen, K.; Xia, T.; Tan, Q.; Wu, B.; Guo, Y.; et al. Mycorrhizal networks interacting with litter improves nutrients and growth for one plant through the vary of N/P ratio under karst soil. *Phyton-Int.J. Exp. Bot.* **2021**, *90*, 701–717. [CrossRef]
63. Wu, Q.; Tang, Y.; Dong, T.; Liao, Y.; Li, D.; He, X.; Xu, X. Additional AM fungi inoculation increase *Populus cathayana* intersexual competition. *Front. Plant Sci.* **2018**, *9*, 607. [CrossRef]
64. Fellbaum, C.R.; Mensah, J.A.; Cloos, A.J.; Strahan, G.E.; Pfeffer, P.E.; Kiers, E.T.; Bücking, H. Fungal nutrient allocation in common mycorrhizal networks is regulated by the carbon source strength of individual host plants. *New Phytol.* **2014**, *203*, 646–656. [CrossRef] [PubMed]
65. Walder, F.; Niemann, H.; Natarajan, M.; Lehmann, M.F.; Boller, T.; Wiemken, A. Mycorrhizal networks: Common goods of plants shared under unequal terms of trade. *Plant Physiol.* **2012**, *159*, 789–797. [CrossRef] [PubMed]
66. Helgason, T.; Fitter, A.H. Natural selection and the evolutionary ecology of the arbuscular mycorrhizal fungi (*Phylum Glomeromycota*). *J. Exp. Bot.* **2009**, *60*, 2465–2480. [CrossRef]
67. Zheng, S.; Guo, S.; Zhang, Y.; Song, X.; Chen, F.; Zhang, J.; Sun, J. Effects of arbuscular mycorrhizal fungi on characteristics of photosynthesis, microbial diversity and enzymes activity in rhizosphere of pepper plants cultivated in organic substrate. *Acta Bot. Boreali-Occident. Sin.* **2014**, *34*, 800–809.
68. Xu, H.; Shao, H.; Lu, Y. Arbuscular mycorrhiza fungi and related soil microbial activity drive carbon mineralization in the maize rhizosphere. *Ecotoxicol. Environ. Saf.* **2019**, *182*, 109476. [CrossRef]
69. Frey, S.D.; Six, J.; Elliott, E.T. Reciprocal transfer of carbon and nitrogen by decomposer fungi at the soil-litter interface. *Soil Biol. Biochem.* **2003**, *35*, 1001–1004. [CrossRef]
70. Verbruggen, E.; Xiang, D.; Chen, B.; Xu, T.; Rillig, M.C. Mycorrhizal fungi associated with high soil N:P ratios are more likely to be lost upon conversion from grasslands to arable agriculture. *Soil Biol. Biochem.* **2015**, *86*, 1–4. [CrossRef]
71. Cavagnaro, T.R.; Bender, S.F.; Asghari, H.R.; Der Heijden, M.G.A.V. The role of arbuscular mycorrhizas in reducing soil nutrient loss. *Trends Plant Sci.* **2015**, *20*, 283–290. [CrossRef]
72. Govindarajulu, M.; Pfeffer, P.E.; Jin, H.; Abubaker, J.; Douds, D.D.; Allen, J.W.; Bucking, H.; Lammers, P.J.; Shachar-Hill, Y. Nitrogen transfer in the arbuscular mycorrhizal symbiosis. *Nature* **2005**, *435*, 819–823. [CrossRef] [PubMed]

forests
MDPI

Article

Diversity Monitoring of Coexisting Birds in Urban Forests by Integrating Spectrograms and Object-Based Image Analysis

Yilin Zhao [1,2,3], Jingli Yan [4,5,*], Jiali Jin [1,2,3], Zhenkai Sun [1,2,3], Luqin Yin [1,2,3], Zitong Bai [1,2,3,6] and Cheng Wang [1,2,3,*]

1 Research Institute of Forestry, Chinese Academy of Forestry, Beijing 100091, China; elyn_zhaoyilin@163.com (Y.Z.); king90emily@gmail.com (J.J.); zksun_caf@caf.ac.cn (Z.S.); yinluqin@caf.cn (L.Y.); baizitong0604@foxmail.com (Z.B.)
2 Key Laboratory of Tree Breeding and Cultivation, National Forestry and Grassland Administration, Beijing 100091, China
3 Urban Forest Research Center, National Forestry and Grassland Administration, Beijing 100091, China
4 School of Agriculture and Biology, Shanghai Jiao Tong University, Shanghai 200240, China
5 Shanghai Yangtze River Delta Eco-Environmental Change and Management Observation and Research Station, Ministry of Science and Technology, Shanghai 200240, China
6 Beijing Institute of Landscape and Traditional Architecture Design and Research Co., Ltd., Beijing 100005, China
* Correspondence: jlyan24@163.com (J.Y.); wch8361@163.com (C.W.); Tel.: +86-21-3420-4780 (J.Y.); +86-10-6288-8361 (C.W.)

Abstract: In the context of rapid urbanization, urban foresters are actively seeking management monitoring programs that address the challenges of urban biodiversity loss. Passive acoustic monitoring (PAM) has attracted attention because it allows for the collection of data passively, objectively, and continuously across large areas and for extended periods. However, it continues to be a difficult subject due to the massive amount of information that audio recordings contain. Most existing automated analysis methods have limitations in their application in urban areas, with unclear ecological relevance and efficacy. To better support urban forest biodiversity monitoring, we present a novel methodology for automatically extracting bird vocalizations from spectrograms of field audio recordings, integrating object-based classification. We applied this approach to acoustic data from an urban forest in Beijing and achieved an accuracy of 93.55% (±4.78%) in vocalization recognition while requiring less than 1/8 of the time needed for traditional inspection. The difference in efficiency would become more significant as the data size increases because object-based classification allows for batch processing of spectrograms. Using the extracted vocalizations, a series of acoustic and morphological features of bird-vocalization syllables (syllable feature metrics, SFMs) could be calculated to better quantify acoustic events and describe the soundscape. A significant correlation between the SFMs and biodiversity indices was found, with 57% of the variance in species richness, 41% in Shannon's diversity index and 38% in Simpson's diversity index being explained by SFMs. Therefore, our proposed method provides an effective complementary tool to existing automated methods for long-term urban forest biodiversity monitoring and conservation.

Keywords: biodiversity monitoring; soundscape ecology; spectrograms; acoustic indices; birds; urban forests; object-based image analysis; novel approach

Citation: Zhao, Y.; Yan, J.; Jin, J.; Sun, Z.; Yin, L.; Bai, Z.; Wang, C. Diversity Monitoring of Coexisting Birds in Urban Forests by Integrating Spectrograms and Object-Based Image Analysis. *Forests* **2022**, *13*, 264. https://doi.org/10.3390/f13020264

Academic Editor: Todd Fredericksen

Received: 6 November 2021
Accepted: 28 January 2022
Published: 8 February 2022

1. Introduction

Biodiversity loss has been a major and challenging problem globally, and is a potential risk factor for pandemics [1]. The ongoing global COVID-19 pandemic has confirmed this concern. The loss of biodiversity has generated conditions that not only favored the appearance of the virus but also enabled the COVID-19 pandemic to surface [2,3]. Biodiversity conservation is, therefore, an urgent global task.

Aiming to produce high-quality habitats and improve regional biodiversity [4], in 2012, Beijing launched its largest decade-long afforestation campaign, the Plain Afforestation Project (PAP), which required building hundreds of gardens and parks in urban areas. To assess the success of PAP, the rapid and effective monitoring of urban biodiversity is key [5], which has led to a need for innovative investigation approaches. However, the need for expert knowledge and the substantial costs in terms of both money and time are major obstacles for any multi-taxa approach based on large-scale fieldwork [6]. Thus, the development of cost-effective and robust tools for monitoring urban forest biodiversity is a pressing need [1].

Operating within the conceptual and methodological framework of ecoacoustics [7], passive acoustic monitoring (PAM) is a promising approach with many advantages, including its availability in remote and difficult-to-reach locations, noninvasiveness, non-observer bias, permanent record of surveys, and low cost [8–11]. In addition, PAM allows for standardized surveys that can provide new insights into sound-producing organisms over enhanced spatiotemporal scales [12]. For example, the global acoustic database Ocean Biodiversity Information System–Spatial Ecological Analysis of Megavertebrate Population (OBIS-SEAMAP) was developed to enable research data commons, and contains more than one million observation recordings from 163 datasets spanning 71 years (1935 to 2005), provided by a growing international network of data users [13].

In terrestrial soundscapes, bird vocalizations are one of the most prominent elements [14] and have been widely used to detect species and monitor and quantify ecosystems [15]. More specifically, acoustic traits have been proved to respond to environmental changes, such as climate change [16], habitat fragmentation [17], vegetation structure, and microclimate [18]. With the emergence of PAM, massive acoustic data have accumulated globally, offering unprecedented opportunities, as well as challenges, for innovative biodiversity monitoring.

A critical challenge in PAM studies is the analysis and handling of very large amounts of acoustic data, especially for programs spanning wide temporal or spatial extents [19]. However, manual analysis is still the primary method for extracting biological information from PAM recordings [20], which typically combines aural and visual inspection of spectrograms [21,22] to achieve graphical representations of acoustic events connected to biophony, geophony, and anthrophony, as well as a general overview of the daily acoustic pattern [21]. When experienced observers are involved, manual analysis is always considered to be the most accurate, but it is time-consuming, costly, frequently subjective, and ultimately fails to be applied across broad spatiotemporal scales [22].

To address the challenges posed by massive data and manual analysis, increasing numbers of studies have been conducted on individual species, and there seems to be a rising interest in the ecological processes of biomes [23,24]. It has been well established that monitoring community acoustic dynamics is key to understanding the changes and drivers of ecosystem biodiversity within the framework of soundscape ecology [25–28]. The burgeoning development of this framework has stimulated research interest in ecological applications of acoustic indices, which have been intensively proposed and tested [29–34].

Unfortunately, automated acoustic analysis remains a difficult subject to study because of the wide variety of information available in each acoustic environment, making it difficult to quickly identify and extract critical ecological information for interpreting recordings [29]. Most existing acoustic indices use simple algorithms to collapse the signal into one domain and quantify the soundscape by summing or contrasting acoustic energy variations [12,24,34], which are intrinsically an extension of the traditional sound pressure and spectral density indices [29,35–38]. Although cheap and fast, this type of analysis leads to massive loss of information, so its eco-efficiency remains controversial. In addition, the difficulty in excluding the interference of noise in order to quantify biophony alone remains a major limitation of the existing indices, which leads to huge bias in the application of these indices in urban areas [33], raising concerns over their applicability [34,37].

Although automated analysis techniques are rapidly improving, software tools still lag far behind actual applications [22,39–42]. We therefore suggest that advancing the theory and practice of soundscape ecology research requires going beyond the limits of the temporal/frequency structure of sound and developing more tools to retain as much ecologically relevant information as possible from recordings, testing our methods in complex urban environments to clarify their robustness.

Remote sensing technology has been broadly used in many applications, such as extracting land cover/usage information. Object-based image analysis (OBIA) has emerged as an effective tool to overcome the problems of traditional pixel-based techniques of image data [43,44]. It defines segments rather than pixels to classify areas, and it incorporates meaningful spectral and non-spectral features for class separation, thereby providing a clear illustration of landscape patterns [43–46]. Owing to its superiority and efficiency [47], OBIA has been utilized in many different areas, such as computer vision [48,49], biomedical imaging [50,51], and environmental scanning electron microscopy (SEM) analysis [52–54]. Just as remote sensing images are numeric representations of the earth surface landscape consisting of water area, forest land, wetlands, etc. [55], spectrograms are visual expressions of collections of various sound components (biophony, geophony and anthrophony). As such, could OBIA provide a novel perspective for extracting bird vocalizations when introducing advanced remote sensing tools in the soundscape field? Could we further digitally summarize vocalization patches and use them as ecologically relevant indicators of acoustic community patterns?

Based on the above hypotheses, an automated bird vocalization extraction method based on OBIA is presented here. We hypothesize that OBIA may allow for the extracting of bird vocalizations from recordings with complex background noise and the representation of long-term acoustic data as numbers describing biophony. From the perspective of community-level soundscape ecology, we are not necessarily concerned with species identification, but with achieving a numerical description of the qualitative patterns of species vocalizations [24]. OBIA enables rapid identification of the number of bird vocalizations while providing multidimensional spectral, morphological, and acoustic traits, unlike other existing methods (whether manual or automatic). Examples of spectral variables include the mean value and standard deviation of a specific spectral band; morphological traits include size, perimeter, and compactness; acoustic traits include song length and frequency information.

In the present paper, we take a first look at how OBIA might provide a new perspective on the current automated acoustic analysis methods and provide a complement to existing acoustic indices that can be used for urban forest biodiversity assessments.

2. Materials and Methods

2.1. Study Area and Data Sets

For the present study, we selected an old urban forest in Zhongshan Park, Beijing (115°24′—117°30′ E, 39°38′—41°05′ N) as our case study area (Figure 1A). As the capital of China and the second-largest city in the world, Beijing is also a major node in the East Asian–Australasian bird flyway [56]. Beijing is a key corridor for birds' spring and autumn migrations, as it is in the ecosystem transition zone from Northeast China to North China. Urban forests in Beijing play an important role in supporting roosting, breeding and other activities of birds, and as a result, they are rich in soundscapes.

Our data were derived from audio recordings continuously obtained during four consecutive sunny, windless days, from 18 to 21 May 2019, provided by three recorders positioned in Zhongshan Park (Figure 1B). Recorders were placed and fixed horizontally at a height of 2 m on healthy growing trees (Figure 1C,D). Auto-recording led to a total of 17,280 min of raw recordings, which were subsequently processed using the AudioSegment function in PYTHON v.3.7.2 and sampled into 15-s clips every 15 min, resulting in a sub-dataset of 1152 15-s clips. A sampling protocol of 15 s was used as it provided a tradeoff between ensuring an effective acoustic survey and a manageable amount of data for processing when there is no standardized protocol [57].

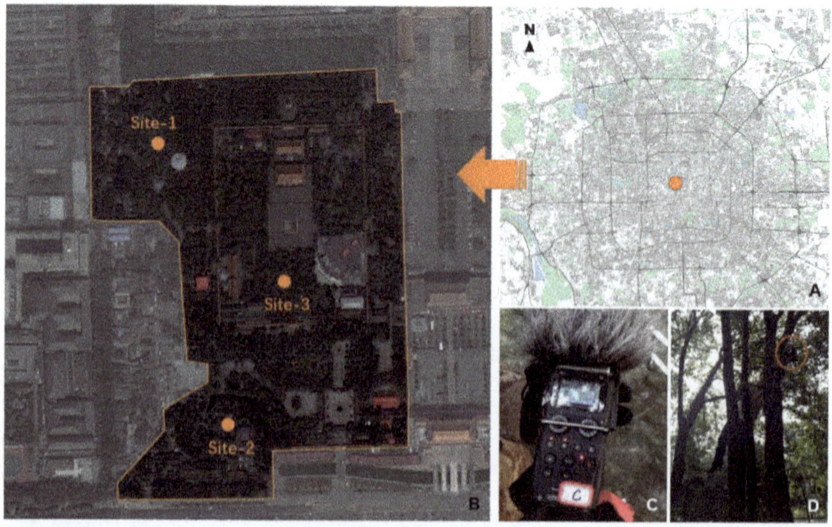

Figure 1. Location of the urban forest (**A**); location of sites used for acoustic recording (**B**); an acoustic recorder (**C**) and its positioning in the urban forest (**D**) are also presented.

Recordings were obtained using Zoom H5 acoustic sensors (Zoom Inc., Tokyo, Japan, System 2.40) with XYH-5 X/Y microphones. Zoom H5 is a commercial digital recording device that has good sound reliability and has been successfully used in other soundscape ecology research [27,58]. Parameters of sensors were set as follows: the sampling rate was 44,100 Hz, bit depth 16 bits, and recording channels two (stereo). Files were saved as non-compressed WAVE files.

Two trained technicians (Yu and Hanchen Huang) manually identified the acoustic events (AEs) in all audio clips for further use in evaluating the reliability and sensitivity of our approach. Over 95% of biological acoustic events (BEs) were from birds; thus, only bird sounds were identified to the species level (List of Bird Species see Table A1), while BEs that were not bird sounds were identified to the family level. For example, cricket sounds were labelled Gryllidae. Because we were unable to distinguish individual animals based on their sounds, technicians identified and counted the total number of BEs for a given species (or family) in each clip. Anthropogenic and geophysical sounds were also counted and classified into AEs. The results of this process were finally confirmed by Hanchen Huang. Richness (S), Shannon's diversity (H') and Simpson's diversity (λ) indices were calculated for each spectrogram (i.e., per 15 s clip) to reflect the diversity of bird species as well as AE and BE types [12].

2.2. Methods

We developed an automated bird vocalization extraction approach based on OBIA, which followed a typical analysis workflow of bird vocalizations with three main steps [59]: preprocessing, automated extraction, and feature calculation. In line with this workflow, the processing details of each step of our approach are described in the following subsections. An overview of the approach is depicted in Figure 2.

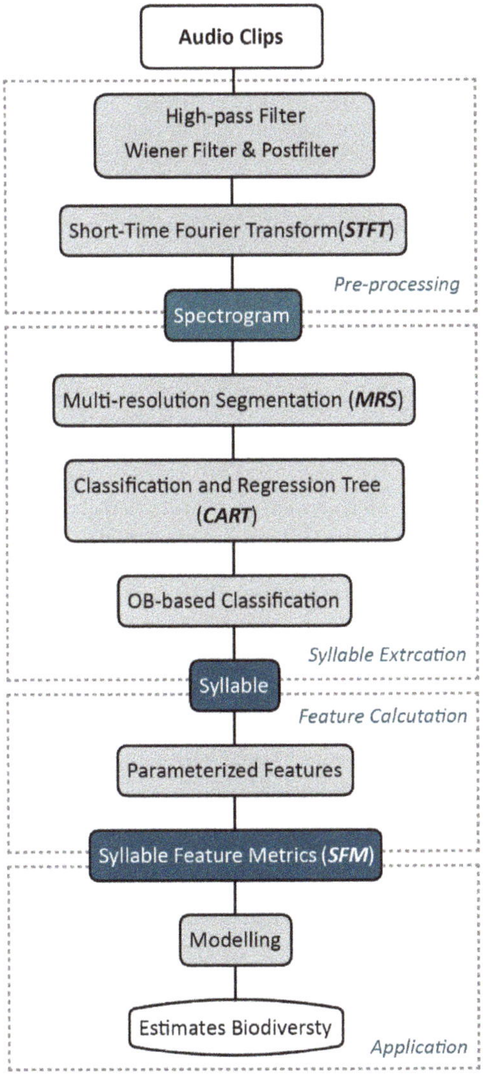

Figure 2. Overall scheme of the proposed approach.

2.2.1. Pre-Processing

1. Audio recordings denoising

The step prior to spectrogram analysis was denoising, and it was carried out to obtain clear vocalization patterns to improve extraction accuracy and minimize false positives [60]. We selected only noise reduction techniques that could perform batch and fast processing, to allow the proposed approach to be more generalizable and to reduce the effects of human operations.

Audio signals are characterized by the presence of higher energy in the low-frequency regions dominated by environmental noise. Therefore, we applied a high-pass filter with a cut-off frequency setting (800 Hz) below the lowest frequency at which bird songs are expected, with a 12 dB roll-off per octave [61–64]. This allowed for the elimination of current and environmental noise, mostly created by mechanical devices such as engines [65,66].

Since medium and high frequencies are the useful part for discriminating between different acoustic events [67], we applied a widely used signal enhancement method to our recordings, the Wiener filter [68–70] and postfilter, to minimize baseline and white noise and improve the quality of bird sound recordings. In field recordings, bird songs are transient but a considerable amount of background noise is nearly stationary. The Wiener filter eliminates this quasi-stationary noise, provided that it approximates a Gaussian distribution [22].

To verify the denoising effect, we randomly chose 80 clips from all datasets (including anthropogenic AEs) that were listened to before and after the denoising process to assess denoising efficacy.

2. Short-time Fourier transformation (STFT)-Spectrogram

Bird vocalizations of songs or calls are complex, non-stationary signals with a great degree of variation in intensity, pitch, and syllable patterns [71]. One of the proven methods for joint time-frequency domain analysis of non-stationary sound signals is STFT [59]. The STFT spectrogram is a two-dimensional convolution of the signal and window function [72]: the X-axis represents time, the Y-axis represents frequency, and the amplitude of a particular frequency at a particular time is represented by its color in the image [73].

Our STFT spectrogram was calculated with a Hamming window of 1024 samples, no zero paddings, and a 75% overlap between successive windows. A peak amplitude of −25 dB was set to standardize the spectrograms [42].

2.2.2. Bird Vocalization Extraction

Generally, bird vocalizations are classified into songs (longer-term) and calls (shorter-term). In the present study, the separation between songs and calls was not considered because both types use syllables as fundamental units [74]. All vocalizations were segmented and classified at the syllable level. To facilitate understanding, a BE refers to either a song (call) in recordings or a syllable in spectrograms [10].

Bird syllable extraction is the key part of the proposed OBIA. All processing steps were conducted in one framework. Firstly, we adopted a multi-resolution segmentation algorithm to segment spectrograms into image objects. Then, we manually selected samples representing a pure spectrum of bird syllables to generate potential extraction features using a Classification and Regression Tree (CART). Finally, to extract bird syllables from background noise, we applied the rulesets established from identified potential features to all spectrograms.

1. Segmentation

Segmentation creates new meaningful image objects according to their spectral properties (Figure 3), including subdividing and merging operations [75]. We applied the MRS algorithm embedded in eCognition Developer™ to generate image objects. This algorithm consecutively merges pixels or existing image objects into larger objects based on relative homogeneity within the merged object [53,54,76]. The process uses three key parameters in the process: scale, shape and compactness [77]. A coarse scale value allows for the forming of larger objects and more heterogeneity, involving more spectral values. Shape defines the influence of color (spectral value) and shapes on the formation of the segments, while compactness defines whether the boundary of the segments should be smoother or more compact [78]. In our segmentation, we employed the "trial and error" method of visual inspection to determine the optimal segmentation parameters [53], as bird syllables and their shapes on spectrograms are easily recognized by human eyes. After several attempts, we finally selected a scale of 40, 0.2 of shape, and 0.5 of compactness to produce segmented image objects.

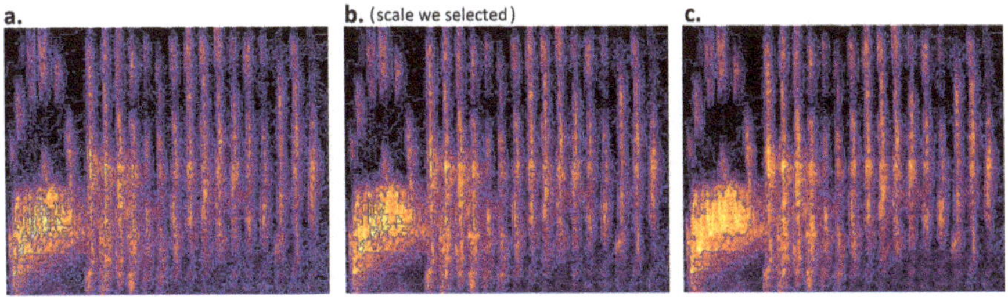

Figure 3. Segmented image objects with three different scales but identical shape and compactness. (**a**) Scale 20 (1657 objects); (**b**) Scale 40 (458 objects); (**c**) Scale 60 (174 objects).

2. Classification

Techniques from machine learning and computational intelligence have been used in bird vocalization analysis [79,80]. In the present study, CART [81] was applied to construct accurate and reliable predictive models for syllable extraction. CART does not require any special data preparation, only a good representation of the problem. Creating a CART model involves selecting input variables and split points on those variables until a suitable tree is trained. To operate CART, we randomly selected 288 syllable samples, one for each of the 288 audio clips, and divided them into training samples (70%, $n = 202$) and testing samples (30%, $n = 86$). Meanwhile, the ten most explanatory object features, including spectral, shape and textural characteristics (Table 1), were identified by calculating the feature variations in syllable samples and other objects, and were imported into CART as input variables. The representation of the CART model is a binary tree derived from recursive binary splitting (or greedy splitting). The binary tree allows for relatively straightforward obtainment of clear classification rules from the model diagram.

Table 1. The summary of the 10 most explanatory object features used in CART.

Feature Name	Description	Rel. Imp
Brightness	Mean value of all image bands	81.24
Shape index	The smoothness of the boundary of an image object	10.85
Area	The area of objects in number of pixels	18.01
Length/width	The ratio of length to width	7.58
Elliptic fit	How well an image object fits into an ellipse	1.36
Hue	Mean of hue, one of three color components	4.97
Saturation	Mean of saturation, one of three color components	23.12
Intensity	Mean of intensity, one of three color components	54.33
GLCM-M	Mean value of GLCM (Gray-level Co-occurrence Matrix)	19.54
GLCM-H	Homogeneity of GLCM (Gray-level Co-occurrence Matrix)	60.49

Rel. imp stands for relative importance; GLCM (Gray-level Co-occurrence Matrix) is a tabulation of how often different combinations of pixel gray levels occur in a scene.

The complexity of a decision tree is defined by the number of splits in each tree. Simpler trees are preferred, as they are easier to understand and less likely to overfit the data. Trees can be pruned to further improve performance. The fastest and simplest pruning method is to work through each leaf node in the tree and evaluate the effect of removing it using a hold-out test set. Leaf nodes are removed only if this results in a decrease in the overall cost function for the entire test set. Node removal is stopped when no further improvements can be made.

We introduced two indicators, relative cost (RC) and rate of change (ROC), to evaluate the performance of the CART model. The value of RC ranges from 0 to 1, with 0 indicating a perfect model with no error and 1 indicating random guessing. Similarly, the value of ROC ranges from 0 to 1, with higher values suggesting better performance. The resulting optimal decision tree consisted of three features: brightness, intensity, and grey level co-occurrence matrix (GLCM) homogeneity (Figure 4). It had an RC of 0.124 and a ROC of 0.979, indicating the reliability of the prediction model. The final step was to apply the ruleset generated from CART to classify all image objects.

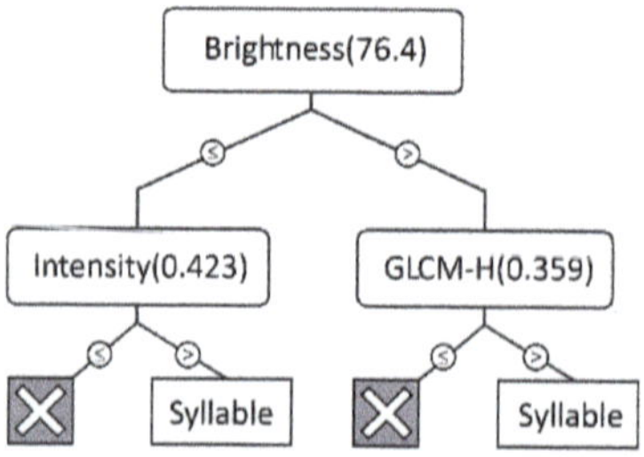

Figure 4. The optimal decision tree resulting from CART. Instances with a split value greater than the threshold (in parentheses) were moved to the right.

2.2.3. Feature Representation of Extracted Syllables

The time-frequency pattern of syllables displayed via spectrograms is a useful representation of species information, which can also characterize bird vocalization patterns [82,83]. Based on the remote sensing framework, where syllables on spectrograms are comparable to landscape patches on maps, our approach could easily calculate and analyze morphological descriptors. Each extracted syllable was characterized by a sequence of feature vectors.

Considering their good performance with respect to feature analysis of syllables in previous studies, instantaneous frequency (IF) [59], and amplitude [84] were adopted in this work. Each parameter was further statistically analyzed to obtain more detailed descriptive values such as maximum (IF_MAX), minimum (IF_MI), mean, and central frequency. The mean instantaneous frequency (IF_MN) is the first moment of the spectrogram relative to the frequency, and it can be calculated using Equation (1):

$$f_i(t) = \frac{\int_{-\infty}^{\infty} f \cdot S(t,f)\mathrm{d}f}{\int_{-\infty}^{\infty} S(t,f)\mathrm{d}f} \quad (1)$$

where $f_i(t)$ is the mean instantaneous frequency at time t, and $S\ (t,f)$ is the spectrogram at frequency f and time t.

In addition to the time-frequency characteristics, we also summarized the landscape metrics that were of practical meaning for syllable patches. Datasets were rasterized in R (https://www.r-project.org (accessed on 28 May 2021)) and imported into FRAGSTATS [85]. For each extracted syllable (patch level), we calculated area, shape index, border index, length, width (duration), etc., to obtain the fundamental spatial character and morphological understanding of each patch. Landscape-level metrics were also further summarized and nondimensionalized in each spectrogram, reflecting time-frequency patterns of acoustic activities, including: (1) NP, the number of syllable patches; (2) CA, the sum of the areas of all patches; (3) SHAPE_MN, mean value of the shape index of each patch; (4) TL, total bandwidth occupancy of all patches, calculated from a transformation of patch length; and (5) TW, total duration of all patches, calculated from a transformation of patch width. To facilitate reading, we abbreviated all these syllable feature metrics as SFMs (Table 2).

Table 2. Summary of the main syllable feature metrics (SFMs).

SFMs	Description in Landscape Ecology	Transformation	Meaning in Acoustics
NP (Patch Number)	NP is a count of all the patches across the entire landscape.	none	Number of acoustic events.
CA (Class Area)	CA is the sum of the areas of all patches belonging to a given class.	none	Proportion of spectrogram covered by acoustic-event patches.
SHAPE_MN	SHAPE_MN equals the average shape index of patches across the entire landscape.	none	Average shape index (complexity of patch shape) of the extracted vocalization syllables.
TL (Total Length)	The sum of the lengths of all patches belonging to a given spectrogram.	×15	Total bandwidth occupancy of acoustic events (Hz).
TW (Total Width)	The sum of the widths of all patches belonging to a given spectrogram.	÷200	The total duration of the acoustic events (s).

2.3. *Statistical Analysis*

All statistical analyses were performed in R version 4.0.5 (R Core Team, Vienna, Austria, 2021).

Because the probability distribution of the raw data failed the Kolmogorov–Smirnov test for normality, transformations were performed using the bestNormalize package [86], which attempts a range of transformations and selects the best one based on the goodness-of-fit statistic to ensure transformations are consistent. It can also remove the effects of order-of-magnitude differences among variables.

2.3.1. Accuracy Assessment

Manual inspection of BEs from audio clips was used as a reference for accuracy assessment. To minimize errors, we marked each syllable patch with a serial number when calculating it to avoid missing or double-counting. We also recorded the time spent by the technician on each spectrogram. We used relative error (RE) as a measurement of accuracy, which is the ratio of the absolute value of the reference value [87]. Specifically, RE was calculated by dividing the number of syllables correctly identified through the automated approach by the total number of BEs identified through manual inspection as a reference.

2.3.2. Correlation Analysis

Correlation analysis (Spearman's rho, $p < 0.01$) was performed between the number of extracted syllable patches and bioacoustic and acoustic events to further verify the reliability of the approach of automated extraction of bird vocalizations. Then, a second Spearman's rho correlation test was performed for the relationships between SFMs and bird species *S*. Non-parametric correlation analysis was selected because not all data were normally distributed despite being transformed.

2.3.3. Modelling

To test the efficacy of SFMs as a biodiversity proxy, we used a random forest (RF) machine learning procedure (randomForest package) [88,89] to predict bird species biodiversity from SFMs calculated from matching recordings.

RF is a meta-estimator and one of the most accurate learning algorithms available. The RF algorithm aggregates many decision trees and combines the results of multiple predictions, while ensuring that the ensemble model makes fair use of all potentially predictive variables and prevents overfitting [90]. In addition, RF accommodates multivariate collinearity among predictors, which is convenient for calculating the nonlinear effects of variables [27]. We used a bootstrapping cross-validation method to select the model structure with the lowest median of mean squared error (MSE) and highest R^2 between tested data and predicted values [12]. MSE was also used to measure the importance of each variable. A higher percentage increase in MSE indicated a greater ability to predict the model [91].

3. Results

3.1. Approach Reliability

With pre-processing denoising, we removed more than 85% (n = 77) of the anthropogenic acoustic events (for a denoising example see Figure 5A,B). The number of syllables identified from the spectrograms varied from 0 to 183 in each spectrogram according to the automatic extraction approach. The REs of the approach ranged from 11.52% to 0.00% across all spectrograms with an average of 6.45% (±4.78%), suggesting that the automated extraction process yielded high accuracy (93.55%) values (for an example of identified syllables see Figure 5).

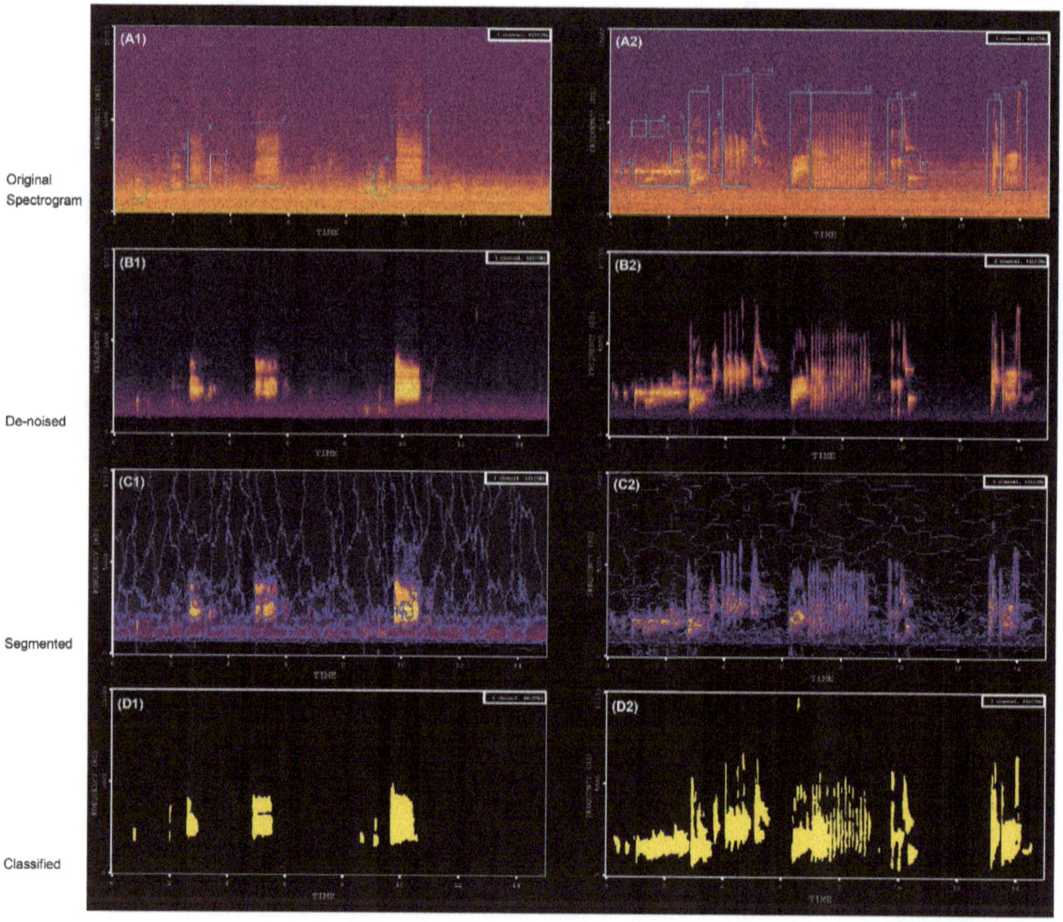

Figure 5. Typical spectrograms of audio scenes during processing procedures. (**A1**,**A2**) is the original spectrogram. (**B1**,**B2**) shows the result of noise reduction: a cleaner recording (although possibly with some artifacts) that is ready to be used as input into segmentation algorithms. (**C1**,**C2**) shows the result of the segmentation procedure and (**D1**,**D2**) shows the classification results; the yellow patches are the extracted syllables.

A high correlation coefficient between the NP values and the number of bio-acoustic events and all acoustic events (r = 0.71, $p < 0.01$; r = 0.89, $p < 0.01$) was observed for the Spearman's rho correlation matrix (Table 3). The CA (area), TL (duration) and TW (frequency) values of extracted syllables were also significantly correlated with bio-acoustic events and all acoustic events, albeit less strongly (Table 3).

Table 3. Spearman's rho correlation matrix ($p < 0.01$; $n = 288$).

	NP	CA	TL	TW
BE	0.71	0.60	0.56	0.63
AE	0.89	0.74	0.71	0.79

These results indicate that this new, automated approach was much more efficient than manual inspection. For manual inspection, on average, it took approximately 27 s of effort to analyze 15-s acoustic data, which was approximately eight times longer than the time required for the automated approach (3.5 s refers to the time taken for the whole process shown in Figure 2, averaged by each 15 s-spectrogram). This is because analysts tend to replay recordings to confirm results [92], and the time spent on loading and annotating vocalizations must also be accounted for. Further, it is expected that the difference in efficiency would be more significant with increasing amounts of data, because spectrograms can be batch processed using our approach.

3.2. SFMs' Correlation with Biodiversity

High correlation coefficients (r > 0.5) between SFMs and bird species richness were found, except for shape index and MIF (avg, min, and max) (Figure 6). Simpson's and Shannon's diversity indices were also correlated with SFMs, albeit less strongly. PN always had the highest correlation with the three diversity indices.

3.3. Prediction of Biodiversity

Random forest regression models confirmed that combinations of SFMs are good predictors of biodiversity (Table 4). Bird species richness was predicted well (R^2 = 0.57) but the acoustic diversity of bird communities was less reliably predicted (Simpson's and Shannon's diversity indices had R^2 of 0.38 and 0.41, respectively). These results suggest that SFMs have great potential for tracking acoustic communities, even in the presence of considerable anthrophony (human-induced noise) in an urban environment [29,93].

Table 4. Mean squared error (MSE) and R^2 of the top models of SFMs that predicted species richness, Shannon diversity, and Simpson diversity in acoustic recording samples.

Response Variable	Model Type [a]	SFM-Covariates	MSE	R^2
Richness	3	PN + CA + TW + TL + Border_len + Width_Pxl + HSI_Transf +	1.47	0.57
Simpson diversity	1	HSI_Tran_1 + Max_pixel_ + Shape_MN + Compactnes + Brightness +	0.30	0.38
Shannon diversity	2	Roundness + Area_Pxl + Border_ind + MIF_min + MIF_max + MIF	0.27	0.41

[a] The number of variables tried at each split. For each number of variables per split from one to six, a new random forest was generated, which was evaluated and chosen both by the error rates in the test set and the out-of-bag OOB error. See Appendix A Materials for the descriptions of variables that are not described in the text.

We used all 14 SFMs as predictors (including soundscape and patch levels) in each RF regression model and investigated the relative contributions of each SFM. These metrics described soundscapes (syllable patches) from different perspectives and levels. Results demonstrated that SFMs effectively explained different pieces of information in acoustic recordings, likely because their unique mathematical properties reflect different dimensions of a soundscape. The number of syllable patches (PN) was the strongest single predictor in the best model found for richness (Figure 7a). In the best models predicting Simpson's and Shannon's diversity, the SFM with the highest importance was the total length of patches (20% and 23% of variance explained, respectively) (Figure 7b,c). It is noteworthy that PN, TL, and PI (Patch Intensity) were always the top three predictors for the three models (the sum of their contributions was 68%, 57%, and 65% in models a, b, and c respectively), suggesting that PN, TL and PI can be considered the most promising SFMs. All other indices exceeded the analytic threshold [63,94], suggesting that they all contributed little to predictive power.

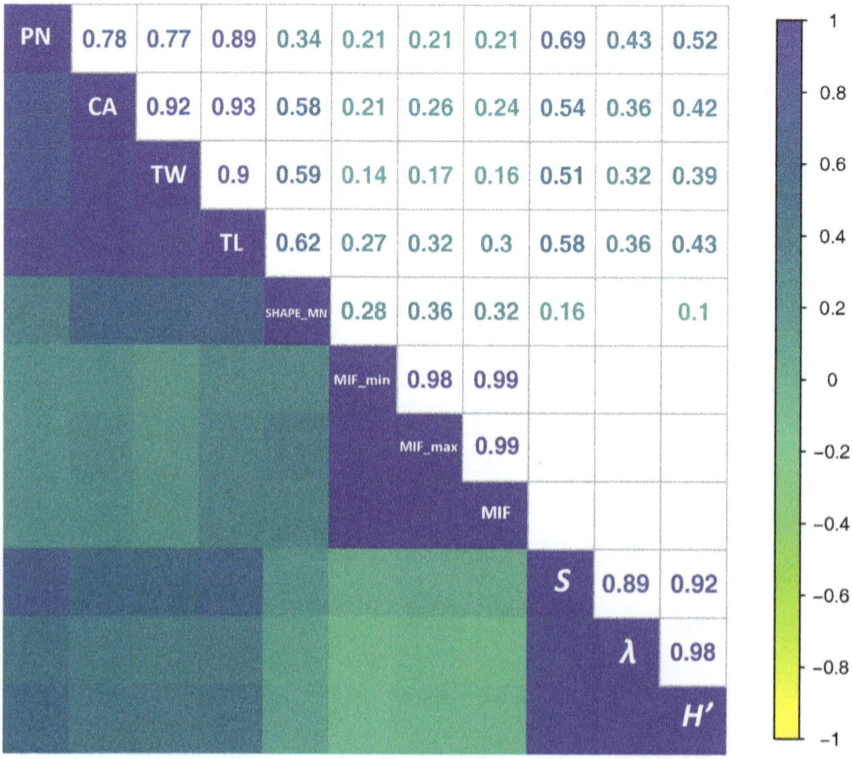

Figure 6. Spearman's correlation coefficient values shown for each relationship in the upper half of the matrix (only showing coefficient values that are statistically significant, $p < 0.001$). The diagonal shows the variables. Labels: NP = the number of syllable patches, CA = the sum of the areas of all patches in the given spectrogram, SHAPE_MN = the average shape index of patches across the entire spectrogram, TW = total duration of all patches in the given spectrogram (s), TL = total bandwidth occupancy of all patches in the given spectrogram (Hz), MIF = the mean instantaneous frequency (Hz), MIF_min = the minimum instantaneous frequency (Hz). MIF_min = the maximum instantaneous frequency (Hz). S = bird species richness, λ = Simpson's Diversity Index, H' = Shannon's diversity index.

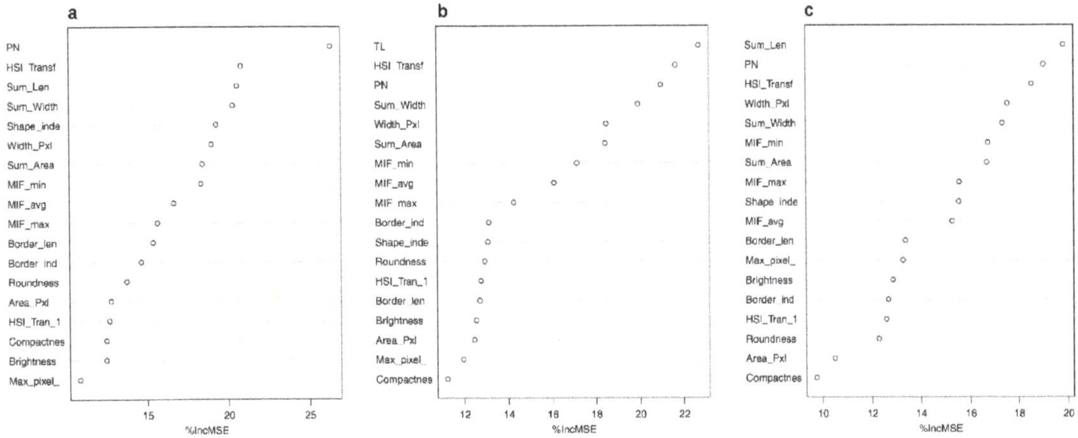

Figure 7. Importance of covariates in random forest models, indicated by mean percent increase in mean squared error (MSE): the final models. (**a**) Richness of bird species, (**b**) Simpson diversity, and (**c**) Shannon diversity. Greater MSE indicates a larger loss of predictive accuracy when covariates are permuted and thus a larger influence in the model. Results are shown for acoustic index covariates, ordered by MSE. See Appendix A for a description of all SFMs.

4. Discussion

When analyzing biophony in urban environments, anthrophony could lead to potential false positives [95]. Low recognition accuracy is often attributed to noise [96], which affects the whole process unless removed initially. Hence, at present, most methods for analyzing biophony in urban environments are trained and tested on relatively low numbers of high-quality recordings that have been carefully selected [22]. This may lead to better results but will limit the generalization of their approach to real field recordings, especially in urban areas with complex acoustic environments. Therefore, in our study, we used only noise reduction, which meant that batch and fast processing could be performed.

According to Spearman's rho correlations, NP was always more strongly correlated with AEs than with BEs, suggesting that SFMs were still somewhat influenced by human-generated noise, even after the denoising process. SFMs were hardly affected by constant-intensity noise (e.g., noise from aircraft or automobile traffic; for an example see Figure 5D: persistent noise was not extracted by the algorithm) [29,33]. In addition to pre-processing filtering, which eliminated most of the noise, CART models minimized the confounding effects of noise and syllables through training samples. The inherent properties of constant-intensity noise are different from those of bird vocalizations and are easily recognized by the model. For example, the brightness (Figure 4) of bird vocalizations was generally greater than 80, while the values for noise were 30–60. Nevertheless, some intermittent human noises, such as car horns and ringtones, might be extracted together with bird vocalizations using our approach. However, we believe that testing the approach in other habitats such as natural forests or biodiversity conservation areas will yield more encouraging results.

By treating spectrograms as images, previous studies have applied image processing techniques to extract bird vocalizations [10,59,60,83,96–98], such as widely used median clipping [41,99,100] and frame- or acoustic event-based morphological filtering [66]. There are plenty of toolboxes available to extract acoustic traits [22], such as central frequency, highest frequency, lowest frequency, initial frequency, and loudest frequency and so on [10,59], which are basically time–frequency characteristics only. To the best knowledge, all these studies aimed to identify or classify one or several bird species specifically. However, when focusing on the entire ecosystem, the species-level approach misses the forest for the trees [101]. Unlike the studies aiming at recognition of one or more species, ours attempted to take a global estimate of the acoustic output of the community. Our results indicated that SFMs are a promising complement to the existing indices working as biodiversity proxies when rapid assessments are required because SFMs were significantly correlated with diversity indices. SFMs allow for the effective interpretation of different pieces of bio-information in recordings, probably because their unique mathematical properties reflect different components of the soundscape, preserving more of the potentially eco-relevant information.

Using the proposed approach, we could collect a series of data including acoustic traits on the time–frequency scale (Figure 8), such as duration and mean, maximum, and minimum frequency of acoustic events, and morphological characteristics of acoustic events (Figure 9), such as the size and shape characteristics for each syllable patch. Such features (i.e., SFMs) may contribute to a more nuanced understanding of the acoustic environment of the study area from multiple perspectives. In Figure 8, the frequency patterns of syllable patches are shown across frequency intervals. The soundscape was dominated by mid-frequency sounds (3–5 kHz); syllable patches were over 50%. Mid-frequency sounds are generally attributed to biophony, especially bird species ranging from larger birds such as the Eurasian magpie (*Pica pica*) to smaller species such as the Oriental reed warbler (*Acrocephalus orientalis*). At the other end of the frequency spectrum, the number of patches within individual high-frequency intervals was quite low (biophony patches in the highest four frequency intervals accounted for 1.5%) and were mainly within 20–22 kHz, which may be attributed to some night-flying moths of the family Noctuidae or to mating calls from grasshoppers. In Figure 9, the average shape index (SHAPE) over 24 h was shown to rise rapidly at dawn chorus to reach the peak of the day, falling rapidly and remaining steady until dusk, when the chorus rose rapidly again, and then fluctuated and fell until the morning. The daily pattern of SHAPE is consistent with previous studies of other acoustic indices [27,31,102,103], thus reflecting a daily activity pattern and highlighting distinct dawn and dusk bird choruses. However, compared to other indices that only focus on sound intensity, SHAPE provides a new perspective on patterns of complexity of bird songs: songs of the dawn and dusk choruses tended to be more complex and elaborate than daytime songs. This is mainly related to defending territory and/or attracting a mate [104].

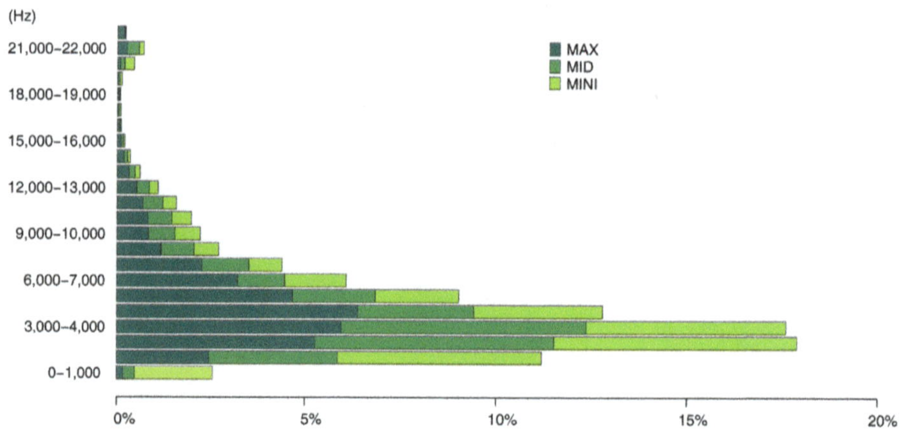

Figure 8. Number of maximum, medium, and minimum frequencies for each syllable patch across 23 1-kHz intervals.

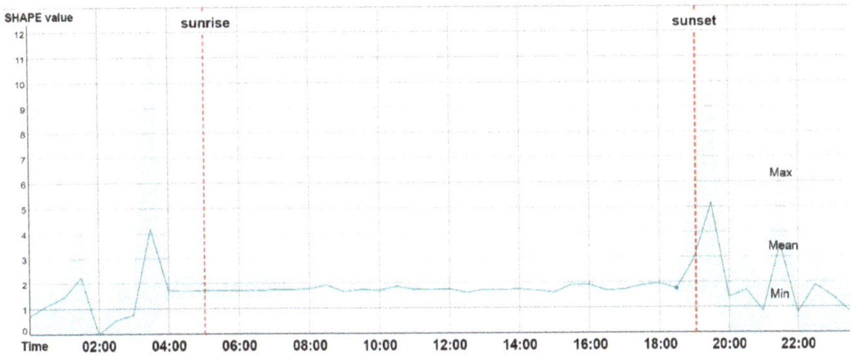

Figure 9. Shape index variation for 24 h. Upper bound is the maximum. Lower bound is the minimum.

Borrowing a framework from landscape ecology, many types SFMs can effectively be used to interpret different aspects of acoustic information and different components of the soundscape, which is presumably attributed to the mathematical properties of SFMs and to the introduction of a spatial concept. SFMs calculated in this study, such as area, compactness, roundness, and shape index of the patches, could be easily generated in eCognition developer. In addition, many existing open-source platforms (e.g., package landscapemetrics in R) or software have integrated huge workflows, which can provide similar functions (e.g., QGIS). Measurement of biophony from multiple dimensions has been considered useful for detecting variations in the behavior and composition of acoustic communities and, as a result, to better monitor their dynamics and interactions with habitats [29]. These results support the possibility that PAM could potentially offer a more comprehensive picture of biodiversity than traditional inspection [63].

Under the high pressure of a surplus of data, and facing the lack of technology, funding and standardized protocols [11], most passive monitoring now lasts one to three years at most [105], while ecosystem conservation and ecological change detection usually require at least ten years. In particular, there is usually a lag period when measuring the benefits of planted forests, as individual trees need to grow and stands need to mature to form a stable structure [4]. Short-term monitoring may lead to a reduction in the quality and reliability of data [106]. This emphasizes the significance of utilizing and applying PAM within the framework of a monitoring strategy, with defined objectives, effective indicators, and standardized protocols [20].

There is developing acknowledgment from governments and related sectors that urban greenery is not monitored adequately to satisfy its crucial roles in biodiversity provisioning and ecosystem support [33,107]. A rich and diverse biophony usually indicates a stable and healthy ecosystem [26]. With its government-led design, planning, and implementation, the in-depth greening project in Beijing has indeed enhanced green space in the plain area, if only considering the total increased amounts of trees and connected urban forest and park patches [108]. However, the large-scale transition between cropland and forest generated by the afforestation process has the potential to lead to original wildlife habitat loss. By long-term monitoring of biodiversity patterns and processes, we can better assess the positive and negative impacts of afforestation projects.

According to our preliminary results, the proposed approach (with high computational efficiency and accuracy) may benefit further research on the rapid assessment and prediction of biodiversity in urban forests, providing an indirect but immediate measurement of bird activity dynamics across enhanced spatio-temporal scales. This would facilitate the application of PAM and the formulation of a standardized sampling protocol. Furthermore, a robust automated approach could support PAM as part of citizen science research. This would benefit developing countries that lack financial budgets, experts, and capacity for massive data processing. Globally, only 5% of PAM studies are conducted in regions of Asia, western Oceania, northern Africa, and southern South America, where some countries still have no record of using PAM [20]. As our approach does not require a priori data, it facilitates the implementation of long-term ecosystem monitoring in developing countries where baseline data are not available.

5. Conclusions

In 2019, the Intergovernmental Science-Policy Platform on Biodiversity and Ecosystem Services (IPBES) warned that the unprecedented deterioration of natural resources was deracinating millions of species and reducing human well-being worldwide. However, global biodiversity loss has not attracted as much public attention as global climate change. To evaluate the diversity of coexisting birds in urban forests and ultimately facilitate the assessment of afforestation benefits in urbanized Beijing, we developed an automated approach to extract and quantify bird vocalizations from spectrograms, integrating the well-established technology of object-based image analysis. The approach could achieve recognition accuracy (93.55%) of acoustic events at much higher efficiency (eight times faster) than traditional inspection methods. In addition, it also provided multiple aspects of acoustic traits, such as quantity, song length, frequency bandwidth, and shape information, which can be used to predict bird biodiversity. In our case, 57% of the variance in bird species richness could be explained by the acoustic and morphological features selected. The proposed soundscape evaluation method sheds light on long-term biodiversity monitoring and conservation during the upcoming global biodiversity crisis.

Author Contributions: Conceptualization, Y.Z., J.Y. and C.W.; data curation, Y.Z. and Z.B.; formal analysis, Y.Z. and J.Y.; funding acquisition, C.W.; investigation, Y.Z. and Z.B.; methodology, Y.Z., J.Y. and J.J.; project administration, C.W.; resources, Z.S., L.Y. and C.W.; software, Y.Z. and J.Y.; validation, Y.Z., J.Y., J.J., Z.S., L.Y. and C.W.; visualization, Y.Z.; writing—original draft, Y.Z.; writing—review and editing, J.Y., J.J. and C.W. All authors have read and agreed to the published version of the manuscript.

Funding: This research was funded by the National Non-Profit Research Institutions of the Chinese Academy of Forestry (CAFYBB2020ZB008).

Institutional Review Board Statement: Not applicable.

Informed Consent Statement: Not applicable.

Data Availability Statement: Not applicable.

Acknowledgments: We thank Hanchen Huang for bird species identification, Junyou Zhang for writing advice, and Shi Xu, Qi Bian for field investigation.

Conflicts of Interest: The authors declare no conflict of interest.

Appendix A

Table A1. List of Bird Species. A total of 30,758 vocalizations pertaining to 33 species were counted during recording sessions, as shown in the table below.

No.	Common Name	Binomial Name	Number of Syllable Patches	%	Predominant Frequency Intervals
1	Common Blackbird	*Turdus merula*	227	23.67049	/
2	Eurasian tree sparrow	*Passer montanus*	158	16.4755	2.3–5 kHz or
3	Azure-winged magpie	*Cyanopica cyanus*	120	12.51303	2–10 kHz
4	Large-billed crow	*Corvus macrorhynchos*	116	12.09593	1–2 kHz
5	Spotted dove	*Spilopelia chinensis*	109	11.36601	1–2 kHz

Table A1. *Cont.*

No.	Common Name	Binomial Name	Number of Syllable Patches	%	Predominant Frequency Intervals
6	Light-vented bulbul	*Pycnonotus sinensis*	55	5.735141	1.5–4 kHz
7	Eurasian magpie	*Pica pica*	46	4.796663	0–4 kHz
8	Common swift	*Apus apus*	42	4.379562	20–16,000 Hz
9	Yellow-browed warbler	*Phylloscopus inornatus*	17	1.77268	4–8 kHz
10	Arctic warbler	*Phylloscopus borealis*	12	1.251303	/
11	Crested myna	*Acridotheres cristatellus*	7	0.729927	/
12	Two-barred warbler	*Phylloscopus plumbeitarsus*	7	0.729927	/
13	Dusky warbler	*Phylloscopus fuscatus*	6	0.625652	/
14	Marsh tit	*Poecile palustris*	5	0.521376	6–10 kHz
15	Grey starling	*Spodiopsar cineraceus*	4	0.417101	above 4 kHz
16	Great spotted woodpecker	*Dendrocopos major*	4	0.417101	0–2.6 kHz
17	Chinese grosbeak	*Eophona migratoria*	4	0.417101	/
18	Barn swallow	*Hirundo rustica*	3	0.312826	/
19	Chicken	*Gallus gallus domesticus*	2	0.208551	5–10 kHz
20	Grey-capped greenfinch	*Chloris sinica*	2	0.208551	3–5.5 kHz
21	Yellow-rumped Flycatcher	*Ficedula zanthopygia*	1	0.104275	/
22	Oriental reed warbler	*Acrocephalus orientalis*	1	0.104275	/
23	Yellow-throated Bunting	*Emberiza elegans*	1	0.104275	/
24	Red-breasted Flycatcher	*Ficedula parva*	1	0.104275	/
25	Carrion crow	*Corvus corone*	1	0.104275	0–8 kHz
26	Dusky thrush	*Turdus eunomus*	1	0.104275	/
27	Black-browed Reed Warbler	*Acrocephalus bistrigiceps*	1	0.104275	/
28	Grey-capped pygmy woodpecker	*Dendrocopos canicapillus*	1	0.104275	4.5–5 kHz
29	Naumann's Thrush	*Turdus naumanni*	1	0.104275	/
30	U1	/	1	0.104275	/
31	U2	/	1	0.104275	/
32	U3	/	1	0.104275	/
33	U4	/	1	0.104275	/

Note: Unidentified species marked as U1-Un.

Table A2. List of other SFMs.

No.	Metric	Description
1	Border_len(Border Length)	The sum of the edges of the patch.
2	Width_Pxl (Width)	The number of pixels occupied by the length of the patch.
3	HSI_Transf	HSI transformation feature of patch hue.
4	HSI_Tran_1	HSI transformation feature of patch intensity.
5	Compactnes	The Compactness feature describes how compact a patch is. It is similar to Border Index but is based on area. However, the more compact a patch is, the smaller its border appears. The compactness of a patch is the product of the length and the width, divided by the number of pixels.
6	Roundness	The Roundness feature describes how similar a patch is to an ellipse. It is calculated by the difference between the enclosing ellipse and the enclosed ellipse. The radius of the largest enclosed ellipse is subtracted from the radius of the smallest enclosing ellipse.

Table A2. *Cont.*

No.	Metric	Description
7	Area_Pxl (Area of the patch)	The number of pixels forming a patch. If unit information is available, the number of pixels can be converted into a measurement. In scenes that provide no unit information, the area of a single pixel is 1 and the patch area is simply the number of pixels that form it. If the image data provides unit information, the area can be multiplied using the appropriate factor.
8	Border_ind (Border index)	The Border Index feature describes how jagged a patch is; the more jagged, the higher its border index. This feature is similar to the Shape Index feature, but the Border Index feature uses a rectangular approximation instead of a square. The smallest rectangle enclosing the patch is created and the border index is calculated as the ratio between the border lengths of the patch and the smallest enclosing rectangle.

References

1. Zhongming, Z.; Linong, L.; Wangqiang, Z.; Wei, L. *The Global Biodiversity Outlook 5 (GBO-5)*; Secretariat of the Convention on Biological Diversity: Montreal, QC, Canada, 2020.
2. World Health Organization. *WHO-Convened Global Study of Origins of SARS-CoV-2: China Part*; WHO: Geneva, Switzerland, 2021.
3. Platto, S.; Zhou, J.; Wang, Y.; Wang, H.; Carafoli, E. Biodiversity loss and COVID-19 pandemic: The role of bats in the origin and the spreading of the disease. *Biochem. Biophys. Res. Commun.* **2021**, *538*, 2. [CrossRef]
4. Pei, N.; Wang, C.; Jin, J.; Jia, B.; Chen, B.; Qie, G.; Qiu, E.; Gu, L.; Sun, R.; Li, J.; et al. Long-term afforestation efforts increase bird species diversity in Beijing, China. *Urban For. Urban Green.* **2018**, *29*, 88. [CrossRef]
5. Turner, A.; Fischer, M.; Tzanopoulos, J. Sound-mapping a coniferous forest-Perspectives for biodiversity monitoring and noise mitigation. *PLoS ONE* **2018**, *13*, e0189843. [CrossRef]
6. Sueur, J.; Gasc, A.; Grandcolas, P.; Pavoine, S. Global estimation of animal diversity using automatic acoustic sensors. In *Sensors for Ecology*; CNRS: Paris, France, 2012; Volume 99.
7. Sueur, J.; Farina, A. Ecoacoustics: The Ecological Investigation and Interpretation of Environmental Sound. *Biosemiotics* **2015**, *8*, 493. [CrossRef]
8. Blumstein, D.T.; Mennill, D.J.; Clemins, P.; Girod, L.; Yao, K.; Patricelli, G.; Deppe, J.L.; Krakauer, A.H.; Clark, C.; Cortopassi, K.A.; et al. Acoustic mon-itoring in terrestrial environments using microphone arrays: Applications, technological considerations and prospectus. *J. Appl. Ecol.* **2011**, *48*, 758. [CrossRef]
9. Krause, B.; Farina, A. Using ecoacoustic methods to survey the impacts of climate change on biodiversity. *Biol. Conserv.* **2016**, *195*, 245. [CrossRef]
10. Zhao, Z.; Zhang, S.; Xu, Z.; Bellisario, K.; Dai, N.; Omrani, H.; Pijanowski, B.C. Automated bird acoustic event detection and robust species classification. *Ecol. Inform.* **2017**, *39*, 99. [CrossRef]
11. Stephenson, P.J. Technological advances in biodiversity monitoring: Applicability, opportunities and challenges. *Curr. Opin. Environ. Sustain.* **2020**, *45*, 36. [CrossRef]
12. Buxton, R.T.; McKenna, M.F.; Clapp, M.; Meyer, E.; Stabenau, E.; Angeloni, L.M.; Crooks, K.; Wittemyer, G. Efficacy of extracting indices from large-scale acoustic recordings to monitor biodiversity. *Conserv. Biol.* **2018**, *32*, 1174. [CrossRef]
13. Halpin, P.N.; Read, A.J.; Best, B.D.; Hyrenbach, K.D.; Fujioka, E.; Coyne, M.S.; Crowder, L.B.; Freeman, S.; Spoerri, C. OBIS-SEAMAP: Developing a biogeographic research data commons for the ecological studies of marine mammals, seabirds, and sea turtles. *Mar. Ecol. Prog. Ser.* **2006**, *316*, 239. [CrossRef]
14. Gross, M. Eavesdropping on ecosystems. *Curr. Biol.* **2020**, *30*, R237–R240. [CrossRef]
15. Rajan, S.C.; Athira, K.; Jaishanker, R.; Sooraj, N.P.; Sarojkumar, V. Rapid assessment of biodiversity using acoustic indices. *Biodivers. Conserv.* **2019**, *28*, 2371. [CrossRef]
16. Llusia, D.; Márquez, R.; Beltrán, J.F.; Benítez, M.; do Amaral, J.P. Calling behaviour under climate change: Geographical and seasonal variation of calling temperatures in ectotherms. *Glob. Chang. Biol.* **2013**, *19*, 2655. [CrossRef] [PubMed]
17. Hart, P.J.; Sebastián-González, E.; Tanimoto, A.; Thompson, A.; Speetjens, T.; Hopkins, M.; Atencio-Picado, M. Birdsong characteristics are related to fragment size in a neotropical forest. *Anim. Behav.* **2018**, *137*, 45. [CrossRef]
18. Bueno-Enciso, J.; Ferrer, E.S.; Barrientos, R.; Sanz, J.J. Habitat structure influences the song characteris-tics within a population of Great Tits Parus major. *Bird Study* **2016**, *63*, 359. [CrossRef]
19. Browning, E.; Gibb, R.; Glover-Kapfer, P.; Jones, K.E. Passive Acoustic Monitoring in Ecology and Conservation. 2017. Available online: https://www.wwf.org.uk/sites/default/files/2019-04/Acousticmonitoring-WWF-guidelines.pdf (accessed on 7 April 2021).

20. Sugai, L.S.M.; Silva, T.S.F.; Ribeiro, J.W.; Llusia, D. Terrestrial Passive Acoustic Monitoring: Review and Perspectives. *BioScience* **2019**, *69*, 15. [CrossRef]
21. Righini, R.; Pavan, G. A soundscape assessment of the Sasso Fratino Integral Nature Reserve in the Central Apennines, Italy. *Biodiversity* **2020**, *21*, 4. [CrossRef]
22. Priyadarshani, N.; Marsland, S.; Castro, I. Automated birdsong recognition in complex acoustic environments: A review. *J. Avian Biol.* **2018**, *49*, jav-01447. [CrossRef]
23. Farina, A. *Soundscape Ecology: Principles, Patterns, Methods and Applications*; Springer: Berlin/Heidelberg, Germany, 2013.
24. Eldridge, A.; Casey, M.; Moscoso, P.; Peck, M. A new method for ecoacoustics? Toward the extraction and evaluation of ecologically-meaningful soundscape components using sparse coding methods. *PeerJ* **2016**, *4*, e2108. [CrossRef]
25. Lellouch, L.; Pavoine, S.; Jiguet, F.; Glotin, H.; Sueur, J. Monitoring temporal change of bird communities with dissimilarity acoustic indices. *Methods Ecol. Evol.* **2014**, *5*, 495. [CrossRef]
26. Pijanowski, B.C.; Villanueva-Rivera, L.J.; Dumyahn, S.L.; Farina, A.; Krause, B.L.; Napoletano, B.M.; Gage, S.H.; Pieretti, N. Soundscape Ecology: The Science of Sound in the Landscape. *BioScience* **2011**, *61*, 203. [CrossRef]
27. Hao, Z.; Wang, C.; Sun, Z.; van den Bosch, C.K.; Zhao, D.; Sun, B.; Xu, X.; Bian, Q.; Bai, Z.; Wei, K.; et al. Soundscape mapping for spa-tial-temporal estimate on bird activities in urban forests. *Urban For. Urban Green.* **2021**, *57*, 126822. [CrossRef]
28. Project, W.S.; Truax, B. *The World Soundscape Project's Handbook for Acoustic Ecology*; Arc Publications: Todmorden, UK, 1978.
29. Pieretti, N.; Farina, A.; Morri, D. A new methodology to infer the singing activity of an avian community: The Acoustic Complexity Index (ACI). *Ecol. Indic.* **2011**, *11*, 868. [CrossRef]
30. Boelman, N.T.; Asner, G.P.; Hart, P.J.; Martin, R.E. Multi-trophic invasion resistance in Hawaii: Bioacoustics, field surveys, and airborne remote sensing. *Ecol. Appl.* **2007**, *17*, 2137. [CrossRef]
31. Fuller, S.; Axel, A.C.; Tucker, D.; Gage, S.H. Connecting soundscape to landscape: Which acoustic index best describes landscape configuration? *Ecol. Indic.* **2015**, *58*, 207. [CrossRef]
32. Mammides, C.; Goodale, E.; Dayananda, S.K.; Kang, L.; Chen, J. Do acoustic indices correlate with bird diversity? Insights from two biodiverse regions in Yunnan Province, south China. *Ecol. Indic.* **2017**, *82*, 470. [CrossRef]
33. Fairbrass, A.J.; Rennert, P.; Williams, C.; Titheridge, H.; Jones, K.E. Biases of acoustic indices measuring biodiversity in urban areas. *Ecol. Indic.* **2017**, *83*, 169. [CrossRef]
34. Ross, S.R.-J.; Friedman, N.R.; Yoshimura, M.; Yoshida, T.; Donohue, I.; Economo, E.P. Utility of acoustic indices for ecological monitoring in complex sonic environments. *Ecol. Indic.* **2021**, *121*, 107114. [CrossRef]
35. Sueur, J.; Pavoine, S.; Hamerlynck, O.; Duvail, S. Rapid acoustic survey for biodiversity appraisal. *PLoS ONE* **2008**, *3*, e4065. [CrossRef]
36. Kasten, E.P.; Gage, S.H.; Fox, J.; Joo, W. The remote environmental assessment laboratory's acoustic library: An archive for studying soundscape ecology. *Ecol. Inform.* **2012**, *12*, 50. [CrossRef]
37. Gibb, R.; Browning, E.; Glover-Kapfer, P.; Jones, K.E. Emerging opportunities and challenges for passive acoustics in ecological assessment and monitoring. *Methods Ecol. Evol.* **2019**, *10*, 169. [CrossRef]
38. Merchant, N.D.; Fristrup, K.M.; Johnson, M.P.; Tyack, P.L.; Witt, M.J.; Blondel, P.; Parks, S.E. Measuring acoustic habitats. *Methods Ecol. Evol.* **2015**, *6*, 257. [CrossRef] [PubMed]
39. Swiston, K.A.; Mennill, D.J. Comparison of manual and automated methods for identifying target sounds in audio recordings of Pileated, Pale-billed, and putative Ivory-billed woodpeckers. *J. Field Ornithol.* **2009**, *80*, 42. [CrossRef]
40. Goyette, J.L.; Howe, R.W.; Wolf, A.T.; Robinson, W.D. Detecting tropical nocturnal birds using auto-mated audio recordings. *J. Field Ornithol.* **2011**, *82*, 279. [CrossRef]
41. Potamitis, I. Automatic classification of a taxon-rich community recorded in the wild. *PLoS ONE* **2014**, *9*, e96936.
42. Ulloa, J.S.; Gasc, A.; Gaucher, P.; Aubin, T.; Réjou-Méchain, M.; Sueur, J. Screening large audio datasets to determine the time and space distribution of Screaming Piha birds in a tropical forest. *Ecol. Inform.* **2016**, *31*, 91. [CrossRef]
43. Blaschke, T. Object based image analysis for remote sensing. *ISPRS J. Photogramm. Remote Sens.* **2010**, *65*, 2. [CrossRef]
44. Blaschke, T. (Ed.) Object Based Image Analysis: A new paradigm in remote sensing? In Proceedings of the American Society for Photogrammetry and Remote Sensing Annual Conferenc, Baltimore, MD, USA, 26–28 March 2013.
45. Benz, U.C.; Hofmann, P.; Willhauck, G.; Lingenfelder, I.; Heynen, M. Multi-resolution, object-oriented fuzzy analysis of remote sensing data for GIS-ready information. *ISPRS J. Photogramm. Remote Sens.* **2004**, *58*, 239. [CrossRef]
46. Johansen, K.; Arroyo, L.A.; Phinn, S.; Witte, C. Comparison of geo-object based and pixel-based change detection of riparian environments using high spatial resolution multi-spectral imagery. *Photogramm. Eng. Remote Sens.* **2010**, *76*, 123. [CrossRef]
47. Burivalova, Z.; Game, E.T.; Butler, R.A. The sound of a tropical forest. *Science* **2019**, *363*, 28. [CrossRef]
48. Hay, G.J.; Castilla, G. Object-based image analysis: Strengths, weaknesses, opportunities and threats (SWOT). In Proceedings of the 1st International Conference on Object-based Image Analysis (OBIA), Salzburg, Austria, 4–5 July 2006; pp. 4–5.
49. Jafari, N.H.; Li, X.; Chen, Q.; Le, C.-Y.; Betzer, L.P.; Liang, Y. Real-time water level monitoring using live cameras and computer vision techniques. *Comput. Geosci.* **2021**, *147*, 104642. [CrossRef]
50. Schwier, M. Object-based Image Analysis for Detection and Segmentation Tasks in Biomedical Imaging. Ph.D. Thesis, Information Resource Center der Jacobs University Bremen, Bremen, Germany.
51. Kerle, N.; Gerke, M.; Lefèvre, S. *GEOBIA 2016: Advances in Object-Based Image Analysis—Linking with Computer Vision and Machine Learning*; Multidisciplinary Digital Publishing Institute: Basel, Switzerland, 2019.

52. Yan, J.; Lin, L.; Zhou, W.; Han, L.; Ma, K. Quantifying the characteristics of particulate matters captured by urban plants using an automatic approach. *J. Environ. Sci.* **2016**, *39*, 259. [CrossRef] [PubMed]
53. Yan, J.; Lin, L.; Zhou, W.; Ma, K.; Pickett, S.T.A. A novel approach for quantifying particulate matter dis-tribution on leaf surface by combining SEM and object-based image analysis. *Remote Sens. Environ.* **2016**, *173*, 156. [CrossRef]
54. Lin, L.; Yan, J.; Ma, K.; Zhou, W.; Chen, G.; Tang, R.; Zhang, Y. Characterization of particulate matter de-posited on urban tree foliage: A landscape analysis approach. *Atmos. Environ.* **2017**, *171*, 59. [CrossRef]
55. Artiola, J.F.; Brusseau, M.L.; Pepper, I.L. *Environmental Monitoring and Characterization*; Academic Press: Cambridge, MA, USA, 2004.
56. Xie, S.; Lu, F.; Cao, L.; Zhou, W.; Ouyang, Z. Multi-scale factors influencing the characteristics of avian communities in urban parks across Beijing during the breeding season. *Sci. Rep.* **2016**, *6*, 29350. [CrossRef] [PubMed]
57. Farina, A.; Righini, R.; Fuller, S.; Li, P.; Pavan, G. Acoustic complexity indices reveal the acoustic commu-nities of the old-growth Mediterranean forest of Sasso Fratino Integral Natural Reserve (Central Italy). *Ecol. Indic.* **2021**, *120*, 106927. [CrossRef]
58. Zitong, B.; Yilin, Z.; Cheng, W.; Zhenkai, S. The public's Perception of Anthrophony Soundscape in Beijing's Urban Parks. *J. Chin. Urban For.* **2021**, *19*, 16, In Chinese. [CrossRef]
59. Pahuja, R.; Kumar, A. Sound-spectrogram based automatic bird species recognition using MLP classifier. *Appl. Acoust.* **2021**, *180*, 108077. [CrossRef]
60. Aide, T.M.; Corrada-Bravo, C.; Campos-Cerqueira, M.; Milan, C.; Vega, G.; Alvarez, R. Real-time bioacous-tics monitoring and automated species identification. *PeerJ* **2013**, *1*, e103. [CrossRef]
61. Linke, S.; Deretic, J.-A. Ecoacoustics can detect ecosystem responses to environmental water allocations. *Freshw. Biol.* **2020**, *65*, 133. [CrossRef]
62. Gasc, A.; Francomano, D.; Dunning, J.B.; Pijanowski, B.C. Future directions for soundscape ecology: The importance of ornithological contributions. *Auk* **2017**, *134*, 215. [CrossRef]
63. Eldridge, A.; Guyot, P.; Moscoso, P.; Johnston, A.; Eyre-Walker, Y.; Peck, M. Sounding out ecoacoustic met-rics: Avian species richness is predicted by acoustic indices in temperate but not tropical habitats. *Ecol. Indic.* **2018**, *95*, 939. [CrossRef]
64. Dufour, O.; Artieres, T.; Glotin, H.; Giraudet, P. *Soundscape Semiotics—Localization and Categorization*; InTech: London, UK, 2013; Volume 89.
65. Karaconstantis, C.; Desjonquères, C.; Gifford, T.; Linke, S. Spatio-temporal heterogeneity in river sounds: Disentangling micro-and macro-variation in a chain of waterholes. *Freshw. Biol.* **2020**, *65*, 96. [CrossRef]
66. De Oliveira, A.G.; Ventura, T.M.; Ganchev, T.D.; de Figueiredo, J.M.; Jahn, O.; Marques, M.I.; Schuchmann, K.-L. Bird acoustic activity detection based on morphological filtering of the spectrogram. *Appl. Acoust.* **2015**, *98*, 34. [CrossRef]
67. Ludeña-Choez, J.; Gallardo-Antolín, A. Feature extraction based on the high-pass filtering of audio signals for Acoustic Event Classification. *Comput. Speech Lang.* **2015**, *30*, 32. [CrossRef]
68. Albornoz, E.M.; Vignolo, L.D.; Sarquis, J.A.; Leon, E. Automatic classification of Furnariidae species from the Paranaense Littoral region using speech-related features and machine learning. *Ecol. Inform.* **2017**, *38*, 39. [CrossRef]
69. Bhargava, S. *Vocal Source Separation Using Spectrograms and Spikes, Applied to Speech and Birdsong*; ETH Zurich: Zurich, Switzerland, 2017.
70. Plapous, C.; Marro, C.; Scalart, P. Improved signal-to-noise ratio estimation for speech enhancement. *IEEE Trans. Audio Speech Lang. Process.* **2006**, *14*, 2098. [CrossRef]
71. Podos, J.; Warren, P.S. The evolution of geographic variation in birdsong. *Adv. Study Behav.* **2007**, *37*, 403.
72. Lu, W.; Zhang, Q. Deconvolutive Short-Time Fourier Transform Spectrogram. *IEEE Signal Process. Lett.* **2009**, *16*, 576.
73. Mehta, J.; Gandhi, D.; Thakur, G.; Kanani, P. Music Genre Classification using Transfer Learning on log-based MEL Spectrogram. In Proceedings of the 5th International Conference on Computing Methodolo-Gies and Communication (ICCMC), Erode, India, 8–10 April 2021; p. 1101.
74. Ludena-Choez, J.; Quispe-Soncco, R.; Gallardo-Antolin, A. Bird sound spectrogram decomposition through Non-Negative Matrix Factorization for the acoustic classification of bird species. *PLoS ONE* **2017**, *12*, e0179403. [CrossRef]
75. Garcia-Lamont, F.; Cervantes, J.; López, A.; Rodriguez, L. Segmentation of images by color features: A survey. *Neurocomputing* **2018**, *292*, 1. [CrossRef]
76. Chen, Y.; Chen, Q.; Jing, C. Multi-resolution segmentation parameters optimization and evaluation for VHR remote sensing image based on mean NSQI and discrepancy measure. *J. Spat. Sci.* **2021**, *66*, 253. [CrossRef]
77. Zheng, L.; Huang, W. Parameter Optimization in Multi-scale Segmentation of High Resolution Remotely Sensed Image and Its Application in Object-oriented Classification. *J. Subtrop. Resour. Environ.* **2015**, *10*, 77.
78. Mesner, N.; Ostir, K. Investigating the impact of spatial and spectral resolution of satellite images on segmentation quality. *J. Appl. Remote Sens.* **2014**, *8*, 83696. [CrossRef]
79. Ptacek, L.; Machlica, L.; Linhart, P.; Jaska, P.; Muller, L. Automatic recognition of bird individuals on an open set using as-is recordings. *Bioacoustics* **2016**, *25*, 55. [CrossRef]
80. Yip, D.A.; Mahon, C.L.; MacPhail, A.G.; Bayne, E.M. Automated classification of avian vocal activity using acoustic indices in regional and heterogeneous datasets. *Methods Ecol. Evol.* **2021**, *12*, 707. [CrossRef]
81. Breiman, L.; Friedman, J.H.; Olshen, R.A.; Stone, C.J. *Classification and Regression Trees*; Routledge: Oxfordshire, UK, 2017.

82. Stowell, D.; Plumbley, M.D. Automatic large-scale classification of bird sounds is strongly improved by unsupervised feature learning. *PeerJ* **2014**, *2*, e488. [CrossRef]
83. Zottesso, R.H.D.; Costa, Y.M.G.; Bertolini, D.; Oliveira, L.E.S. Bird species identification using spectro-gram and dissimilarity approach. *Ecol. Inform.* **2018**, *48*, 187. [CrossRef]
84. Bai, J.; Chen, C.; Chen, J. (Eds.) *Xception Based Method for Bird Sound Recognition of BirdCLEF*; CLEF: Thessaloniki, Greece, 2020.
85. McGarigal, K.; Cushman, S.A.; Ene, E. FRAGSTATS v4: Spatial Pattern Analysis Program for Categorical and Continuous Maps. Computer Software Program Produced by the Authors at the University of Massachusetts, Amherst. 2012. Available online: http://www.umass.edu/landeco/research/fragstats/fragstats.html (accessed on 25 June 2021).
86. Peterson, R.A.; Peterson, M.R.A. Package 'bestNormalize'. 2020. Available online: https://mran.microsoft.com/snapshot/2020-04-22/web/packages/bestNormalize/bestNormalize.pdf (accessed on 5 April 2021).
87. Van Loan, C.F.; Golub, G. *Matrix Computations (Johns Hopkins Studies in Mathematical Sciences)*; Johns Hopkins University Press: Baltimore, MD, USA, 1996.
88. Liaw, A.; Wiener, M. Classification and regression by randomForest. *R News* **2002**, *2*, 18.
89. ColorBrewer, S.R.; Liaw, M.A. *Package 'randomForest'*; University of California: Berkeley, CA, USA, 2018.
90. Chakure, A. Random Forest Regression. *Towards Data Science*, 12 June 2019.
91. Breiman, L. Random forests. *Mach. Learn.* **2001**, *45*, 5. [CrossRef]
92. Wimmer, J.; Towsey, M.; Roe, P.; Williamson, I. Sampling environmental acoustic recordings to determine bird species richness. *Ecol. Appl.* **2013**, *23*, 1419. [CrossRef] [PubMed]
93. Krause, B.; Bernard, L.; Gage, S. *Testing Biophony as an Indicator of Habitat Fitness and Dynamics. Sequoia National Park (SEKI) Natural Soundscape Vital Signs Pilot Program Report*; Wild Sanctuary, Inc.: Glen Ellen, CA, USA, 2003.
94. Ishwaran, H.; Kogalur, U.B.; Gorodeski, E.Z.; Minn, A.J.; Lauer, M.S. High-dimensional variable selection for survival data. *J. Am. Stat. Assoc.* **2010**, *105*, 205. [CrossRef]
95. Brumm, H.; Zollinger, S.A.; Niemelä, P.T.; Sprau, P. Measurement artefacts lead to false positives in the study of birdsong in noise. *Methods Ecol. Evol.* **2017**, *8*, 1617. [CrossRef]
96. Jancovic, P.; Kokuer, M. Bird Species Recognition Using Unsupervised Modeling of Individual Vocaliza-tion Elements. *IEEE/ACM Trans. Audio Speech Lang. Process.* **2019**, *27*, 932. [CrossRef]
97. Zsebők, S.; Blázi, G.; Laczi, M.; Nagy, G.; Vaskuti, É.; Garamszegi, L.Z. "Ficedula": An open-source MATLAB toolbox for cutting, segmenting and computer-aided clustering of bird song. *J. Ornithol.* **2018**, *159*, 1105. [CrossRef]
98. Potamitis, I. Unsupervised dictionary extraction of bird vocalisations and new tools on assessing and visualising bird activity. *Ecol. Inform.* **2015**, *26*, 6. [CrossRef]
99. Lasseck, M. Bird song classification in field recordings: Winning solution for NIPS4B 2013 competition. In Proceedings of the Neural Information Scaled for Bioacoustics (NIPS), Lake Tahoe, NV, USA, 10 December 2013; p. 176.
100. Lasseck, M. Large-scale Identification of Birds in Audio Recordings. In Proceedings of the Conference and Labs of the Evaluation Forum (CLEF), Sheffield, UK, 15–18 September 2014; p. 643.
101. Servick, K. Eavesdropping on ecosystems. *Science* **2014**, *343*, 834. [CrossRef]
102. Sueur, J.; Farina, A.; Gasc, A.; Pieretti, N.; Pavoine, S. Acoustic Indices for Biodiversity Assessment and Landscape Investigation. *Acta Acust. United Acust.* **2014**, *100*, 772. [CrossRef]
103. Gage, S.H.; Axel, A.C. Visualization of temporal change in soundscape power of a Michigan lake habitat over a 4-year period. *Ecol. Inform.* **2014**, *21*, 100. [CrossRef]
104. Gil, D.; Llusia, D. The bird dawn chorus revisited. In *Coding Strategies in Vertebrate Acoustic Communication*; Springer: Berlin/Heidelberg, Germany, 2020; Volume 7, p. 45.
105. Stephenson, P.J. The Holy Grail of biodiversity conservation management: Monitoring impact in pro-jects and project portfolios. *Perspect. Ecol. Conserv.* **2019**, *17*, 182. [CrossRef]
106. Stephenson, P.J.; Bowles-Newark, N.; Regan, E.; Stanwell-Smith, D.; Diagana, M.; Höft, R.; Abarchi, H.; Abrahamse, T.; Akello, C.; Allison, H.; et al. Unblocking the flow of biodiversity data for decision-making in Africa. *Biol. Conserv.* **2017**, *213*, 335. [CrossRef]
107. Abolina, K.; Zilans, A. Evaluation of urban sustainability in specific sectors in Latvia. *Environ. Dev. Sustain.* **2002**, *4*, 299. [CrossRef]
108. Jin, J.; Sheppard, S.R.; Jia, B.; Wang, C. Planning to Practice: Impacts of Large-Scale and Rapid Urban Af-forestation on Greenspace Patterns in the Beijing Plain Area. *Forests* **2021**, *12*, 316. [CrossRef]

MDPI
St. Alban-Anlage 66
4052 Basel
Switzerland
www.mdpi.com

Forests Editorial Office
E-mail: forests@mdpi.com
www.mdpi.com/journal/forests

www.ingramcontent.com/pod-product-compliance
Lightning Source LLC
Chambersburg PA
CBHW040311270926
51850CB00007BC/1023
* 9 7 8 3 0 3 6 5 8 4 4 2 3 *